UPROOTING URBAN AMERICA

This book is part of the Peter Lang Education list.
Every volume is peer reviewed and meets
the highest quality standards for content and production.

PETER LANG
New York • Washington, D.C./Baltimore • Bern
Frankfurt • Berlin • Brussels • Vienna • Oxford

UPROOTING URBAN AMERICA

multidisciplinary perspectives on race, class and gentrification

edited by
HORACE R. HALL,
CYNTHIA COLE ROBINSON
& AMOR KOHLI

PETER LANG
New York • Washington, D.C./Baltimore • Bern
Frankfurt • Berlin • Brussels • Vienna • Oxford

Library of Congress Cataloging-in-Publication Data

Uprooting urban America: multidisciplinary perspectives on race, class and gentrification / edited by Horace R. Hall, Cynthia Cole Robinson, Amor Kohli.
pages cm
Includes bibliographical references.
1. Urbanization—United States. 2. Sociology, Urban—United States.
3. United States—Social conditions—1945– I. Hall, Horace R.
II. Robinson, Cynthia Cole. III. Kohli, Amor.
HT123.U67   305.0973'091732—dc23   2014000352
ISBN 978-1-4331-2257-6 (hardcover)
ISBN 978-1-4331-2256-9 (paperback)
ISBN 978-1-4539-1270-6 (e-book)

Bibliographic information published by **Die Deutsche Nationalbibliothek**.
Die Deutsche Nationalbibliothek lists this publication in the "Deutsche Nationalbibliografie"; detailed bibliographic data is available on the Internet at http://dnb.d-nb.de/.

The paper in this book meets the guidelines for permanence and durability of the Committee on Production Guidelines for Book Longevity of the Council of Library Resources.

© 2014 Peter Lang Publishing, Inc., New York
29 Broadway, 18th floor, New York, NY 10006
www.peterlang.com

All rights reserved.
Reprint or reproduction, even partially, in all forms such as microfilm, xerography, microfiche, microcard, and offset strictly prohibited.

Printed in the United States of America

Table of Contents

Foreword: On Politics, Property and Wealth ix
    William H. Watkins

Introduction: Understanding Gentrification and the Recolonization
    of U.S. Urban Space 1
    Horace R. Hall

Part One: Urban Renewal and the Marriage of Civic Government and Free Market

Chapter One: Relocating Gentrification: The Working Class, Science and
    the State in Recent Urban Research 13
    Loïc Wacquant

Chapter Two: Hurting or Helping: Gentrification and African
    American Neighborhoods in Chicago 23
    Adrienne Holloway

Chapter Three: The Shrinkage Machine: Race, Class and the Renewal
    of Urban Capital 41
    Emily Rosenman, Samuel Walker and Elvin Wyly

Part Two: Reorienting Gentrification: Space, Equity and Voice

Chapter Four: Sustainable Urban Development and Environmental Gentrification: The Paradox Confronting the U.S. Environmental Justice Movement 77
Daniel Faber and Shelley McDonough Kimelberg

Chapter Five: Visualizing Change: Using Technology and Participatory Research to Engage Youth in Urban Planning and Health Promotion 93
Antwi Akom, Aekta Shah and Aaron Nakai

Chapter Six: Reframing Spatial Inequality: Youth, Photography and a Changing Urban Landscape 107
Stuart Greene, Kevin Burke and Maria McKenna

Part Three: Is There a Class-Conscious, Racially Sensitive Doctor in the House?

Chapter Seven: Training Physicians for the Demographics of the 21st Century: The Importance of Diversity and Cultural Competency 131
Donald A. Barr

Chapter Eight: Gentrification and Health: Patterns of Environmental Risk 143
Russell Lopez

Chapter Nine: Residential Segregation: Trends and Implications for Conducting Effective Community-Based Research to Address Ethnic Health Disparities 157
T. Henry Akintobi, Ronald Braithwaite and Anika Dodds

Part Four: The Customer Is Always Right: Democracy under Threat in Public Schools

Chapter Ten: Topsy-Turvy: Education at the End of Empire 173
William Ayers

Chapter Eleven: Cultural Studies in Dark Times: Public Pedagogy and the Challenge of Neoliberalism 185
Henry A. Giroux

Chapter Twelve: Disparity, Austerity and Public Schooling in the United States: Why Quentin Can't Read 207
Sue Books

Chapter Thirteen: School Activism and the Production of Urban Space in Atlanta, Georgia 227
Katherine Hankins and Elizabeth Egan Henry

Part Five: The Groundswell: Community Responses to Urban Renewal

Chapter Fourteen: "History Still Matters": Leveraging Historicity in
    Struggles to Control Space 245
    Miranda Martinez

Chapter Fifteen: Gentrification as Class Politics 259
    Judith N. DeSena

Chapter Sixteen: Foreclosure Crisis and the Role of Community
    Organizing in a U.S. Latino Community 273
    James Jennings

Chapter Seventeen: Community Parading and Symbolic
    Expression in Post-Katrina New Orleans 287
    Diane Grams

Afterword: Things Have Fallen Apart but We Are Planning to Stay 301
    Mindy Thompson Fullilove

Contributors 307

Foreword

On Politics, Property and Wealth

WILLIAM H. WATKINS

Uprooting Urban America moves beyond the more narrowly conceptualized works now in print. The editors have clearly glimpsed the new social order through a broad interdisciplinary assemblage of writings by well-qualified and thoughtful scholars. In the introduction, Hall writes:

> This book examines salient demographic changes facing the United States in the 21st century, with a focus on the social phenomena of gentrification, the suburbanization of poor and minority groups, increased cultural competency in fields of healthcare and education, and the growing necessity for community organizing.

He goes on to note that this anthology is building on critical discourses pointing to the colonization of urban spaces, housing redevelopment, and privatization of healthcare and education. If you are not deeply affected by the essays you are about to read, then you had better check your pulse.

This is a book about the state of our being. The inescapable message is that oppression and abuse breed resistance and protest. It turns our attention to politics. Politics is an umbrella concept covering power, race, equity, displacement, and human interaction. This edited volume is about hope and imagination. As churched Black folks say, "trouble don't last always."

The editors were astute in locating demographics as central to their inquiry. Property, politics, and culture are inextricably connected. They are at the heart of our socioeconomic existence. I would like to look at the ubiquitous impact of property to help foreground the new social order.

SPACE AND THE CULTURE OF PROPERTY: DEEP ROOTS

Space is an overused term in today's world. It has both personal and societal connotations. Often abstracted, the concept of space is deeply rooted in notions of property and power. The fetish of property is central to class society and prominent in Western thought and, perhaps, all human history.

The land, the oceans, the skies, and all that lies beneath and above have long been contested. Mother Nature's bounty providing sustenance and beauty is now commodified, fought over, and perhaps forever destroyed. The concept of *property* has been inextricably connected to human existence, commercial interaction, and partisan interests. The meaning of property now extends far beyond natural resources, reaching to nearly everything that can be possessed. Now schools have been drawn into this discourse.

Questions surrounding property seem endless. Who owns what? Where are the boundaries? How are public and private property defined? How are those definitions changing? Concepts of property and property rights were theorized long ago and have evolved over time.

DISCOURSES ON PROPERTY: THEOLOGICAL AND PHILOSOPHICAL ORIGINS

Notions of property appear in the Bible and extend back to the Creation. John Battle's (2008) biblical scholarship offers insights into the theological origins of property. He references the Creation and the Old Testament, where God is cast not only as the redeemer but also as creator. (S)He is presented as the owner of all people and things. Battle writes that the Old Testament presents the world as God's possession by right of his "creation and sustaining providence." God's sovereign rights over the earth and its inhabitants are demonstrated by events such as the Flood and the destruction of Sodom and Gomorrah. Battle concludes that the Old Testament establishes God's possession of the earth and its people.

By the 11th century, concepts of "natural rights" were explored (Machan, 2002). Private property came to be seen by many as a natural right. Among the ancient Greek philosophers, Plato's communal notions of property were countered by Aristotle in *The Politics* (2010).

In chapters 3 and 4, Aristotle discusses the family or household, starting with its property, including slaves. Property, for him, includes *possessions* and *instruments*, which Aristotle distinguishes. Possessions are connected to human activity, and instruments are means to the production of artifacts, the products being either possessions or instruments—i.e., means either to further production or to human action. Artisans or employees in farming or industry are human beings who are means to production; slaves are human beings who are possessions, means to action, along with domestic servants and secretaries. The slave "wholly belongs" to the master.

The father of classical liberalism, John Locke (1632–1704), wrote extensively on questions of property, positing that to preserve the public good, the central function of government must be the protection of private property. Initially, Locke suggested the earth and everything on it belonged to all of us in common; among perfectly equal inhabitants, all have the same right to make use of whatever they found and could use.

In his major writings (Locke, 1690), he underscored the "moral" dimension of property. He drew a connection between acting freely and responsibly as moral agents and having the right to private property. He defended the institution of the right to private property as well as the way property might be assigned. He theorized that individuals who toiled on the land and improved its productivity acquired a property interest in the result. In the *Second Treatise* (1690) he argued that our bodies and their movements are our own; therefore whenever we use our own effort to improve the natural world, the resulting products belong to us as well. He wrote (1690):

> The labour of his body, and the work of his hands, we may say, are properly his. Whatsoever then he removes out of the state that nature hath provided, and left it in, he hath mixed his labour with, and joyned to it something that is his own, and thereby makes it his Property.

His refined arguments contributed to the platform for Western social thought and capitalist ownership.

Building on the theoretical platform advanced by Locke, Founding Father James Madison took a leading role in theorizing governance, power, property, and structure for the new territory that became the United States. He believed an arrangement was required that could satisfy both those who favored strong central government as well as others wanting strong states rights. Rejecting the feudal autocracy of Europe, Madison charted a course of "federalism" where the central government would be superior to state and local governing bodies while allowing great latitude to the states. The inalienable rights would be protected at all levels. Property was one of those rights. His writings (1792) joined property to government.

> Government is instituted to protect property of every sort; as well that which lies in the various rights of individuals, as that which the term particularly expresses. This being the end of government, that alone is a just government, which impartially secures to every man, whatever is his own.

Madison associated property with consciousness, choice, and safety. It is as if property were God given and an organic part of humanness. Property was also inextricably associated with freedom. Property had to be protected at all cost. He defined property (1792) as:

> that dominion which one man claims and exercises over the external things of the world, in exclusion of every other individual. In its larger and juster meaning, it embraces every thing to which a man may attach a value and have a right; and which leaves to every one else the like advantage. (pp. 266–68)

Madison (1792) concluded: "If the United States means to obtain or deserve the full praise due to wise and just governments, they will equally respect the rights of property, and the property in rights...."

In ensuing years, the United States expanded its boundaries through purchase and capture. Colonialism and imperial conquest elevated questions of property to national, even international, status. Modern capitalism complicated issues as profit and advantage became increasingly attached to property questions. Trademarks, patents, and copyrights were designed to protect and make exclusive the right to property protection. Fiercely competitive globalized capitalism has pushed notions of intellectual property into that discourse.

PROPERTY IN THE 21ST CENTURY

Property is a major dividing line in society. Property is linked to affluence, power, and privilege. Class divisions are seen as those between the propertied and the propertyless. Property is at the heart of politics and the life or death struggle for survival faced by many throughout the world.

Demanding the reorganization of society, neo-liberalism (Harvey, 2005) has elevated notions of property to new importance. Its market-oriented view advances capitalism and politics to a new level where, masked by the promise of opportunity, the usurpation, reconfiguration, and concentration of wealth and property enriches the very few and impoverishes many. A return to medieval property relations is evidenced as throughout the world partitions are being constructed to keep the "rampaging hordes" and "rabble" out of valued and protected spaces. The neo-liberal free market ideology re-divides the world.

Parenthetically, I note that the 21st-century global neo-liberals in the United States have, quietly and unofficially, embraced diabolical and extreme new military policies involving sovereignty, war, and property. Pursuing war in the Middle East, the American generals and corporatists no longer honor national boundaries as theatres of war. They now treat the *entire world* as the battlefield. Illegal incursions into Pakistan and drone attacks in Yemen and Syria suggest that any nation and any government *anywhere* can be disrupted or overthrown as part of their war.

PROPERTY, POLITICS, AND EDUCATION

As this book will demonstrate, the spacial and political landscape of the United States is being reconfigured. Having grown accustomed to liberal democracy, welfarism, and a safety net, the nation is now experiencing the terror of neo-liberal politics and economics. Public spaces are now usurped by private interests. Public schools, traditionally public property, are being reconceptualized as cities are being re-walled.

As an educator I am interested in how "walls," wealth, and property issues impact education. Gentrification and school "reform" are now inextricably connected (Lipman, 2011) as the re-structuring of urban spaces proceeds rapidly. Universal education, a staple of liberal democracy in the United States, is under attack (Watkins, 2012). Public schooling, as we know it, will not likely survive. Let us move to understanding the neoliberal re-imagining of public education by first contextualizing the transformation of public to private spaces.

PRIVATIZING PUBLIC SPACE

President George W. Bush's (2004) call for an "ownership society" mouthed a sales pitch that contributed to the banks and lending agencies engaging in apocalyptic mortgage scams that nearly undermined the entire economic order. The appeal to property ownership was difficult to resist.

Rights of property have evolved over time. Property under feudalism was fundamentally land or landed estates. This was the base of real property or real estate, a term that usually refers to a property that generates income for the owner without the owner having to do the actual work of the estate. In Europe, agrarian landed property typically consisted of a manor, several tenant farms, and some privileged enterprises. Property was inextricably connected to control. Control of property today has shifted from control by aristocracy and nobility to individual and corporate control.

Capitalism and the Industrial Revolution shifted the paradigm of property away from land (and bodies—e.g., American slavery—ownership/value of the body, often seen as less than human—refer to Locke) onto tools, machines, and the instruments of production, which represented the greatest source of profits. In Europe and North America, capitalism also brought about monumental changes in land occupation as workers moved to industrial venues where production was centralized. The urban construct altered the dynamics of property, ownership, and space.

By the 19th century, definitions and practices of space became more narrowly rendered. Public space or "commons" were now shared and created for open usage. Private space was individually or corporately owned. Physical arrangements became socially and legally constructed. Pursuant to what we now call zoning, areas were identified for production or housing or recreation and entertainment. Organization, limitations, and behaviors were imposed by law to create acceptable public conduct and prevent, for example, drunkenness, loitering, and "unlawful" assemblies.

The advancement of industrial production and evolution of cities in the late 19th and early 20th centuries further redefined space arrangements. Parks and open spaces were now accompanied by thoroughfares, walkways, bridges, walls, and tunnels. A two-tiered urban environment was created where the privileged class was able to live apart from the proletariat. Midway through the 20th century, space allotment became more refined with the advent of the suburb, which allowed the more privileged, mostly white population to live, be educated, and shop within planned and comfortable isolation, away from the city's less fortunate. Late 20th-century gentrification has altered, perhaps reversed, the suburban construct, as the inner city is now being reclaimed by the middle classes.

URBAN SPACE AND SCHOOLS

Urban school "reform" is profoundly impacted by transformed urban spaces. Mike Davis (1990, 2006) among others, has written extensively about changing urban spaces and the accompanying dynamics of power and race. Downtown areas in major metropolitan areas are the epicenter of space reorganization. Their spaces are policed by private security guards instructed to exclude "undesirables" such as people of color, the poor, and the homeless as well as uninvited activity, such as loitering. Political protests are problematic for the downtown core area where valuable assets and the seat of local government usually reside.

"Good" schools are central to the plans for gentrification, as white middle-class parents will not tolerate substandard schools. Gentrification and privatization are wedded. The privatization of public space also represents a more fundamental elite agenda. Privatization of the public realm substitutes the private corporation for public institutions as the repository of ownership, trust, legitimacy, and common identity in our society.

The corporation now replaces the community, as it dominates the economic, political, and spacial order. The privatization of public space is part of a pattern, which includes the privatization and corporate invasion of public schools and the proposed privatization of Social Security, housing, healthcare, and other institutions indispensable to the needs of the people.

HEALTHCARE

While healthcare occupies limited space in this volume, the editors and authors recognize its importance in this demographic and political moment. Healthcare festers as one of the nation's most explosive issues. Millions of people walk around with decaying teeth, uncorrected eyesight, and in throbbing pain. It is where the inequities in our everyday lives become magnified. If you are a little sick, they charge you a little. If you are very sick, they charge you a fortune. It is capitalism at its worst. How can the United States be the only industrial country without state-sponsored healthcare?

America is a sad story of quality healthcare for the affluent and minimal service for the poor and propertyless. Essays in this volume enlighten us about salient issues in this field, including physician shortages and cultural insensitivity among others. Those of us who have had to endure long stretches without health insurance must conclude that any society that commodifies healthcare must be taken to task.

PUBLIC SPACE, PEOPLE, DEMOCRACY, AND PROTEST

(Re)Situating people in America is connected to larger social, economic, and political issues. Commerce in urban areas relies on the social imprisonment of poor people of color to staff the service industry. Traditional living space configurations, that is, ghettos and barrios, are being altered in some cities, while the demand for service jobs remains. Housing the working poor has become a thorny issue as many are driven out of the central cities.

The "urban renaissance" has become the urban nightmare. As public amenities shrink with the closing of libraries and playgrounds, the neglect of parks, and the desolation of increasingly dangerous streets, public resources are being diverted for developments that further privatize public space and subsidize new exclusive enclaves (Davis, 1990; 2006).

Beyond the obvious segregating of people by race and class, the power politics of public space aims to limit, redefine, and disempower citizen participation. In fact, democracy itself is at issue in the ongoing contest of the peoples' rights versus private interests. The transformation of public space into corporate preserves is an attempt by powerful elites to erase from our minds a consciousness of ourselves as people who own the government. The disappearance of public space renders people less and less able to communicate with their fellow citizens. Instead of the government fearing the people as it should, people fear the government and big wealth. Apologists counter-claim that spaces are "public" as long as anyone can enter.

The owners of wealth pursue changes in government, law, and the State that will enable them to hold on to and increase what they have in the face of increased economic instability and the threat of social and political challenge. Those who must work to live desperately hang on to what jobs they can get and hope for better times. The epochal shift from industrialization, with production and distribution based on wage labor, to a system based on laborless digitized production is at the root of the social and political changes affecting individuals, organizations, institutions, and governments—the world as a whole.

With economic, social, and political power, the capitalist class is struggling over the process of making the changes in society and government that will shore up its ability to make a profit and retain its increasingly unstable control. Ironically, each attempt to institute stability introduces further instability into the system. Government bailouts encourage risk and anger the populace. Printing money stimulates the economy (although less than hoped) but risks inflation. Wars send a message of power and support the military-industrial complex but foster instability at home and abroad. We are entering into a period in which the economic foundation of society is conflicted. Automated production polarizes society with unheard-of wealth at one pole and unheard-of poverty at the other.

CONCLUSION

The poverty–wealth gap has never been as apparent as it is now, given the potential of abundance made possible by the microchip. Anyone can walk into a Costco or Sam's Club and find commodities stacked to the ceiling. The problem of scarcity has been solved. Now we need the wherewithal to obtain these goods. Worldwide, the richest 0.5% owns over a third of the world's wealth. Data from tax returns show that the top 1% of households in the U.S. received 9% of all pre-tax income in 1976. By 2008, the top 1% share had more than doubled to 21%. (See www.incomeinequality.org)

Within this process speculative capital exacerbates economic instability and, thus, social and political instability. It accelerates the increasing polarity between wealth and poverty and intensifies the severity of the global crisis.

The growing economic and social inequality spells danger for the ruling class, making authoritarianism a necessity for them. The ruling class cannot continue to allow the growing number of propertyless to make decisions that affect their wealth and privileges. We must pay close attention to the merger of the political state and the corporations. The political aspects of repression are in place with such laws as "Stand Your Ground," the Patriot Act, and the National Defense Authorization Act alongside a social culture that disregards the cheapening of life, suffering, and displacement.

REFERENCES

Aristotle (2010, August 28). *Politics: A treatise on government: A powerful work by Aristotle*. Available from http://www.amazon.com/Politics-Treatise-GovernmentPowerfulAristotle/dp/1453780173/ref=sr_1_1?s=books&ie=UTF8&qid=1299535951&sr=1-1#reader_1453780173

Battle, J. A. (2008). Property Rights and Responsibilities in the Old Testament. *WRS Journal, 15*(1), 15–27. Retrieved from http://www.wrs.edu/Materials_for_Web_Site/Journals/15-1_Feb-2008/Battle--Property_Rights_Responsibilities_in_OT.pdf

Bush, G. W. (2004). *President's remarks to National Association of Home Builders*. Columbus, Ohio: October 2.

Davis, M. (1990). *City of quartz: Excavating the future in Los Angeles*. London, UK: Verso.

Davis, M. (2006). *Planet of slums*. London, UK: Verso.

Hall, H. R., Robinson, C. R., & Kohli, A. (2014). *Uprooting urban America: Multidisciplinary perspectives on race, class and gentrification*. New York, NY: Peter Lang.

Harvey, D. (2005). *A brief history of neoliberalism*. New York, NY: Oxford University Press.

Lipman, P. (2011). *The new political economy of urban education: Neoliberalism, race, and the right to the city*. New York, NY: Routledge.

Locke, J. (1690). *Two treatises of government, the second treatise*. Retrieved from http://www.constitution.org/jl/2ndtreat.htm

Machan, T. R. (2002). *The right to private property*. Stanford, CA: Hoover Institution on War, Revolution and Peace, Stanford University. Retrieved from http://www.hoover.org/publications/monographs/27164

Madison, J. (1792). Property. *Federalist papers*. Chapter 16, 266–68. doi: 2329 or http://press-pubs.uchicago.edu/founders/documents/v1ch16s23.html

Meyer, E. (2011). http://articles.chicagotribune.com/2011-02-08/news/ct-met-aclu-surveillance-cameras-20110208_1_s

Watkins, W. H. (Ed.). (2012). *The assault on public education*. New York, NY: Teachers College Press.

Introduction

Understanding Gentrification and the Recolonization of U.S. Urban Space

HORACE R. HALL

Shifts in America's socioeconomic geography have been documented since the 1960s, showing the reversal of White flight and a reshaping of a nation through 2050. Evidenced is the splaying divide between underprivileged citizens and those of the wealthier stratum. As state and local governments continue to scale back social services that impact health and well-being, how will disenfranchised groups fair in this expanding market-driven global society? Who will have unfettered civilian access and who will have their rights boldly compromised? Under the interminable doctrine of Manifest Destiny, whose welfare and security will matter most as we progress through the second decade of this century and beyond?

The rise of demographic shifts has been most palpable in recolonized urban sectors. In cities like New York, Boston, Chicago, Atlanta and Los Angeles, urban renewal—by way of housing (location, variety, use and reuse) and business development—has been used to reduce concentrated poverty and slow the migration of middle-class families to the suburbs by enticing and keeping these groups in central metropolises (Feagin & Parker, 1990; Freeman, 2006; Thompson, 2001). Gentrification, as one component in the complex engine of urban renewal, has served as a powerful tool in shaping demographic landscapes within the United States and abroad (Harvey, 2009; Smith, 1996, 2002).

Nationwide, gentrification has shown to alter the social and economic fabric of communities by restoring depleted infrastructures with thriving industries and high-income households and retail services—restaurants, grocery and clothing stores, galleries and other businesses, which accommodate wealthy consumers (Hyra, 2008; Solnit, 2001). Gentrified areas are transformed into "new communities" (structurally and socially) primarily through private-market expenditures, which function to regenerate local tax bases and improve public goods and services (Millington, 2011; Wittberg, 1992).

Debatable outcomes of gentrification include the "pushing out" and "keeping out" of urban poor and minority occupants in order to add to the attractiveness of city living (Smith, 2002). Researchers argue that, while some of these tenants are retained in metropolitan spaces, larger proportions are resettled in suburban and rural regions (Atkinson & Bridge, 2005; Marcuse, 1999). Thus, gentrification partly reduces inner-city poverty by removing minority and low-wage earners from the urban core, rather than equalizing the distribution of social wealth. Moreover, this process has been criticized for shrinking resources (i.e., small businesses, jobs and services) for underserved groups and intensifying polarization between displaced citizens and affluent ones, cultivating a new kind of segregation based on class (DeSena, 2009; Hanna, 2010; Zukin, 2011).

So what are the implications of corporatizing forces on American public life? Being that the process of gentrification typically involves an initial disinvestment and then subsequent reinvestment of a given area, looking at services and opportunities within the public domain—housing, employment, healthcare and schooling— we observe multiple and interrelated consequences.

In neighborhoods undergoing revitalization, newly arriving, high-income earners inflate property and business costs (Harvey, 2009). Low-income tenants and homeowners, whose wages remain flat in the present economy, are consequently faced with the burden of paying for affluent amenities or being "priced out" of their households. With affordable accommodations, in some cities, being limited in supply, low-waged earners can lose their bids for homes to highly paid service sector workers (Ball & Vincent, 2007; Zukin, 2011). Left over is housing at the luxury end of the real estate spectrum, leading poor and working-class citizens to relocate in cheaper urban, suburban or rural dwellings (Hyra, 2008; Kennedy & Leonard, 2001).

Individuals and families who struggle financially to stay in the inner city tend to use public or private aid in the form of housing choice vouchers (HCVs), project-based subsidizing programs (Section 8), rental increase exemptions for seniors or public housing. In spite of the monetary assistance that enables them to stay, it does come at a cultural cost. Occupants can have their established community ethos and social networks fragmented in the renewal process (Crump, 2003; Hartung & Henig, 1997; Small, 2002). For those urban residents reassigned to suburban or rural areas, displacement is more than a geographical "shell game." In addition to sacrificing community culture and social capital, they may also incur significant stress from being relocated a great distance away from family and friends (Heller, 1982). Whether relocation is voluntary or not, citizens migrating to culturally and ethnically dissimilar neighborhoods may also incur culture shock—that is, feeling conflict between their own cultural identity and the dominant sociocultural framework of their new surroundings (Fullilove, 2004).

Part and parcel to housing displacement are employment viability and spatial mismatch. Over the past few decades, manufacturing industries have been decentralizing and moving along suburban beltways and more remote peripheries—if not the United States altogether (Whitford, 2005). This has shifted job prospects away from poor and working-class city workers as well as low-waged earners relocated in job-scarce suburban or rural sectors. In both cases, higher income, job-rich spaces can be difficult to access (where vehicle ownership is low) due to insufficient transit services between home and work locations (Dieleman, 2011; Midgley, Tracy, & Livermore, 1999).

While gentrification is theoretically intended to moderate spatial mismatch, we find that working-class African Americans are more geographically isolated from employment in high job-sprawl areas, despite region, city size and their percentage in metro population (Krysan, & Reynolds, 2002). The correlation among these factors also holds true for Puerto Rican and Mexican Americans, but to a lesser extent—nearly half to African Americans, respectively (Kirkland, 2008). All three groups, however, are disproportionately impacted by spatial mismatch when compared to Whites. Even though gentrification is perceived mostly as a class-dividing phenomenon, high concentrations of low-income and working-class workers, living within major U.S. cities, are minority groups (Freeman, 2006; Martinez, 2010).

Without question, where people work, go to school, eat, play and live has tremendous bearing on their overall health and well-being. Despite ongoing demographic changes in U.S. cities, there is still a paucity of information that outlines how socioeconomic alterations impact health outcomes. We do, however, recognize previous research indicating that places with more people of lower socioeconomic status (SES) tend to have higher rates of health problems: more birth defects; greater infant mortality; shorter life expectancy; higher cancer rates and incidence of asthma, diabetes and cardiovascular disease; as well as mental health issues (Barr, 2008; Braithwaite & Taylor, 2001). Environmental and consumer equity is of particular import here, as low-income people of color have historically been segregated from access to quality healthcare, whether in the public sphere or private domain with health maintenance organizations (HMOs).

For decades, land use configurations and community designs have shaped built environments that produce and perpetuate health disparities (Osypuk & Acevedo-Garcia, 2010). Where housing and industrial sites are launching poorly planned or underresourced infrastructure modification, vulnerable groups (e.g., children, women, the elderly and low-SES minorities) living within those vicinities can be overexposed to environmental hazards such as toxic waste, noise pollution and poor air quality (Corburn, 2009). Also distinctive to these settings are inadequate medical facilities or an absence of them, requiring residents to journey long distances to receive appropriate services (Barr, 2008). Urban renewal has the potential to eliminate these structural barriers, but only if space and economic capital are set aside to do so.

Obstacles to quality healthcare are protracted further when low-SES and minority populations encounter health and social service providers who have limited understanding of the problems and needs of culturally diverse groups. Over the last 20 years, cultural competency training has been at the forefront of health provider education, given multicultural shifts—via gentrification and immigration—in U.S. demographics (Rundle, Carvalho, & Robinson, 2002). Resources for this training continue to be essential for confounding some of the underlying assumptions and biases practitioners might hold toward their patients with respect to cultural values and expectations, language and communication styles, technology accessibility (online services) and differing opinions on ability versus disability (White & Chanoff, 2011).

Many of the social and economic concerns observed in healthcare, housing and employment are also reflected in the field of public education. As learning institutions labor over financial resources, some states and districts have been looking to the private sector for economic capital to help improve school efficiency and educational quality. Commercial involvement within public education has resulted primarily in the form of "chain store" charter schools and educational maintenance organizations (EMOs). As some of these institutions are partly subsidized through local property taxes, they siphon already scarce capital away from public schools, even as the necessity for services can be identical across districts (Hankins, 2007). Limited resources, coupled with low academic performance, places public schools in the precarious position of being shut down. Students attending closing schools are often transferred (within the same district) into a cache of overcrowded ones or may opt to enroll into a charter or EMO (André-Bechely, 2005; Lipman, 2004).

As schools tend to gentrify along with their neighborhoods, there is a visible change in student demographics with respect to race and class stratification. Suburban areas that were large receptacles for White middle-class flight in the 1960s are confronted by a reverse pattern of newly arriving minority and low-income children into neighborhood schools (Welch, Sigelman, Beldsoe, & Combs, 2001). Reminiscent of the White flight phenomenon, some suburban parents have been electing to send their children to private institutions or public ones outside of areas designated for relocated urban folks. In other instances, suburban families are moving out entirely, either to the city or other regions, and taking their social and economic assets with them (Brantlinger, 2003; Welch et al., 2001).

As in suburban areas, social avoidance also occurs within metropolitan spaces, yet with a distinct inverted trend. There are those affluent and White gentrifiers who choose to keep their children out of neighborhood schools until there are signs of improved institutional quality (Brantlinger, 2003; DeSena & Ansalone, 2009). Restructuring of this nature usually entails overhauling or shutting down a low-achieving school or, in other cases, building a new facility from scratch. In the course of gentrification, low-income African American and Hispanic students (who make up the vast majority of the student body in city schools) might attend these newer institutions, but only momentarily. Ultimately, they will be mandated to leave if their families relocate to other parts of the city or state.

With poor and low-income minorities gradually being resegregated and amassed into urban spaces, as well as smaller suburban and rural pockets, public schools are increasingly being faced with meeting the educational and social needs of a diverse student body. In major U.S. cities, Black and Latino/a students make up over 80% of the total student population. And that same percentage of youth is qualified for federal free and reduced lunches. Conversely, we find that more than 70% of teachers nationwide are White and middle class (Kopetz, Lease, & Warren-Kring, 2006). Given this institutional composition, White educators are likely to teach students who are of a different race, ethnicity or SES than their own. Education scholars have noted that some of the more inexorable issues confronting teachers presently relate to language barriers, knowledge of student backgrounds and value systems and managerialist paradigms of accountability that involve excessive curriculum standardization (Capps et al., 2005; Forrest & Alexander, 2004; Giroux, 2001).

The last feature of gentrification that this volume explores is community responses. In light of social and economic disparities intensified by urban renewal and the privatization of public space, engaging community members negatively impacted by these phenomena proves vital. Policies and decisions related to land use configuration, housing, healthcare, education, industry and transportation mapping are largely made by affluent parties and venture capitalists. Seldom are their determinations reflective of the concerns of an entire community, but rather in the best interest of realtors, bankers, developers, architects and politicians (Durington, Maddox, Ruhf, Gass, & Schwermer, 2009; Harvey, 2009).

Arguably, gentrification and urban renewal are processes that, by and large, systematically disenfranchise low-income and poor citizens, as they deny them voice in our participatory democracy. The ostracism of these groups, undoubtedly, perpetuates class distinctions and an imbalance of power that maintain the status quo in both our society and world. Public agencies and grassroots community organizations that realize the ethical standard of public dialogue and authentic participation have been working together to bring fairness and inclusion to the process of urban renewal. Some of the strategies they are employing include the following: educating community members of the pros and cons of gentrification and their rights, contesting zoning changes reserved for private investors, developing community land trusts and co-ops, organizing tenants to prevent the deterioration or demolition of public and subsidized housing and using technology to highlight issues of environmental injustice, racism and classism (Smock, 2003; Theodore & Martin, 2007).

Uprooting Urban America is an edited volume that comprises junior and senior scholars from multiple disciplines. The first goal of this anthology is to examine and assess the ways in which American society has been managing current and anticipated changes in social geographies. There are presently several books that look at underlying policy and market causes of unequal distributions of income and opportunities—for example, *The Gentrification Reader* by Lees, Slater and Wyly (2010), *Social Justice and the City* (2009) by Harvey and *The Right to the City: Social Justice and the Fight for Public Space* (2003) by Mitchell. *Uprooting Urban America* parallels these works, yet offers a more comprehensive analysis of gentrification vis-à-vis housing, healthcare services, education and community organizing.

Previous literature exploring urban renewal typically highlights theories or solutions or a combination of both. Hence, the second goal of this work is to add to existing theories and contemporary research findings, while also presenting innovative strategies for contesting many of the socioeconomic challenges posed by privatization and spatial inequality. Conceptual frameworks, community endeavors and best practices for contending with the aforementioned are discussed by contributing authors throughout each section.

The third and final objective of *Uprooting Urban America* is based on the understanding that its audience will include a range of professionals, practitioners and college-level students from the disciplines of education, psychology, sociology, cultural studies, public policy, urban planning and healthcare and human services. With that in mind, this volume serves to compel readers to genuinely reflect on, if not alter, their perceptions of marginalized groups as they engage in research and practice (and the implications of both) moving further into the 21st century. As more than mere academic text, the essays presented here are meant to inform and arouse us to politically regressive forces that are destabilizing the public sphere and undermining the legal and human rights of not only poor and low-income minorities but all U.S. inhabitants.

REFERENCES

André-Bechely, L. (2005). Public school choice at the intersection of voluntary integration and not-so-good neighborhood schools: Lessons from parents' experiences. *Educational Administration Quarterly, 41,* 267–305.

Atkinson, R., & Bridge, G. (2005). *Gentrification in global context: The new urban colonialism*. London, UK: Routledge.
Ball, S., & Vincent, C. (2007). Education, class fractions and the local rules of spatial relations. *Urban Studies, 44*(7), 1175–1189. Retrieved from http://usj.sagepub.com/content/44/7/1175
Barr, D. A. (2008). *Health disparities in the United States: Social class, race, ethnicity, and health*. Baltimore, MD: The Johns Hopkins University Press.
Braithwaite, R. L., & Taylor, S. E. (Eds.). (2001). *Health issues in the Black community* (2nd ed.). San Francisco, CA: Jossey-Bass.
Brantlinger, E. (2003). *Dividing classes: How the middle class negotiates and rationalizes school advantage*. New York, NY: Routledge.
Capps, R., Fix, M., Murray, J., Ost, J., Passel, J. S., & Herwantoro, S. (2005). *The new demography of America's schools: Immigration and the No Child Left Behind Act*. Washington, DC: The Urban Institute.
Corburn, J. (2009). *Toward the healthy city: People, places, and the politics of urban planning*. Cambridge, MA: The MIT Press.
Crump, J. R. (2003, May). The end of public housing as we know it: Public housing policy, labor regulation and the US city. *International Journal of Urban and Regional Research, 22*(1), 179–187.
DeSena, J. (2009). *The gentrification and inequality in Brooklyn: New kids on the block*. Lanham, MD: Rowman & Littlefield.
DeSena, J. N., & Ansalone, G. (2009, September). Gentrification, schooling and social inequality. *Educational Research Quarterly, 33*(1), 61–76.
Dieleman, F. M. (2011). Modeling residential mobility; a review of recent trends in research. *Journal of Housing and the Built Environment, 16*(3–4), 249–265.
Durington, M., Maddox, C., Ruhf, A., Gass, S., & Schwermer, J. (2009). Civic engagement and gentrification issues in metropolitan Baltimore. *Metropolitan Universities, 20*(1), 101–114.
Feagin, J., & Parker, R. (1990). *Building American cities: The urban real estate game* (2nd ed.). Upper Saddle River, NJ: Prentice Hall.
Forrest, M., & Alexander, K. L. (2004, Fall/Winter). The influence of population demographics: What does it mean for teachers and teacher education? *Journal of Family and Consumer Sciences Education, 22*(2), 67–73.
Freeman, L. (2006). *There goes the "hood": Views of gentrification from the ground up*. Philadelphia, PA: Temple University Press.
Fullilove, M. (2004). *Root shock: How tearing up city neighborhoods hurts America, and what we can do about it*. New York, NY: Ballantine.
Giroux, H. A. (2001). *Theory and resistance in education: Towards a pedagogy for the opposition*. Westport, CT: Bergin & Garvey.
Hankins, K. B. (2007, March). The final frontier: Charter schools as new community institutions of gentrification. *Urban Geography, 28*(2), 113–128.
Hanna, R. R. (2010). *Afraid of the darks: The gentrification of Shaw*. Retrieved from http://www.lulu.com
Hartung, J. M., & Henig, J. R. (1997, January) Housing vouchers and certificates as a vehicle for deconcentrating the poor. *Urban Affairs Review, 32*(3), 403–419.
Harvey, D. (2009). *Social justice and the city: Geographies of justice and social transformation*. Athens: The University of Georgia Press.
Heller, T. (1982). The effects of involuntary residential relocation: A review. *American Journal of Community Psychology, 10*(4), 471–492.

Hyra, D. S. (2008). *The new urban renewal: The economic transformation of Harlem and Bronzeville*. Chicago, IL: The University of Chicago Press.

Kennedy, M., & Leonard, P. (2001). *Dealing with neighborhood change: A primer on gentrification and policy choices*. Washington, DC: Brookings Institution.

Kirkland, E. (2008). What's race got to do with it? Looking for the racial dimensions of gentrification. *Western Journal of Black Studies, 32*(2), 18–30.

Kopetz, P. B., Lease, A. J., & Warren-Kring, B. Z. (2006). *Comprehensive urban education*. Boston, MA: Pearson.

Krysan, M., & Reynolds, F. (2002). The residential preferences of blacks: Do they explain persistent segregation? *Social Forces, 80*(3), 937–980.

Lees, L., Slater, T., & Wyly, E. (Eds.) (2010). The gentrification reader. New York, NY: Routledge.

Lipman, P. (2004). *High stakes education: Inequality, globalization, and urban school reform*. New York, NY: Routledge.

Marcuse, P. (1999). Comment on Elvin K. Wyly and Daniel J. Hammel's "Islands of decay in seas of renewal: Housing policy and the resurgence of gentrification": Gentrification resurgence. *Housing Policy Debate, 10*(4), 789–797.

Martinez, M. J. (2010). *Power at the roots: Gentrification, community gardens, and the Puerto Ricans of the Lower East Side*. London, UK: Lexington Books.

Midgley, J. O., Tracy, M. B., & Livermore, M. M. (1999). *The handbook of social policy*. Thousand Oaks, CA: Sage.

Millington, G. (2011). *"Race," culture and the right to the city: Centres, peripheries, margins*. London, UK: Palgrave Macmillan.

Mitchell, D. (2003). *The right to the city: Social justice and the fight for public space*. New York, NY: Guilford Press.

Osypuk, T. L., & Acevedo-Garcia, D. (2010, November). Beyond individual neighborhoods: A geography of opportunity perspective for understanding racial/ethnic health disparities. *Health & Place, 16*(6), 1113–1123.

Rundle, A., Carvalho, M., & Robinson, M. (Eds.). (2002). *Cultural competence in health care: A practical guide* (2nd ed.). San Francisco, CA: Jossey-Bass.

Small, M. L. (2002, July). Culture, cohorts, and social organization theory: Understanding local participation in a Latino housing project. *American Journal of Sociology, 108*(1), 1–54.

Smith, N. (1996). *The new urban frontier: Gentrification and the revanchist city*. New York, NY: Routledge.

Smith, N. (2002). New globalism, new urbanism: Gentrification as global urban strategy. *Antipode, 34*(3), 427–450.

Smock, K. (2003). *Democracy in action: Community organizing and urban change*. New York, NY: Columbia University Press.

Solnit, R. (2001). *Hollow city: Gentrification and the eviction of urban culture*. New York, NY: Verso.

Theodore, N., & Martin, N. (2007, August). Migrant civil society: New voices in the struggle over community development. *Journal of Urban Affairs, 29*(3), 269–287.

Thompson, H. A. (2001). *Whose Detroit? Politics, labor, and race in a modern city*. Ithaca, NY: Cornell University Press.

Welch, S., Sigelman, L., Beldsoe, T., & Combs, M. (2001). *Race and place: Race relations in an American city*. New York, NY: Cambridge University Press.

White, A. A., III & Chanoff, D. (2011). *Seeing patients: Unconscious bias in health care*. Cambridge, MA: Harvard University Press.

Whitford, J. (2005). *The new old economy: Networks, institutions, and the organizational transformation of American manufacturing.* New York, NY: Oxford Press.
Wittberg, P. (1992). Perspectives on gentrification: A comparative review of the literature. *Research in Urban Sociology, 2,* 17–46.
Zukin, S. (2011). *Naked city: The death and life of authentic urban places.* New York, NY: Oxford University Press.

PART ONE

Urban Renewal AND THE Marriage OF Civic Government AND Free Market

What is the social and cultural cost of urban renewal? This question might best be answered by turning to the neighborhood of Bronzeville—a predominantly African American community located near downtown Chicago. Proclaimed the "Black Metropolis" at the beginning of the 20th century, Bronzeville served as home for African American citizens of virtually every socioeconomic status between 1910 and 1930. Just after World War II, however, the area witnessed decades of reduced capital investment, as local businesses shut down and property owners and renters migrated to other sectors of the city. By the 1980s, this once flourishing community had become deeply tarnished by abandoned buildings, failing public schools, gun and gang violence and other social problems relative to residential poverty. It was not until the mid-1990s that the area became a site for redevelopment. The process was strikingly earmarked between 1998 and 2007 with the demolishment of the Robert Taylor Homes—then one of the nation's largest public housing projects. In its stead are now a series of low-rise, mixed-income apartments, refurbished community facilities and a small number of retail and commercial businesses. Presently, one can drive through Bronzeville and take notice of the district's expanded police station, a Starbucks coffee shop, lakefront townhomes starting at $250,000 and preliminary construction for a future Walmart store.

Today, Bronzeville is benefitting, to some degree, from economic recovery, with new homeowners and local businesses gradually seeing increases on their returns from property assets. Yet, there are those longtime occupants, specifically renters and senior citizens, who are being "squeezed out" and replaced by higher-paying customers. Therein lies the rub. Some of these residents not only value and respect the changes to their neighborhood but also have been working for decades to render such improvements. Staying in Bronzeville represents more than simply wanting to delight in the new amenities, as the existing space still comprises family members, social networks and a history of lived experiences. Again, what is the social and cultural price tag attached to urban renewal? The essays in the following section address this query more discursively through the lens of our "choice-driven" society, examining the reality of those priced-out and displaced, as well as the strategies of government agencies and free-market enterprises currently reshaping urban landscapes.

HRH

CHAPTER ONE

Relocating Gentrification

The Working Class, Science and the State in Recent Urban Research

LOÏC WACQUANT

Tom Slater's (2006) provocative essay on "The Eviction of Critical Perspectives from Gentrification Research" is a timely wakeup call for scholars of class, space and politics in the city. It points to a surprising twist and troublesome trend in recent studies of gentrification, whereby the takeover of working-class districts by middle- and upper-class residents and activities is increasingly presented wholesale as a collective good if not boon. By focusing narrowly on the practices and aspirations of the gentrifiers through rose-tinted conceptual glasses, to the near-complete neglect of the fate of the occupants pushed aside and out by urban redevelopment, this scholarship parrots the reigning business and government rhetoric that equates the revamping of the neoliberal metropolis as the coming of a social Eden of diversity, energy and opportunity. But Slater's diagnosis of the facets and causes of what might be termed *the gentrification of gentrification research* after the close of the Fordist-Keynesian era does not go deep enough and, as a result, his plea for "reclaiming the term from those who have sugarcoated what was not so long ago 'a dirty word'" (p. 737) risks falling short of its aim on both the scientific and the policy fronts.[1]

The shift from the acidic denunciation to the glib celebration of gentrification, the elision of the displacement of the established residents of the inner city of lower socioeconomic standing, the bland focus on "social mixing" and euphemistic invocation of "residentialization" are not isolated developments peculiar to the study of neighborhood upgrading. They partake of a broader pattern of *invisibility of the working class* in the public sphere and social inquiry over the past two decades. This literal and figurative effacing of the proletariat in the city is reinforced by the *growing heteronomy of urban research*, as the latter becomes ever more tightly tethered to the concerns and outlook of city rulers and correspondingly unmoored from self-defined and self-propelled theoretical agendas. And both tendencies in turn reveal, confirm and abet the *shifting role of the state* from provider of social support for lower-income populations to supplier of business services and amenities for middle- and upper-class urbanites—chief among them the cleansing of the built environment and the streets of the physical and human detritus wrought by economic deregulation and welfare retrenchment so as to make the city over into a pleasant site of and for bourgeois consumption (Wacquant, 2009). I take each of these issues in turn, with a view toward sharpening as well as amplifying Slater's exhortation to critical reflexivity in gentrification research.

THE VANISHING OF THE WORKING CLASS IN THE PUBLIC SPHERE AND URBAN RESEARCH

Any rigorous study of gentrification would seem *ex definitionis* to hold together the trajectories of the lower-class old-timers and of the higher-class newcomers battling over the fate of the revamped district, since this class nexus forms the very heart of the phenomenon (Glass, 1964; Lees, Slater, & Wyly, 2008). Yet Slater (2006) reported that the social and spatial dislocations caused by neighborhood upgrading have virtually vanished from recent research, and he argued that "the reason why displacement itself got displaced" is essentially "methodological" (p. 748). But the physical absence of those dislodged from the neighborhood hardly explains why researchers do not widen their observational scope to capture their peregrinations through urban space or resort to different methodologies (such as tracking a panel of former inhabitants or drawing out extended life stories) to document the housing turbulence brewing at the bottom of the urban structure. Certainly, these "methodological" obstacles did not stump the founding generation of gentrification scholars! The empirical evaporation of working-class exiles from the literature on renovated urban quarters is not the result of defective research techniques: It mirrors the objective fragmentation of the industrial working class, in the historic incarnation in which we have known it during the long century (1870–1970) of industrialism, climaxing with the conjoint maturation of the Fordist production regime and the Keynesian state and its correlative marginalization in the political and intellectual fields.

With deindustrialization and the shift to deregulated service employment, the spread of mass unemployment and work instability and the universalization of schooling as means of access to even unskilled jobs, the unified and compact working class that occupied the front stage of history until the 1970s has shriveled, splintered and dispersed. Together with low-level service employees, workers continue to compose a majority of the active population of most advanced countries (Marchand & Thélot, 1997; Wright, 1997), but their morphology has been remade by deepening divisions of skills, employment status and reproduction strategy, as well as by spatial scattering. Many working-class households have left public housing estates, entered tract housing or migrated outside the city in search of cheaper lodging. More crucially, these morphological changes have been accompanied by *collective demoralization and symbolic devaluation* in civic and scientific debate, as unions declined and left parties moved rightward. The educated middle classes and the knights of the financial, cultural and technological sectors that drive neoliberal capitalism now occupy both the cultural and the electoral center; their views and aspirations dominate public discussion and orientate the actions of politicians and government alike. There are workers, to be sure, but the working class as such is unfashionable, inscrutable, unnoticed if not invisible.[2]

Instead of tracking the mechanisms and modalities of *class decomposition* and its spatial correlates with the zest with which they tackled class consolidation and conflict in an earlier era, to plumb how deproletarianization and casualization are moulding the emerging urban proletariat of the turn of the century, researchers have turned away from it. Accordingly, the classic studies of "traditional working-class neighborhoods" dissected by Topalov (2003) have disappeared, to be replaced by inquiries into ethnicity and segregation, on the one hand, and urban poverty and street crime on the other. For every book on a lower-class district focusing on social structure and everyday life among workers (such as Kefalas, 2003; Schwartz, 1990 there are a dozen centering on racial isolation, ethnic tension and cultural succession (e.g., Hartigan, 1999; Sharman, 2006; Small 2004; Wilson & Taub, 2006), and another dozen on immigration, violence and the underground economy (Bourgois, 1995; Lepoutre, 2005; Smith, 2005; Venkatesh 2006). At the foot of the metropolitan order, the language of class has been supplanted by the tropes of the "underclass" in the United States and "exclusion" in Western Europe, wherever working-class neighborhoods have undergone involution, and by the theme of "regeneration" and "renaissance" in those areas taken over by higher classes migrating back into the dualizing city. When gentrification researchers ignore the tribulations of working-class residents displaced by rising rents, shrinking housing options and state policies supporting business development and middle-class settlement, they are only following the general pattern of class blindness by urban researchers even as class inequalities sharpen before their very eyes.[3]

The Growing Heteronomy of Urban Inquiry

The breakup of the industrial working class is not the only cause for its virtual vanishing from social inquiry and for the reincarnation of its established territories in the figures of the "ethnic ghetto" and the defamed district of vice and violence

(Wacquant, 2008). Another major factor at play here is the growing subservience of urban research to the concerns, categories and moods of policymakers and opinion-makers.

Twenty-odd years ago, inquiries into class and culture in the city were stamped by the battles of the theoretical schools vying for intellectual dominance: human ecology, Marxism, Weberian political economy and an insurgent culturalist current fed by identity theories, feminism and postmodernism (Hayden, 1986; Logan & Molotch, 1987; Walton, 1990). But in the new climate of political disenchantment and state withdrawal spawned by the concurrent collapse of the Soviet Union and rising hegemony of neoliberalism, intellectual radicalism receded and then seceded from reality. The "false promises of Marxism" and "mirage of the cultural turn"—to borrow the words of Michael Storper (2001)—left a gaping theoretical void, which was quickly filled by the prosaic attractions of conducting research on topical issues and the press of finding funding.[4] Nowadays, urban research is guided primarily by the priorities of state managers and the worries of the mainstream media. A panorama of recent sociological investigations into "The Texture of Hardship" in the American metropolis opens on this note:

> The decade that spans 1995–2005 saw many new avenues of research develop among qualitative sociologists interested in poverty. Welfare reform, passed by the U.S. Congress in 1996, turned attention to the world of low-wage work as it began to dawn on journalists and researchers alike that poverty was as much, if not more, a matter of inadequate earnings as it was a consequence of welfare dependency. Employers emerged as actors whose expectations, normative orientations, and cultural distance from the low-wage world play a powerful role in the sorting of job seekers into success stories and excluded failures. Researchers gave renewed attention to patterns of family formation among the poor. (Newman & Massengill, 2006, p. 423)

"Welfare reform turned attention…": This capsule speaks volumes about how political developments and the funding bandwagons they create drive the intellectual agenda. In the 1980s, the "underclass" had come to monopolize the attention of American researchers by seducing first philanthropic foundations and then journalists and policymakers with its loathsome moral and racial connotations (Katz, 1989). After 1996, it was summarily dismissed overnight from the scholarly stage without remonstration to make room for studies of the valiant low-wage earners making the transition from "welfare to work," the families that support them, the employers that rush or balk at hiring them and the bureaucrats who supervise their career (contrast, for instance, Jencks & Peterson, 1991, with Hays, 2003). In the European Union, Brussels's Targeted Socio-Economic Program on Exclusion and Integration similarly drew researchers away from the study of mass unemployment and its spatial impact toward the new bureaucratic problematic of "exclusion" and "integration." In France, the Netherlands, Germany and Belgium, political tensions around postcolonial immigration and the deterioration of public housing have fueled a wave of studies and policy evaluation programmes on "neighborhood mixing," "community-building" and crime-fighting centered on working-class neighborhoods but studiously avoiding the socioeconomic underpinnings of urban degradation, in keeping with the design of politicians to deploy territory, ethnicity and insecurity as screens to obscure the desocialization of wage labor and its impact on the life strategies and spaces of the emerging proletariat (Wacquant 2006).

So when gentrification researchers offer rosy accounts of neighborhood "renewal" as an urban "solution" to the ills of sociospatial decay, in lockstep with the views of government and business elites, they walk in good company: Theirs is only an acute case of the common malady of heteronomy that afflicts with growing virulence large sectors of social research in general and urban research in particular.[5]

THE STATE AS HOMEMAKER AND STREET-CLEANSING AGENCY

It is revealing that the 26-volume *International Encyclopedia of Social & Behavioral Sciences* (2004), edited by Neil Smelser and Paul Baltes (2004), does not contain an entry on "gentrification." The term appears as a subtheme under the entries "Neighborhood Revitalization and Community Development" and "Neighborhood," where one reads as follows:

> Within the group of *public–individual partnerships,* the most salient phenomenon is gentrification, the entrance of upper middle-class residents into low-income neighborhoods, usually in the vicinity of vibrant central business districts of cities. In spite of the frequently found and often denounced consequence of displacement of poor veteran residents, gentrification processes have been encouraged by city councils in Europe and America, especially through enabling regulations and tax deductions. Another process in this group is housing and neighborhood upgrading by incumbent residents. Local people invest their own resources in improving their living environment, and often manage to receive at least some assistance from voluntary and public bodies. Last but not least is neighborhood upgrading by immigrants.... What is common to the three processes is that they usually start as spontaneous private investments, which later receive some support from local public agencies. Together they are changing parts of our old cities and neighborhoods. (Carmon, 2004, p. 10493)

I cite this article, not only because it confirms Slater's main thesis, but because, while it mentions the role of public agencies, it vastly underestimates its timing, scope and effects. It is high time students of gentrification recognize that the primary engine behind the (re)allocation of people, resources and institutions in the city is the state.

Slater (2006, pp. 746–747) listed among the causes of the eviction of critical perspectives on gentrification "the resilience of theoretical squabbles" that have stalled and sterilized debate. But the ritualized opposition between Neil Smith's economic explanation and David Ley's culturalist take that Slater recapitulates as the dominant theoretical trope of gentrification research is problematic for what it leaves out: politics, policy and the state. The "rent-gap thesis" favored by neo-Marxist analysts, the "cultural distinction" approach adopted by neo-Weberian or postmodernist scholars (who invoke the phraseology of Bourdieu just as quickly as they disregard his theoretical principles) and the globalization thesis inspired by Saskia Sassen all leave out the crucial role of the state in producing not only space but the space of consumer and producers of housing. Logan and Molotch (1987) were right to insist that place is not a regular commodity but a battleground between use and exchange value. But they did not go far enough in their specification of the parameters of that battle and, in keeping with the national U.S. common sense, they grossly underestimated the weight of Leviathan in it. Pierre Bourdieu ([2001] 2005, pp. 30–31) has shown in *The Social Structures of the Economy* that housing is "the product of a double social construction, to which the state contributes crucially," by shaping the universe of builders and sellers via fiscal, banking and regulatory policies on the economic side and by moulding the dispositions and capabilities of house buyers (including the propensity to rent or buy) on the social side.

This double state structuring of the housing "market" is then trebled by the political steering of urban and regional planning, however weak its agencies may be. For, as Tedd Gurr and Desmond King (1987, p. 4) reminded us two decades ago, "those who hold and use state power *can* allow the fate of cities to be determined mainly in the private economy, but that is a matter of public choice rather than iron necessity." The weight of the central and local state is all the more decisive in lower-class neighborhoods, insofar as workers and the poor are most dependent on public provision to access social rented housing (Harloe, 1995). But the role of the state in gentrification hardly stops at building and distributing housing or shaping the pool of home buyers: It extends to the gamut of policies that impact urban living, from infracture maintenance to schooling and transportation, to the provision of cultural amenities and policing. Without the campaigns of aggressive policing of the streets fostered by the rolling out of the penal state in and around neighborhoods of relegation over the past decade (Herbert, 2006; Wacquant, 2008), the middle classes could not have moved into the inner city and gentrification would not have grown beyond the sprinkling of "islands of revitalization within seas of decay" (Carnon, 2004, p. 10493). More generally, the historic shift from the Keynesian state of the 1950s to the neo-Darwinist state of the *fin de siècle*, practicing economic liberalism at the top and punitive paternalism at the bottom, entails a sea change in the political framing of neighborhood upgrading. Here the literature on gentrification surveyed by Slater reproduces for those districts the general tendency of public policy to *invisibilize the urban poor*, either by dispersing them (as with the demolition and deconcentration of public housing) or by containing them in reserved spaces (stigmatized districts of perdition and the expanding prison system to which they are preferentially linked).

To build better models of the changing nexus of class and space in the city, we need to do much more than renew the critical spirit that animated the pioneers of gentrification research out of a feeling of intellectual loyalty and reverence for their political engagement: We need to relocate gentrification in a broader and sturdier analytic framework. First, we must revive and revise class analysis to capture the (de)formation of the postindustrial proletariat and inscribe the evolution of "revitalized" districts within the overall structures of social and urban space and their linked makeovers. Second, we must better resist the seductions of the prefabricated problematics of policy and advance research agendas sporting greater separation from the imperatives of city rulers and carrying a higher theoretical payload. And third, we must give pride of place to the state as generator of sociospatial inequality in the dualizing metropolis. For, much like the fate of neighborhoods of relegation that fester at the bottom of the system of places that compose the metropolis (Wacquant, 2008, pp. 283–284), the trajectory of gentrified districts in the 21st century is *economically underdetermined and politically overdetermined*. It behooves us, then, to restore the primacy of the political in our efforts to analytically dissect and practically redirect the social transformation of the neoliberal city.

NOTES

1. A more elaborate argument than can be offered here due to space limitations would discuss the analytic and the political moments separately and then proceed to link them. The reasoning would be the same with gentrification as is deployed to draw out the implications of "urban polarization from below" for social theory and public policy (Wacquant 2007, pp. 247–256).
2. For a brutal contrast with the centrality of the working class in an earlier era of research inspired by Marxism, read Katznelson's (1992, pp. 203–256) analytic account of how "Working Classes Map the City."
3. Since 2000, this journal has published all of one article featuring the words "working class" in its title (Watt, 2006), out of a grand total of nine with the word "class" (most often qualified by "middle" and its equivalent). Interestingly, among the most frequent terms in titles are global/ization, governance, exclusion, scale, social movement, network, enterprise and ethnicity, namely, the staple ingredients of "neoliberal newspeak" (Bourdieu & Wacquant 2001).
4. Milicevic's (2001) analysis of the "de-radicalization" of the New Urban Sociology of the 1960s and 1970s could be extended, with suitable qualifications (moving from the level of personal interactions to that of the structure of positions in the intellectual field) from Britain to France and the United States.
5 Of course, the tug of war between autonomy and heteronomy is itself traversed with tensions and contradictions that would need to be mapped out. But the pendulum had definitely swung

in favor of the latter. In France, for instance, sociologists of the city have moved from the "critical pole" to the "technical pole" and from academic to professional orientations (Lassave 1997, pp. 23–29): Whereas the structuralist generation defined itself by systematically subverting state demands, the current cohort has largely accompanied and even anticipated them. The mutation of the work of Jacques Donzelot, from Foucauldian critic of state discipline to advocate of the state as "animator" of neighborhood sociability, is emblematic of this collective drift.

REFERENCES

Bourdieu, P. (2005). *The social structures of the economy.* Cambridge, UK: Polity Press. (Original work published 2001).
Bourdieu, P., & Wacquant, L. (2001). Neoliberal newspeak. *Radical Philosophy, 105,* 2–5.
Bourgois, P. (1995). *In search of respect: Selling crack in El Barrio.* New York, NY: Cambridge University Press.
Carmon, N. (2004). Neighborhood: General. In N. J. Smelser & P. Baltes (Eds.), *International encyclopedia of the social & behavioral sciences* (pp. 10488–10494). Oxford, UK: Elsevier.
Glass, R. (1964). *London: Aspects of change.* London, UK: McGibbon and Kee.
Gurr, T. R., & King, D. (1987). *The state and the city.* Chicago, IL: The University of Chicago Press.
Harloe, D. (1995). *The people's home? Social rented housing in Europe and in America.* Oxford, UK: Basil Blackwell.
Hartigan, J. (1999). *Racial situations: Class predicaments of whiteness in Detroit.* Princeton, NJ: Princeton University Press.
Hayden, D. (1986). *Redesigning the American dream: Gender, housing, and family life.* New York, NY: W.W. Norton.
Hays, S. (2003). *Flat broke with children: Women in the age of welfare reform.* New York, NY: Oxford University Press.
Herbert, S. (2006). *Citizens, cops, and power: Recognizing the limits of community.* Chicago, IL: The University of Chicago Press.
Jencks, C., & Peterson, P. E. (Eds.). (1991). *The urban underclass.* Washington, DC: Brookings Institution.
Katz, M. B. (1989). *The undeserving poor: From the war on poverty to the war on welfare.* New York, NY: Pantheon.
Katznelson, I. (1992). *Marxism and the city.* Clarendon, UK: Oxford University Press.
Kefalas, M. (2003). *Working-class heroes: Protecting home, community, and nation in a Chicago neighborhood.* Berkeley: University of California Press.
Lassave, P. (1997). *Sociologists and urban research in contemporary France.* Toulouse, France: Presses Universitaires du Mirail.
Lees, L., Slater, T., & Wyly, E. (2007). *Gentrification.* London, UK: Routledge.
Lepoutre, D., with Isabelle Cannoodt. (2005). *Memories of Immigrant Families.* Paris, France: Odile Jacob.
Logan, J. R., & Molotch, H. L. (1987). *Urban fortunes: The political economy of place.* Berkeley: University of California Press.
Marchand, O., & Thélot, C. (1997). Work in France, 1800–2000. Paris, France: Nathan.
Milicevic, A. S. (2001). Radical intellectuals: What happened to the new urban sociology? *International Journal of Urban and Regional Research, 25*(4), 759–783.

Newman, K. S., & Massengill, R. P. (2006). The texture of hardship: Qualitative sociology of poverty, 1995–2005. *Annual Review of Sociology, 32*, 423–446.

Schwartz, O. (1990). *The private world of workers. Men and women of the north.* Paris, France: Presses Universitaires de France.

Sharman, R. L. (2006). *The tenants of East Harlem.* Berkeley: University of California Press.

Slater, T. (2006). The eviction of critical perspectives from gentrification research. *International Journal of Urban and Regional Research, 30*(4), 737–979.

Small, M. (2004). *Villa Victoria: The transformation of social capital in a Boston barrio.* Chicago, IL: The University of Chicago Press.

Smelser, N. J., & Baltes, P. (Eds.). (2004). *International encyclopedia of social & behavioral sciences.* Oxford, UK: Elsevier.

Smith, R. (2005). *Mexican New York: Transnational lives of new immigrants.* Berkeley: University of California Press.

Storper, M. (2001). The poverty of radical theory today: From the false promises of Marxism to the mirage of the cultural turn. *International Journal of Urban and Regional Research, 25*(1), 155–179.

Topalov, C. (2003). "Traditional working-class neighborhoods": An inquiry into the emergence of a sociological model in the 1950s and 1960s. *Osiris, 18,* 213–233.

Venkatesh, S. A. (2006). *Off the books: The underground economy of the urban poor.* Cambridge, MA: Harvard University Press.

Wacquant, L. (2006). "Ghetto, banlieues, state: Reasserting the primacy of the political." *Nouveaux Regards, 33,* 62–66.

Wacquant, L. (2008). *Urban outcasts: A comparative sociology of advanced marginality.* Cambridge, UK: Polity Press.

Wacquant, L. (2009). *Punishing the poor: The new government of social insecurity.* Durham, NC: Duke University Press.

Walton, J. (1990). Urban sociology: The contribution and limits of political economy. *Annual Review of Sociology, 19,* 301–320.

Watt, P. (2006). Respectability, roughness and "race": Neighbourhood place images and the making of working-class social distinctions in London. *International Journal of Urban and Regional Research, 30*(4), 776–797.

Wilson, W. J., & Taub, R. (2006). *There goes the neighborhood: Racial, ethnic, and class tensions in four Chicago neighborhoods and their meaning for America.* New York, NY: Knopf.

Wright, E. O. (1997). *Class counts: Comparative studies in class analysis.* Cambridge, UK: Cambridge University Press.

CHAPTER TWO

Hurting or Helping

Gentrification and African American Neighborhoods in Chicago

ADRIENNE HOLLOWAY

INTRODUCTION

While current debate on the extent of segregation in the city of Chicago continues to be unresolved among scholars and the general public, consensus does exist on the effect that public policy has on determining an area's residential composition (Glaeser & Vigdor, 2012; Seng & Caruso, 2013). Early public policy, implemented in Chicago, aided in the formation of a city of vastly segregated neighborhoods. Though recent data suggest that Chicago is less segregated than in the 1900s, current public policy, particularly public housing policy, if left unchecked, is quite capable of reestablishing Chicago as a high conglomeration of neighborhoods segregated by race and income (Forman & Krysan, 2008; Seng & Caruso, 2013), accomplished, as in past efforts, through public policy implemented in the spirit of redevelopment and positive neighborhood change.

During the early 20th century, distinct White neighborhoods evolved on the north side of Chicago near the financial, cultural and social heart of the city, while the Great Migration brought nearly 500,000 African Americans to the south and west sides of the downtown area, near manufacturing and industrial sites. Public housing was also developed along fortified residential segregated patterns. Working-class White households had access to public housing in predominately White neighborhoods and African American working-class families accessed public housing in largely African American neighborhoods. Over time, the profile of public housing residents shifted from working-class families to poor families. The compounded effect of racially segregated residential patterns, growing poor communities and race-based housing policy implementation initiated the creation of African American–dominated public housing communities in urban Chicago.

By the end of the 20th century, Chicago's neighborhood configuration was firmly segregated by race and income, to some extent, fortified by decades of misguided federal and local housing policy (Hirsch, 1998; Massey & Denton, 1988; Polikoff et al., 2009; von Hoffman, 1996). In response to failed public housing policy, new initiatives were developed in an attempt to reverse negative trends of poverty concentration, disinvestment and economic decline. Within this context, this chapter explores neighborhood change effects of specific initiatives, namely the federal HOPE VI program and the Chicago Housing Authority Plan for Transformation (PFT), on four Chicago neighborhoods (Douglas, Near North Side, Near West Side and East Garfield Park). Analyses will elucidate how these housing redevelopment programs are resulting in positive neighborhood change for some communities and their respective families but not in others. Consequently, implications of policy outcomes are discussed within the context of their affect on long-standing community residents.

NEIGHBORHOOD CHANGE: A WORKING DEFINITION

To examine the outcomes of the HOPE VI and PFT programs relative to Chicago's Douglas, Near North Side (NNS), Near West Side (NWS) and East Garfield Park (EGP) neighborhoods, a working definition of neighborhood change is warranted. Neighborhood change, within the context of gentrification, is understood to be the movement of middle-class families into revitalized neighborhoods that were previously depressed, neglected and poor (Schaffer & Smith, 1986; Wyly & Hammel, 1999). Further, communities with sought-after amenities such as access to transportation and entertainment venues are ripe for gentrification (Wyly & Hammel, 1999). Thus, proponents of gentrification emphasize its ability to "reverse [the] economic and social decline" (Schaffer & Smith, 1986, p. 349) plaguing inner-city neighborhoods by attracting families with financial resources that can be leveraged through highlighting community amenities.

Contrary viewpoints argue that gentrification facilitates the displacement of long-term residents who weathered the storm of economic decline (Freeman, 2005; Lees & Ley, 2008; Levy, Comey, & Padilla, 2006; Smith, 2001). For instance, scholars contend that an influx of middle-income persons into gentrifying neighborhoods increases property values, influencing an uptick in rental prices (Freeman, 2005; Levy et al., 2006). Accordingly, low-income residents are forced to leave their neighborhoods in order to locate affordable housing elsewhere (Levy et al., 2006). Thus, one challenge of gentrification is striking a balance between economic stability in historically neglected neighborhoods and housing affordability, particularly for low-income persons. Research has determined this to be a difficult balance to attain (Kennedy & Leonard, 2001).

Unfortunately, cities across the country are failing to achieve such a delicate balance (Bates, 2013; Henneberger, 2008; Paul, 2011), including Chicago. While it is appropriate to acknowledge Chicago leadership for taking action to improve severely neglected communities, it is also necessary to confront leadership when the results of such actions are likely to hurt people who have long suffered from its neglect. Investment in communities to produce neighborhood change must benefit the population who sustained years of divestment. In this particular instance, the African American community is of concern.

Neighborhood change, stimulated by way of public housing redevelopment activities, has a disproportionate impact on the African American community for two fundamental reasons. First, early U.S. public housing developments, as discussed in the next section, were predominately occupied by African Americans. Combined with the high African American resident population, public housing developments were also overwhelmingly built in African American neighborhoods. Consequently, when policy requires the demolition of public housing and the development of mixed-income housing, more African American households will lose their housing units and subsequently may not be able to return to revitalized neighborhoods.

Second, African American–community residents who reside in unsubsidized housing located within targeted redevelopment districts may be subjected to higher rental costs or property taxes due to increased property values resulting from mixed-income housing development and gentrification. Accordingly, they too may find it financially challenging to remain in their respective communities. Such forced relocation causes economic, social and political marginalization of African American households as it dismantles social networks, disrupts political and economic activity and weakens cultural influences and contributions (Hunter, 2011; Tavernise, 2011). Therefore, establishing if not balancing the economic stability and housing affordability elements of redevelopment activity requires (public) housing policy to be both complicated in its approach and sensitive to the needs of neighborhood constituents to avoid gentrification that displaces and disempowers African American residents. Historically, public housing policy has been devoid of these characteristics.

PUBLIC HOUSING POLICY AND NEIGHBORHOOD CHANGE

Public housing programs have strategically changed the landscape of urban neighborhoods for decades. Beginning with the Housing Division of President Roosevelt's Public Works Administration (PWA), 51 low-rise public housing projects comprised of over 25,000 units were developed in (segregated) low-income, working-class communities across the country (Cam, 1939; Hunt, 1997). Consequently, prevailing supposition of early public housing policy contended that housing developed within segregated neighborhoods would decrease blight and stabilize communities.

Moving forward in the drive to eliminate neighborhood blight, the federal government passed the Housing Act of 1949 where Title I of the act facilitated the elimination of blight, through Urban Renewal funding, by authorizing the use of eminent domain to purchase and demolish dilapidated structures. Displaced families were often forced to locate housing in other, usually poorer, neighborhoods (Groberg, 1965). Title III of the Housing Act financed the development of over 800,000 units of public housing, where the physical design shifted from bungalow style to multi-unit high-rise development, as a way to conserve funds. As a result, large public housing communities were developed in primarily poor and racially segregated neighborhoods (von Hoffman, 1996). Collectively, Title I and III housing development activity of the Housing Act of 1949 contributed to forced relocation, class and racial segregation and the progressive decline of public housing communities for decades. Evidence of the negative impact of these policies continues to exist in the current landscape of Chicago's poor neighborhoods.

Fast forward 50 years, however, and you find a new shift in public housing strategy, from owning and managing pubic housing units to developing mixed-income communities and overseeing sizeable Housing Choice Voucher (HCV) portfolios. Accepting that negligible public housing maintenance activity and growing concentrations of poverty contributed to the decline of public housing neighborhoods, the U.S. Department of Housing and Urban Development (HUD) authorized the HOPE VI program, a significant investment of federal resources, to reverse the status of public housing neighborhoods across the country.

HOPE VI PROGRAM

HOPE VI evolved as a result of recommendations from the National Commission of Severely Stressed Housing that was charged with developing a plan that would eliminate distressed public housing by 2000 (Abt Associates, Forsburg, Popkin, & Locke, 1996). What resulted was a public housing poverty-reduction program, implemented through redevelopment plans that included demolition of densely populated and severely distressed public housing complexes and the construction of several smaller and affordable housing complexes in sufficient number to house temporarily displaced families in targeted communities. Early programmatic rule changes altered implementation strategies where replacement housing was accomplished either through the issuance of an HCV[1] or a new unit (limited in number) in mixed-income communities with low-rise garden apartments and single-family (town) homes (Wexler, 2001). Thus, impoverished communities that were once the home of towering, sprawling public housing high- and low-rise developments in cities like Chicago[2] leveraged HOPE VI funds to demolish subsidized housing units.

Operating under the philosophy that poverty deconcentration and socioeconomic diversity can convert disinvested and dilapidated communities into economically stable spaces, the HOPE VI program altered the socioeconomic fabric of targeted communities. To facilitate demolition, residents of targeted housing sites are relocated to private market housing using HCV. Through its use of HCV, HOPE VI effectively diffuses low-income and poor households across a wider geography and contributes to demographic shifts of both sending and receiving communities. HOPE VI also caused additional spillover effects in other communities. Positive effects such as increased property values and commercial investment and negative spillover effects, like rising rental prices and resident displacement, occurred in HOPE VI contiguous communities as well (Goetz, 2011).

Early research of HOPE VI relocation suggests that impacted residents using HCV, on average, moved to either racially segregated or impoverished urban neighborhoods or are clustered in less poor but still tenuous neighborhoods (Comey, 2007; Goetz, 2010). For instance, displaced African American households tended to relocate to nearby disadvantaged neighborhoods (Oakley & Burchfield, 2009) that contained underperforming schools (Popkin, 2006) and limited employment opportunities aimed at increasing self-sufficiency (Clampet-Lundquist, 2006). Collectively, these neighborhood deficiencies contribute to the perpetuation of the marginalization and disempowerment of African American households.

The Chicago Plan for Transformation (PFT)

In response to the federal government's shift in public housing strategy that launched HOPE VI, the Chicago Housing Authority (CHA), developed the Chicago Plan for Transformation (PFT). PFT is an aggressive model of dismantling public housing and changing neighborhood landscapes through the creation of mixed income communities in emerging neighborhoods capable of leveraging private investment. Approved by HUD in 2000 and extended in 2008, CHA's PFT secured a commitment of over $1.6 billion in HUD funding to redevelop or rehabilitate approximately 25,000 units of public housing by the end of Fiscal Year 2018 (CHA, 2013). To date, PFT is 85% complete, having redeveloped or built 21,432 units of affordable and market-rate housing (CHA, 2013). Unfortunately, at the completion of PFT, CHA will have eliminated approximately 13,000 predominantly family-sized public housing units (replaced with HCVs; Vale & Graves, 2010). The decrease in the supply of public housing units increases the reliance on private landlords to rent their respective units to public housing families.

Over the last 10 years, HOPE VI and PFT, combined, have produced drastic changes in the aesthetics and composition of former public housing–dominated neighborhoods, and though welcomed to a degree by community residents, ongoing implementation of HOPE VI and PFT has revealed some concerns. For instance, CHA depends on HCVs to be a conduit to disperse former place-based subsidized public housing tenants to receiving communities located throughout the city of Chicago and suburban communities across the metropolitan area. Relocation to receiving communities should be temporary, where families eventually return to revitalized communities at the completion of HOPE VI activity to avoid social network disruption and feelings of isolation in new communities. However, redevelopment activity spans many years and low-income families using HCVs are compelled to consider the value of uprooting and returning to their originating neighborhoods. The cost associated with having to again sever network ties and develop new relationships in redeveloped communities, to unenroll children from schools and enroll them in new schools that require them to forge new friendships and adjust to a new school environment, and a whole host of additional relocation tasks, may likely be higher than families deserving of a new redeveloped housing unit/neighborhood are willing to pay.

Moreover, research contends that while HOPE VI and HCV households receive the right to return, many do not because of resident choice (as discussed earlier), inadequate relocation services, poor lines of communication or stringent readmission screening criteria (National Housing Law Project, 2002; Jones & Popke, 2010; Marquis & Ghosh, 2008). Further complicating the displaced household's relocation decision equation are specific HOPE VI/PFT program objectives that mandate public housing authorities to decrease public housing unit density in redevelopment sites. Consequently, redeveloped sites do not have sufficient public housing units to accommodate all relocated residents if they desire to return to their original communities. Thus, as we explore recent HOPE VI neighborhood change, it is essential to also assess the extent to which HCV households have successfully returned to these communities.

Another concern is whether communities which have suffered from sustained racial segregation, disinvestment and concentrated poverty, attain similar neighborhood change outcomes compared to public housing communities that were in lesser states of decline. Research contends that HOPE VI/PFT site selection favored high-quality or gentrifying neighborhoods that support mixed-income development (Keating, 2000; Swope, 2001). In this vein, another probing question is whether the four Chicago highlighted communities experienced the same level of neighborhood change as a result of public housing policy intervention by way of HOPE VI and PFT.

FOUR CHICAGO COMMUNITIES

The Douglas, NWS, NNS and EGP neighborhoods are highlighted for their distinct historical (race and income) composition, significant public housing presence (comprised mostly of African American families) and respective role in HOPE VI/PFT. In addition, these communities have suffered from decades of disinvestment and neglect. Despite this, collectively, these communities contained over 20,000 units of public housing and were considered by CHA to have strong potential to leverage private investments due to their respective proximity to city amenities.

The Douglas neighborhood, defined politically as Community Area 35, is notable for accommodating the first University of Chicago in 1860 and being part of the contiguous area known in the 1920s as the "Black Metropolis." Douglas is also infamous for being the location of sprawling public housing complexes developed with PWA and Title III funds. Built public housing communities included low-rise and high-rise complexes totaling nearly 10,000 public housing units in this 1.67-mile community.

Ida B. Wells, built in 1941, was the first public housing project developed exclusively for African Americans. It was comprised of 32 row houses and garden-style apartment buildings of 1,662 units (see Table 1). However, in the 1960s, public housing design changed from being predominately low-rise buildings to high-rise public housing developments. Both Stateway Gardens and the Robert Taylor Homes, built in the Douglas neighborhood, complied with the prevailing public housing design strategy of this era. Stateway Gardens, an eight-building, high-rise development, and Robert Taylor Homes, a conglomeration of 28 sixteen-story high-rise apartments, were built along a two-mile stretch just south of Chicago's central business district. Containing nearly 4,500 low-income units, Taylor represented the epitome of concentrated poverty in the Douglas community.

EGP, home of the Garfield Park Conservatory, is Community Area 27. In the 1800s, EGP was a vibrant, prosperous area. However, urban renewal funding used to create the Eisenhower expressway changed the landscape of the neighborhood by compelling development-impacted (African American) residents of the South and NWS communities to relocate to EGP. Not as densely populated with public housing developments as Douglas, EGP had a large share of public housing units, including Maplewood Courts, Harrison Courts and Rockwell Gardens, totaling more than 1,500 public housing units in this two-square-mile community. Unlike Douglas, EGP was a community defined by its moderate African American population but, similar to Douglas, it was an impoverished community.

The NWS, Community Area 28, began as a neighborhood divided along ethnic, racial and economic lines; it was home to many of the city's elite and simultaneously to many of its poor. Similar to EGP, NWS suffered as a result of the urban renewal projects that displaced swaths of residents. In addition, the uprising that followed Martin Luther King's assassination in 1968 destroyed acres of land, contributing to an increase of poverty in the area. The construction of Henry Horner Homes in 1959, a 1,761-unit high-rise development in the NWS, further deepened the racial and economic split of this community.

Wealth was always present in NNS, Community Area 8. It was recognized for its expensive residential strip on the east and home to the McCormick family and feared for its impoverished west side. Beginning in 1942 and ending in 1962, the CHA built over 3,600 public housing units across Frances Cabrini Homes, Cabrini Extension and William Green Homes, which were collectively known as Cabrini-Green. Deferred maintenance, poverty concentration and high levels of gang-related criminal activity contributed to the rapid decline of this community, despite its proximity to economically vibrant neighborhoods.

Prior to the implementation of PFT in 2000, the Douglas and EGP communities had high African American populations (85% and 97%, respectively) while the NNS and NWS communities had a considerably lower African American population (19% and 53%, respectively; U.S. Census Bureau, 2000). A parallel range existed relative to the percentage of impoverished families within these communities, where the NNS had a low poverty percentage (14%) and Douglas had a high percentage (36%), and the NWS and EGP had approximately 30% of their respective families living below poverty (U.S. Census Bureau, 2000). These data confirm that, while all aforementioned public housing communities are part of CHA's HOPE VI/PFT initiatives, they were not mirror images of each other with respect to race and poverty concentration. EGP and Douglas appear to suffer from high racial and poverty concentration compared to NNS and NWS. Therefore, the selection of these communities allows for an interesting comparison of preliminary neighborhood change outcomes related to public housing policy intervention for communities in dissimilar states of decline.

PUBLIC HOUSING POLICY AND FOUR CHICAGO COMMUNITIES: ASSESSING NEIGHBORHOOD CHANGE

How have the Douglas, EGP, NNS and NWS Chicago neighborhoods changed during the implementation of HOPE VI and PFT? For this analysis, I updated the Nathalie P. Voorhees Center (Voorhees) gentrification index[3] by adding 2010 census and 2005–2009 American Community Survey (ACS) data to the previously established 1970–2000 census data for each Chicago community Area. The resulting database allowed for a robust analysis of five decades of neighborhood demographic data. Voorhees created a scoring mechanism for the index that identified six neighborhood typologies[4] that defined the level of change that had occurred in each community area over the selected time frame. Analysis 1 includes analyses that determine whether the four highlighted Chicago communities have changed over time, and if so, in which direction (positive or negative), with a particular focus on the time frame that corresponds with HOPE VI and PFT.

To assess the ability of displaced households to access the four Chicago highlighted communities, post-HOPE VI and PFT activity, I retrieved Illinois HCV data for the years 2000 and 2007 from the HUD Multifamily Tenant Characteristics System. This analysis, labeled as Analysis 2, allows for a nuanced examination of access to changing neighborhoods by measuring the extent to which private developers and landlords in PFT communities are leasing to Housing Choice Voucher (HCV) households.

Analysis 1

Based on the original Voorhees index (1970–2000 analysis), the NNS and NWS were labeled Type 4 communities, indicating that overall both communities experienced positive neighborhood change with respect to gentrification. Index data and associated scoring found that the NNS community began its process of positive change in 1980 and the NWS in 1970. Key demographic shifts, relative to gentrification, across both communities over the 40-year period include a decline in the number of families living below poverty and an increase in the respective communities' median family income and median home value (see Table 2). The most notable shift among these was change in median home value. The median home values for NNS and NWS increased by 1,496% and 1,149% between 1980 and 2000, respectively. Price escalation of this magnitude raises concern as to whether an outcome of PFT redevelopment activity is a constricted housing market that prices out the HCV households.

Douglas and EGP were identified by the original Voorhees index as stagnant, poor communities that experienced no considerable change between 1970 and 2000. Table 2 presents Douglas's and EGP's scores against NNW and NNS. Housing value differences were slightly remarkable between Douglas and EGP, with a difference of $100,000 in median home value in 2000. More significant is the 145% difference in median home value between EGP and NNS. In this instance, the concern is whether certain PFT communities, like NNS and NWS, were better positioned to capitalize on an infusion of resources, enabling them to realize gains in neighborhood change that far exceed the outcomes of their peers. For the median income and families in poverty variables, Douglas appears to be at par with NWS, ahead of EGP and far behind NNS. The original Voorhees index was able to capture some affects related to HOPE VI activity. However, the lion's share of HOPE VI and PFT activity occurred post-2000. The following analyses based on recent census data reveal the extent to which redevelopment activity requires an adjustment in gentrification classification for each neighborhood.

Updated index analyses (see Table 2) reveal that conditions within the NNS neighborhood continued to improve to where it is currently defined as a Type 1 stable, upper-middle-class community. With the exception of the *percentage of elderly* and *children* data values, movement in all other gentrification factors between the last two decades exhibited positive change. Conversely, there was no substantial change in gentrification scores over the five decades under study to warrant an adjustment in the respective typology classification for the remaining three communities.

The NWS continues its process of positive gentrification, improving on several index factors during the last decade. For instance, the median home values in the NWS increased by 126%, while the number of families living in poverty decreased by 44%. If neighborhood change continues in this direction, the NWS will be firmly placed in the Stable Middle Income typology classification in the near future. EGP and Douglas, however, continue to be struggling, poor communities. EGP has made some improvements along the index with a huge 250% increase in median home value and a modest 11% increase in median family income. Nonetheless, consistent and long-term resource investment in EGP is required if substantial neighborhood change is to be attained. Currently, EGP is still firmly placed in the extreme poverty category.

Douglas's typology classification remains "poor," notwithstanding its enhanced index scores in 2010 compared to 2000. In addition to improvements realized across several gentrification factors, home values and incomes, specifically, increased by a large margin (136% and 76%, respectively), while the percentage of families living in poverty decreased by 48%. Though not as challenged as EGP, Douglas has had a long history of disinvestment and neglect as well. Examining the five decades of index data for Douglas suggests that the community is on a steady path toward positive neighborhood change and should be able to attain Stable Middle Income typology status, possibly by the completion of PFT.

As mentioned previously, CHA public housing in each of the highlighted communities was largely comprised of poor African American households earning approximately 50% to 80% of median income (HUD, 2013). Thus, HOPE VI/PFT redevelopment activity that targets CHA communities will disproportionately affect its dominant residents, poor African American families as discussed earlier. Index analyses suggest this phenomenon to be true. With the exception of the NNS, the remaining communities had high African American populations in 1970 that were relatively reduced each subsequent decade (Douglas's African American population grew slightly between 1980 and 1990 but resumed the patterned decline in 2000). Focusing on percentage change between 2000 and

2010, the reduction of the African American population was most pronounced in the NWS relative to the NNS, Douglas and EGP (21%, 8%, 11%, and 6%, respectively). While the African American population was decreasing between 2000 and 2010, the White population was increasing for all highlighted communities (see Table 2). Poverty was lessened as well, if for only two of the four highlighted communities (NWS: 43% reduction; Douglas: 44% reduction). Yet, family income climbed for each of the communities. It is logical to conclude that HOPE VI/PFT housing demolition activity aided in reducing the number of African American and, in some instances, poor households, while mixed-income redevelopment increased the size of the White and higher-income populations of communities formerly dominated by public housing.

Analysis 2

Using 2000 HCV per capita data (1,000 households) for the four highlighted communities analyzed against the original Voorhees index revealed that both the NNS and the NWS contained lower HCV households per capita compared to the remaining two communities. The NNS's 1.66 and the NWS's 4.24 per capita totals were significantly lower than the EGP's 33.09 and the Douglas's 14.05 per capita totals (see Table 2). Bearing in mind that both the EGP and Douglas community populations were lower in 2000 than the populations for NNS and NWS, and both housed more HCV households than did NNS and NWS, it is probable that EGP and Douglas experienced HCV household concentration. In other words, from an income perspective, there were more low-income HCV households concentrated in the two neighborhoods that were typed as poor than in the two neighborhoods that exhibited considerable positive change.

To determine whether the HCV household concentration trend persisted beyond 2000, I analyzed 2007 HCV per capita data against the upgraded gentrification index. Both the NNS and NWS communities experienced an increase in their respective HCV household populations from 1.6 and 4.2 to 6.6 and 12.5. It must be noted that the populations of both communities increased by approximately 8,000 persons, which may account for the increase in their respective HCV households.

The Douglas and EGP communities also realized rather generous increases in their respective HCV households. EGP had the greatest per capita change, increasing from 33 HCV households to 87 households. Douglas increased to 44 HCV households per capita from 14 households. As for population shifts, both EGP and Douglas experienced a loss in their respective community populations by 314 persons and 8,482, respectively. Thus, HCV household concentration persists for the poorest of the highlighted communities.

NEIGHBORHOOD CHANGE: WHERE DO WE GO FROM HERE?

Highlighting the Douglas, EGP, NWS and NNS communities facilitates the telling of a story—a story of how the presence of public housing is not an automatic and absolute death sentence to an entire community. Data presented in the preceding analyses suggest that some PFT neighborhoods are transforming by way of increase per capita income levels and racial integration, while other neighborhoods continue to exhibit economic and racial segregation. For instance, despite being the location of the previous Cabrini-Green housing development, the NNS has benefitted by redevelopment efforts. A similar statement can be made for the NWS. Yet, research contends that neighborhood redevelopment efforts that successfully leverage advantaged spatial geography and proximity to wealth are more likely to secure additional private investment that furthers positive neighborhood change (Keating, 2000). This may explain the unequal housing policy outcomes for the NNS and NWS compared to Douglas and EGP.

While redevelopment advocates may view neighborhood change within communities formally dominated by public housing as positive outcomes for HOPE VI and PFT, other scholars have raised concerns with the disparate affect redevelopment activity has had on African American and poor public housing residents. Evidence presented here suggests that African Americans are moving out of and White households are moving into redeveloped neighborhoods, thus confirming scholarly literature (Marquis & Ghosh, 2008; Oakley & Burchfield, 2009). Such race-based displacement can be tempered if displaced African American and poor households were moving into higher-quality neighborhoods using HCVs. The data presented here suggests the contrary for the highlighted communities, yet confirms research contending that HCV households move to poor, segregated urban areas (Comey, 2007; Marquis & Ghosh, 2008; Oakley & Burchfield, 2009). Substantially more HCV households were found to live in the Douglas and EGP communities that had higher percentages of poor and African American families than in the NNS and NWS neighborhoods. In other words, Douglas and EGP are still poor, largely African American neighborhoods despite the tearing down of high numbers of public housing units.

While the CHA recognized the need to address horrendous living conditions of residents living in Chicago public housing, the true challenge for CHA, however, is to create communities that will not further disadvantage poor people by forcing them from their homes into existing or newly created low-income communities. Analyses contained within are limited by scope of geography and access to refined, granular data, thus they cannot make wholesale claims as to the success or failure of CHA's housing policy initiatives. Nonetheless, this research presents potential concerns with preliminary redevelopment outcomes of the HOPE VI and PFT housing programs. Future investigation, conducted well after the completion of PFT in 2018, will produce a more definitive determination as to whether CHA successfully changed the economic mix of targeted neighborhoods while also benefitting long-standing residents of affected communities. Will HOPE VI and PFT achieve the elusive balance of a broad economic mix of new and former households that reside in racially diverse revitalized communities throughout the city of Chicago, or will Chicago experience the similar fate of other HOPE VI cities and redevelop poor communities into segregated exclusive spaces for an economically advantaged and racially specific clientele? Time will tell.

Table 1. Public Housing Developments in Highlighted Communities.

Public Housing Development	Community	Year Built	Building Type	# Units	Year Demolished	PFT Planned Redevelopment
Ida B. Wells	Douglas	1941	Garden Apts; Row House	1654	2002–2011	Oakwood Shores
Ida B. Wells Extension	Douglas	1941	Garden Apts; Row House	376	2002–2011	Oakwood Shores
Dearborn Homes	Douglas	1950	Midrise	800	2010 Renovation	Reducing to 660 units
Prairie Courts	Douglas	1952	Row house; High Rise	343	2003	South Park Plaza
Darrow Homes	Douglas	1961	High Rise	480	2000	Oakwood Shores
Stateway Gardens	Douglas	1961	High Rise	1644	1997	Park Boulevard
Robert Taylor Homes	Douglas	1961	High Rise	4415	2007	Legends South
Madden Park Homes	Douglas	1970	Midrise; High Rise	450	2000	Oakwood Shores
Rockwell Gardens	East Garfield Park	1959	High Rise	1136	2006	Jackson Square
Maplewood Courts	East Garfield Park	1952	High Rise	140	2006	Jackson Square
Harrison Courts	East Garfield Park	1952	High Rise	140	2006	Jackson Square
Francis Cabrini Homes	Near North Side	1942	Low Rise	586	1997–2011	Parkside
Cabrini Extension	Near North Side	1957	High Rise	1925	1997–2011	Parkside
William Green Homes	Near North Side	1962	High Rise	1096	1997–2011	Parkside
Henry Horner	Near West Side	1969	High Rise	1761	2008	West Haven

Table 2. Select Gentrification Index Variables and Housing Choice Voucher Per Capita.*

	1970	1980	1990	2000	2009/2010		1970	1980	1990	2000	2009/2010
	Near North Side						Near West Side				
# Gentrification Factors**	2	6	10	9	11	# Gentrification Factors*	-11	-7	-3	-3	6
Population	70,269	67,167	62,842	72,811	80,484	Population	78,703	57,305	46,197	46,419	54,881
% White	58.1	62.7	71.0	69.2	72	% White	16.3	12.8	18.8	25.3	42
% Black	37.1	32.5	23.1	19.1	11	% Black	72.2	74.1	66.5	52.9	32
Median Family Income	$11,274	$23,395	$62,317	$93,934	$100,382	Median Family Income	$6,012	$7,535	$10,268	$32,822	$74,026
Median House Value	$50,100	$200,100	$506,300	$625,692	$467,775	Median House Value	$12,800	$27,900	$121,900	$204,411	$287,400
% Family Below Poverty	18.2	26.2	18.1	13.6	17	% Family Below Poverty	31.7	48.9	52.3	30.1	17
HCV/Capita	%	%	%	1.66	6.6	HCV/Capita	%	%	%	4.24	12.5
	East Garfield Park						Douglas				
# Gentrification Factors*	-9	-9	-9	-10	-8	# Gentrification Factors*	-5	-5	-5	-6	-1
Population	52,185	31,580	24,030	20,881	20,567	Population	41,276	35,700	30,652	26,470	18,238
% White	0.7	0.6	0.7	1.1	3	% White	9.3	9.7	5.2	6.6	10

Table 2. Continued

	East Garfield Park						Douglas			
	1970	1980	1990	2000	2009/2010		1970	1980	1990	2000	2009/2010
% Black	98.0	98.3	98.5	97.2	91	% Black	87.6	86.2	91.2	85.5	73
Median Family Income	8,457	9,682	14,301	27,185	30,064	Median Family Income	$6,260	$8,578	$10,577	$25,720	$60,742
Median House Value	$13,500	$22,800	$39,700	$98,699	$345,322	Median House Value	$17,000	$25,900	$113,000	$208,449	$365,956
% Family Below Poverty	29.5	40.3	45.6	28.7	35	% Family Below Poverty	27.1	42.6	49.4	36.4	19
HCV/Capita	∅	∅	∅	33.09	87.3	HCV/Capita	∅	∅	∅	14.05	44.25

*Publically available Voorhees Center gentrification index data for 1970–2000 was accessed from http://www.uic.edu/cuppa/voorheesctr/Gentrification%20Index%20Site/Main%20Neighborhood%20Change%20Revised.htm. Index analyses of 2009/2010 gentrification data was computed by author.

**Gentrification Factor numbers (based upon Voorhees Center index convention) are determined by comparing neighborhood variable data against city of Chicago variable data. If a variable was "positively associated with gentrification" and the community area received a higher value than the city, it received a score of +1; if it received a lower value than the city, it received a –1. Variable factor scores are summed to arrive at the # of gentrification factors.

NOTES

1. The HCV provides rental assistance to low-income households that earn between 30% and 80% of median income who use the housing vouchers to secure private-market rental housing across the country.
2. Between 1994 and 2001, CHA received HOPE VI grants for Cabrini-Green, Henry Horner Homes, Robert Taylor Homes, Ida B. Wells and Rockwell Gardens.
3. The publically accessible Voorhees index included data on the following variables: median family income, percentage of families below poverty, median house value, percentage of owner-occupied housing, race/ethnicity, percentage of children ages 5–19, percentage of elderly (65+), percentage of managers and professionals (of all workers), percentage of adults with college education, percentage of children enrolled in private school, and percentage of female-headed households with dependent children under age 18.
4. See http://www.uic.edu/cuppa/voorheesctr/Gentrification%20Index%20Site/Main%20Neighborhood %20Change%20Revised.htm for the definitions of all typologies.

REFERENCES

Abt Associates, Fosburg, L. B., Popkin, S. J., & Locke, G. P. (1996). *A historical and baseline assessment of HOPE VI.* Washington, DC: U.S. Department of Housing and Urban Development.

Bates, L. (2013). Gentrification and displacement study: Implementing an equitable inclusive development strategy in the context of gentrification: City of Portland. Retrieved from http://www.portlandoregon.gov/bps/article/454027

Cam, G. (1939). United States government activity in low cost housing, 1932–1938. *The Journal of Political Economy, 47*(3), 357–378.

Chicago Housing Authority. (2013). Plan forward: Communities that work. Retrieved from http://www.thecha.org/pages/plan_forward/2677.php

Clampet-Lundquist, S. (2006). Moving over or moving up? Short-term gains and losses for relocated HOPE VI families. *Journal of Policy Development and Research, 7*(1), 57–80.

Comey, J. (2007). *Hope VI'd and on the move* (Brief No. 1). Washington, DC: Metropolitan Housing and Communities Center, The Urban Institute.

Forman, T., & Krysan, M. (2008). Racial segregation in metropolitan Chicago. *Institute of Government and Policy Affairs Policy Forum, 20*(3), 1–5.

Freeman, L. (2005). Displacement or succession? Residential mobility in gentrifying neighborhoods. *Urban Affairs Review, 40,* 463–491.

Glaeser, E., & Vigdor, J. (2012). The end of the segregated century: Racial separation in America's neighborhoods, 1890–2010. Retrieved from http://www.manhattan-institute.org/pdf/cr_66.pdf on December 11, 2013.

Goetz, E. G. (2010). Better neighborhoods, better outcomes? Explaining relocation outcomes in HOPE VI. *Cityscape: A Journal of Policy Development and Research, 12*(1), 5–31.

Goetz, E. G. (2011). Where have all the towers gone? The dismantling of public housing in US cities. *Journal of Urban Affairs, 33*(3), 267–287.

Groberg, R. P. (Winter, 1965). Urban renewal realistically reappraised. *Law and Contemporary Problems and Prospects, 30,* 212–229.

Henneberger, J. (2008, March 27). Study the shifting geographic nature of low-income neighborhoods in urban and suburban areas, focusing on factors that contribute to economic displacement of low income homeowners and factors that contribute to beneficial redevelopment of distressed neighborhoods. Testimony before the Texas Senate Intergovernmental Relations Committee. Retrieved from http://www.texashousing.org/blogfiles/s_test_gentrification.pdf

Hirsch, A. R. (1998). *The making of the second ghetto: Race and housing in Chicago 1940–1960*. Chicago, IL: The University of Chicago Press.

Hunt, D. B. (1997). Public housing in America: Lost opportunities. *Reviews in American History, 25*(4), 637–642.

Hunter, M. (2011, Autumn). Gentrification and marginalization of people of color. *The (Sub)-Urban Scene*. The John Perkins Center. Retrieved from https://spu.edu/depts/perkins/about/perspective/2011-autumn/gentrification-and-marginalized-people-of-color.asp

Jones, K. T., & Popke, J. (2010). Re-envisioning the city: Lefebvre, HOPE VI, and the neoliberalization of urban space. *Urban Geography, 31*(1), 114–133.

Keating, L. (2000). Redeveloping public housing: Relearning urban renewal's immutable lessons. *Journal of the American Planning Association, 66*(4), 384–396.

Kennedy, M., & Leonard, P. (2001). *Dealing with neighborhood change: A primer on gentrification and policy choices*. Washington, DC: Brookings Institution.

Lees, L., & Ley, D. (2008). Introduction to special issue on gentrification and public policy. *Urban Studies, 45*(12), 2379–2384.

Levy, D. K., Comey, J., & Padilla, S. (2006). *In the face of gentrification: Case studies of local efforts to mitigate displacement*. Washington, DC: The Urban Institute.

Marquis, G. P., & Ghosh, S. (2008). Housing opportunities for people everywhere (HOPE VI): Who gets back in? *The Social Science Journal, 45*, 401–418.

Massey, D. S., & Denton, N. A. (1988). The dimensions of residential segregation. *Social Forces, 67*(2), 281–315.

National Housing Law Project, Poverty & Race Research Action Council, Sherwood Research Associates, Everywhere and Now Public Housing Residents Organizing Nationally Together. (2002). *False HOPE: A Critical Assessment of the HOPE VI Public Housing Redevelopment Program*. Oakland, CA: National Housing Law Project.

Oakley, D., & Burchfield, K. (2009). Out of the project, still in the 'hood: The spatial constraints on public-housing residents' relocation in Chicago. *Journal of Urban Affairs, 31*(5), 589–614.

Paul, B. (2011, February 22). Affordable housing policies may spur gentrification, segregation. *Gotham Gazette*. Retrieved from http://www.gothamgazette.com/index.php/topics/4282-affordable-housing-policies-may-spur-gentrification-segregation

Polikoff, A., McConnell, E. H. II, Brown, J., Hill, K., Pomper, K., Underhill, N.,... & Falk, J. (2009). *The third side: A mid-course report on Chicago's transformation of public housing*. Chicago, IL: Business and Professional People for the Public Interest.

Popkin, S. J. (2006). The HOPE VI program: What has happened to the residents? In L. Bennett, J. L. Smith, & P. A. Wright (Eds.), *Where are poor people to live? Transforming public housing communities* (pp. 68–90). Armonk, NY: M. E. Sharpe.

Schaffer, R., & Smith, N. (1986). The gentrification of Harlem? *Annals of the Association of American Geographers, 76*(3), 347–365.

Seng, M. P., & Caruso, F. W. (2013). Segregation in the Chicago metropolitan area: Some immediate measures to reverse this impediment to fair housing. *John Marshall Law School Fair Housing*

Legal Support Center. Retrieved from http://www.jmls.edu/fairhousing/pdf/2013-chicago-segregation-study.pdf on December 11, 2013.

Smith, J. L. (2001, May 17). *Mixing it up: Public housing redevelopment in Chicago.* Paper presented at the Area-Based Initiatives in Contemporary Urban Policy conference, Copenhagen, Denmark. Retrieved from http://www.sbi.dk/eura/workshops/papers/workshop1/smith.pdf

Swope, C. (2001, May). Rehab refugees. *Governing Magazine.* Retrieved from http://www.governing.com/topics/public-justice-safety/Rehab-Refugees.html on December 11, 2013.

Tavernise, S. (2011, July 18). A population changes, uneasily. *The New York Times,* p. A9.

U.S. Census Bureau. (2000). Summary File 3. Profile of selected social, housing and economic characteristics: Chicago, IL: 2000 Census.

U.S. Department of Housing and Urban Development (HUD). (2013). Income limits. Retrieved from http://www.huduser.org/portal/datasets/il/il13/index.html

Vale, L. J., & Graves, E. (2010). *The Chicago Housing Authority's plan for transformation: What does the research show so far?* Cambridge: Massachusetts Institute of Technology, Department of Urban Studies and Planning.

von Hoffman, A. (1996). High ambitions: The past and future of American low-income housing policy. *Housing Policy Debate, 7*(3), 423–446.

Wexler, H. J. (2001). HOPE VI: Market means/public ends: The goals, strategies, and midterm lessons of HUD's Urban Revitalization Demonstration Program. *Journal of Affordable Housing, 10*(3), 195–233.

Wyly, E. K., & Hammel, D. J. (1999). Islands of decay in seas of renewal: Housing policy and the resurgence of gentrification. *Housing Policy Debate, 10*(4), 711–771.

CHAPTER THREE

The Shrinkage Machine

Race, Class and the Renewal of Urban Capital

EMILY ROSENMAN, SAMUEL WALKER AND ELVIN WYLY[1]

INTRODUCTION

The arc of postindustrial urban landscapes surrounding the Great Lakes—also known as the rust belt—was hit especially hard by the financial crisis that began in 2007. Decades of de-industrialization and "White flight" migration shaped by uneven urban development had already undermined Detroit, Michigan, Cleveland, Ohio, and a galaxy of smaller industrial cities before the bills for the American credit binge finally came due. The economic base of rust belt cities was dismantled by the financialization of the American (and global) economies beginning in the 1970s, as capital was withdrawn from domestic industrial production in favor of transnational corporate investment and a wide array of speculative financial instruments (Krippner, 2012). Real estate, a central "fictitious form of capital" (Harvey, 2012, p. 28) whose pseudocommodity form has played a starring role as a spatial fix in centuries of crises of capital accumulation, was financialized in ways that created a partial but dangerous new industrial base for rust belt urbanism. With new practices and policies in housing finance, "'home' became...intricately woven" (Newman, 2012, p. 236) into financial circuits, as secondary-market securitization transformed "local housing" into "an electronic instrument" (Sassen, 2012, p. 74). As financial institutions developed new techniques to maximize accumulation through financial and transactional means while offloading the risks, predatory practices flourished in communities marginalized by race and class (Aalbers, 2009; Wyly, Atia, Foxcroft, Hammel, & Phillips-Watts, 2006; Wyly, Moos, & Hammel, 2012). "In the new economy," Newman (2012, p. 236) concluded, "mortgage loans are the modern-day widget" for the "postindustrial financialization industry," and "financial institutions needed to produce more widgets to fuel expansion." Many of these widgets were produced by deceiving lower-middle-class households in America's burgeoning Sunbelt suburbs, but many also came from the racialized capitalist innovations of predatory lending in de-industrialized rust belt cities. Racially marginalized communities that had long faced *exploitative exclusion* from good credit became lucrative targets for *stratified inclusion* into transnational circuits of bad, risky credit. After the collapse, rust belt neighborhoods have been devastated by widespread vacancy, abandonment, and deterioration—undermining house prices and land values, in turn eviscerating the local tax base as the ongoing global crisis works its way through dynamic urban systems of inequality. The wave of foreclosures has been especially severe in Cleveland (see Figure 1).

Faced with these problems, shrinking cities in the United States have adopted numerous strategies in their attempts to stem and reverse economic decline. Since the 1970s, efforts to revitalize struggling downtowns have taken many forms, including prominent mixed public and private projects to attract outside capital (stadiums and casinos) and tax incentives for business relocation. Generally, these actions have occurred in a national context of neoliberal reforms that first cut social services and institutions and then began to roll out neoliberalized modes of urban governance and regulation (Peck & Tickell, 2002). In an environment of interurban competition, entrepreneurial urban governments compete with each other to gain the preconditions of growth: public money, population, private investment, and development resources. Unlike the elite interest-based growth strategies described in Molotch's (1976) influential "growth machine" thesis, however, in the current moment, what can more aptly be defined as a "shrinkage machine" mentality has emerged as the policy consensus for the new, postcrisis preconditions of urban economic growth. The shrinkage machine policy regime marks a departure from a long-standing connection between population and economic growth. In the aftermath of the housing and financial crisis, a new consensus has emerged: since population growth can no longer be anticipated in rust belt cities, economic growth must be pursued in new ways. Crucially, land management strategies are being rapidly altered to conform to the new "realities" of urban shrinkage.

Harking back to the "planned shrinkage" strategy developed for the South Bronx in the mid-1970s (see Stone, 1978), rust belt cities have pursued drastic measures designed to tackle their surplus—and rapidly deteriorating—housing stock. A major part of such plans involves the demolition of abandoned foreclosed homes. While demolitions of tax-foreclosed homes have been occurring in the rust belt for decades (Hackworth, 2014), the recent crisis has led to a growing effort to tackle the vacant or abandoned structures that are a product of widespread mortgage foreclosure. For example, officials of Ohio's Cuyahoga County, of which Cleveland is the largest city, have destroyed over 3,200 foreclosed homes since 2008 (Center on Urban Poverty and Community Development, 2012) and are actively seeking state, federal, and private-sector funding to demolish more (c.f. Atassi, 2012, 2013; Bulava, 2011). Not surprisingly, these demolitions follow the general outlines of the landscapes of destruction wrought by predatory capital and subsequent foreclosures (see Figure 2). The corollary of demolitions in Ohio and other states is land banking, in which cities or counties work to acquire vacant land so as to exert some measure of oversight and planning regarding its eventual reuse. Both demolitions and land banking are generally portrayed as commonsense, apolitical solutions to managing currently value-less land with the anticipation of future demand and renewed urban economic growth. This neutral, technocratic framing, however, belies significant class and race implications—such as when Jim Rokakis (2013), former chairman of the Cuyahoga County Land Bank and current advocate of Ohio-wide land banking, claims that many Cleveland neighborhoods, specifically many high-poverty areas inhabited by an African American majority, are "primed for a demographic turnaround."

This chapter considers two postcrisis managerial strategies—land banking and the demolition of abandoned foreclosed homes—through the lens of these strategies' connections to past urban renewal policies, which had highly uneven effects in terms of race and class. Given Cleveland's comparatively well-developed postcrisis land banking strategies, which are fast becoming a policy example for other cities, we use the city as a case study to draw parallels between three themes present in both the contemporary moment and urban renewal policies of the past: the role of governance and public oversight, the justification and means of land appropriation by the state, and the use of neighborhood typologies in these policies. These arguments lead us to a critique of present-day managerial land banking that recalls the unequal results of similar economic-revitalization techniques of the not-so-distant past. While avoiding predictions regarding the outcome of the still-new postcrisis demolition and land banking policies, we argue that remembering the legacy of past rounds of discriminatory, state-led land redistribution is important for planners, researchers, and city residents and activists. This is especially crucial in a climate of neoliberal urban governance that involves more and more private sector "partners" with ties to the business and real estate sectors.

POSTCRISIS LAND MANAGEMENT STRATEGIES IN THE RUST BELT

Many local governments have used land banking to intervene when market processes culminate in property abandonment. As Alexander (2011) and Hackworth

(2014) have noted, there is a long history of land banking in the rust belt and these attempts at land market intervention are not one-dimensional. Furthermore, the state's role in land management can vary substantially. Hackwork identifies three general categories of land abandonment regimes, elucidating how various ideological projects influence policy responses. Following Harvey (1989), Hackworth classifies one set of responses as "managerial," "those designed to regulate [private capital] development by placing conditions on it" (Hackworth, 2014, p. 25). Cities adopting a managerial stance often change legislation to allow faster and more punitive fines to be levied against negligent or absentee property owners, use eminent domain to reclaim abandoned property, and carry out reforms on the tax revision process, such as limiting who can buy foreclosed properties or giving the city or land bank priority to purchase auctioned property. In this model, the final goal is often the demolition of abandoned structures and land-holding by a (quasi) governmental entity. Land is banked in anticipation of future development, in an attempt to increase current land values (by creating scarcity), and to combat the negative price effects that foreclosures inflict on occupied homes in and adjacent to high-vacancy neighborhoods.

The second category of land abandonment intervention is what Hackworth categorizes as "market-first," which attempts to entice current owner-occupants or potential investors (both firms and individuals) to invest in blighted neighborhoods. This form of land governance has resulted in popular "side lot" programs, in which current owner-occupiers can purchase abandoned or vacant lots adjacent to their property for a discounted rate. Another policy falling under the market-first umbrella is buyer and renovation assistance, often provided to buyers with low incomes or poor credit histories (Hackworth, 2014, pp. 65–66).

The third land abandonment policy regime is what Hackworth calls "market only." While a land market that functions with no government intervention is mostly a neoliberal fantasy, cities like Detroit do provide stark examples of reforms designed to reduce the state's role in establishing a controlled market for abandoned or vacant properties. Market-only policy reforms include accelerating the tax reversion process so that properties are returned to the market more quickly, lowering or eliminating taxes to reduce delinquency, and holding no-minimum bid auctions in which the government and land banks do not participate (Hackworth, 2014, p. 29). In Detroit, these extreme free-market reforms have unleashed rampant speculation that has de-stabilized housing sector recovery (Akers, 2013; see also MacDonald, 2011).

Hackworth suggests that market-only land abandonment regimes are less socially beneficial than the managerial approach. Like many policy analysts (e.g., Alexander, 2011), he situates Cleveland and its land banking authority, the Cuyahoga County Land Reutilization Corporation (CCLRC), as a relatively successful use of a land bank to curtail speculation, while portraying

Detroit's free-market approach as leading to "the erosion of market conditions" (Hackworth, 2014, p. 2). While acknowledging that such neoliberal, "free-market" land management regimes are economically and socially detrimental to high-abandonment neighborhoods, we argue that managerial land banking policies can also have negative—but perhaps less obvious—effects, especially when the race and class implications of these policies are considered. Using Cleveland as a case study, we analyze connections between earlier eras of slum removal and the current regime of foreclosure demolition and land banking. Through this analysis we contribute to planning and housing policy literatures on both urban renewal and today's era of "planned shrinkage" (Gratz, 2010). Though the postcrisis land management policies detailed here are overwhelmingly new, we situate them historically in an effort to ascertain the effects of changing land governance in today's economically declining rust belt cities.

URBAN RENEWAL AND THE EXPANSION OF "PLANNED SHRINKAGE"

In discussing urban renewal, we refer both to the so-named era of postwar slum clearance, public housing construction and inner city economic investment, as well as to subsequent, more market-based interventions in the urban housing market. We follow Carmon (1999) in conceptualizing urban renewal as an ongoing process rather than a bracketed historical period. However, government-led urban renewal must also be read through wider, multiscalar economic conditions and development factors. Post-1970s trends toward entrepreneurial urban governance and a global expansion of finance capital are particularly salient to what some scholars have deemed a new era of urban renewal policies (c.f., Fainstein, 2005; Goetz, 2011). While 1940s–1960s urban renewal was primarily a federal policy enforced at the urban level, subsequent rounds have been decidedly multiscalar; the influence of globalized capital is such that transitioning "underutilized" land into future investment fodder is both a global and local interest, with the federal level largely filling a finance, not policy, role (Hyra, 2012).[2] While urban renewal's interested parties have shifted, we identify two common and enduring factors: first, the focus on residential land clearance in the name of future (higher-value) economic investment, and second, exacerbated social and economic inequalities as a result of such policies.

In each round of urban renewal, policymakers have used demolitions in an effort to address social and economic problems and catalyze neighborhood change. The post-2007 crisis period has seen accelerated demolition trends, with a remarkable expansion of "planned shrinkage" throughout the American metropolitan system. The flood of foreclosures has installed "foreclosure demolition" as a standard policy template in rust belt cities, whose targets have evolved while still reflecting the historical roots of urban renewal. Two shifts are apparent. First, the social problem targeted by demolition policies has shifted from the shortcomings of *actually existing* residents (which the urban renewal of previous eras often branded as the root of neighborhood decline) to the risks associated with *potential* residents of buildings slated for demolition. Rather than a problem of existing poverty concentration, high-foreclosure neighborhoods are now framed in terms of their potential use by criminals, drug dealers, and squatters—a potential that was vividly illustrated to a global audience with an image of a Cleveland foreclosure that won the 2008 World Press Photograph of the Year (see Figure 3). Empty houses, not existing social conditions, are now framed as the primary problem for the state to address in a context where the planner's motto of "highest and best use" has collapsed along with land values—the doctrine was dismantled with the steel mills, forced to leave town, or stripped of anything salable before being razed to the ground.

Second, whereas urban renewal and HOPE VI were paired with rehousing programs (however imperfect) for people displaced by demolition, foreclosure demolitions are wholly focused on building clearance without social remediation for the people who lost these homes. Given popular associations between empty foreclosed homes, crime, and sagging property values, demolitions are increasingly demanded by middle-class and wealthy neighborhoods in Cleveland (cf. Narkin, 2011). The present-day politics of demolition targeting thus also align with historical examples, in which demolitions were especially common in slums or public housing projects adjacent to more valuable real estate (Goetz, 2011; Metzger, 2000). In the following section, we examine three congruencies between the current expansion of planned shrinkage regimes and past policies that reflected and reinforced systemic urban inequalities.

GOVERNANCE

Perhaps one of the most striking contrasts between the urban renewal programs of the middle of the 20th century and today's land bank-led managerial regimes involves governance—specifically, decision-making processes and the role of the state. Today's managerial regimes reflect the general historical trajectory through which urban governance has shifted from the state-led urban renewal projects of the mid-20th century to the era of public-private partnerships and the privileging of private capital in roll-out neoliberal urban development since the 1980s (Brenner & Theodore, 2002; Harvey, 1989). The urban renewal policies of the 1940s and 1950s were characterized by direct government involvement in land redistribution and housing access (Hirsh, 1983/1998). Government decisions and attempts to redraw racial and class boundaries within cities reshaped urban forms and demographic patterns. Redevelopment legislation churned out discursive connections between the physical deterioration of housing stock (blight) and the health and morals of residents (Weber, 2002). In this era, government discourse emphasized the use-values of slum residents through well-publicized plans to relocate people into new, "modern" low-income housing, even though these programs also subsidized the clearance of poor neighborhoods for private redevelopment and profit. Beginning in the 1970s, however, globalizing capitalism and a governance style more oriented toward urban entrepreneurialism and economic growth redirected municipal governments' role in redevelopment. Government's new role was to assist the private market by identifying blight and facilitating private-sector "reclamation" of land. The state no longer carried out land transfers directly, focusing instead on enabling private investment where the differences between current and potential land rents yielded potential for revalorization and growth (Weber, 2002). Governments now use a variety of zoning, financial, and tax mechanisms to steer urban redevelopment without intervening directly. Also central to this new style of governance is the formation of extragovernmental entities, such as urban redevelopment agencies and land banks, which shifts land governance outside of direct municipal oversight.

The specific context in which land banks emerge is vitally important to their structure and outcomes. The Cuyahoga County Land Reutilization Corporation (CCLRC) was formed in 2009 after the 2008 Ohio Land Bank Act, but has its roots in the 1976 establishment of the Cleveland Land Bank, which was created to handle the large number of tax foreclosures in the city inventory (Hackworth, 2014, p. 33). The CCLRC is part of a regional plan to manage the increasing number of abandoned properties resulting from mortgage foreclosure. In order to achieve an "active" model in which land is quickly returned to private and "productive" use, rather than simply held for future development, Ohio legislators decided to legally separate the CCLRC from county and municipal government. Thus, the CCLRC is granted its own staff and funding, separate from local government. Additionally, it can coordinate a regional strategy among municipalities, acquire REO (real estate owned) properties,[3] contract building and property maintenance to third parties, and bear legal liability on its inventory, thereby relieving local government of many financial, legal, and administrative responsibilities (Fitzpatrick, 2009).

For these reasons, the CCLRC was established as a quasigovernmental entity: a nonprofit corporation with the ability to levy fines, collect fees, and coordinate the management of parcels within participating municipalities. These features have made the CCLRC a highly effective land bank; it has acquired and demolished thousands of structures and banked this newly vacant land. We probe this "success story" further, however, to analyze how this land governance structure balances the interests of private capital and declining Cleveland neighborhoods.

An initial question is why the state legislature thought it necessary to make the CCLRC a private legal entity. The CCLRC's 2009 six-month business plan stresses the importance of market-based approaches to dealing with the foreclosure crisis:

> Existing structures lack entrepreneurial capabilities and efficiencies. Indeed, unlike traditional governmental bodies that necessarily operate with bureaucracies which must comply with unnecessary regulations, the CCLRC is uniquely designed to act much like a private corporation but with a public purpose. Flexibility and efficiency are its hallmarks and are critical to its ability to respond to the demands of the market as well as solve problems in a timely manner. (CCLRC, 2009, p. 13)

This discourse echoes the words used by Robert Moses to justify a new kind of public authority in New York with "freedom from the red tape" so that we can "function freely and efficiently" outside the "old-line governmental agencies" (Caro, 1974, p. 633); the logic used to *build* the growing cities of 20th-century American industrial modernity now is retooled to dismantle Cleveland and other cities of permanent de-industrialization. This short-term logic envisions the creation of "flexible" and de-/unregulated private actors (at the local scale) as the only way to recover from a crisis caused by "flexible" and de-/unregulated predatory financial practices (at extralocal scales). This is a textbook illustration of the "fail forward" logic of austerity neoliberal urbanism (Peck, 2012).

Additionally, the idea of removing "unnecessary regulation" is not just talk: the CCLRC has been granted immunity from certain "environmental regulations and orders, permits, licenses, variances, or plans approved or issued under any such regulations" (Fitzpatrick, 2009, p. 137). Cleveland demolition officials have publicly announced their commitment to properly handling the removal of asbestos from demolished homes, but they have also complained that the strict Environmental Protection Agency regulations on home demolition will use up too much of scarce demolition dollars (The Plain Dealer Editorial Board, 2012). Finally, the funding structure of the CCLRC is dictated in many ways by the neoliberal political atmosphere of the Ohio legislature: as noted by Hackworth (2014), most land banks must be self-funding to justify their existence. The CCLRC is funded primarily through the collection of interest on delinquent real estate tax, not from any primary tax sources. However, in the climate of competitive entrepreneurial funding for public services, the land bank has also been aggressive in pursuing grants and competing for federal government money such as Neighborhood Stabilization funds from the Department of Housing and Urban Development[4] and, more recently, nearly

$12 million from a mortgage settlement between the federal government and several large financial institutions (Atassi, 2012). Cleveland's sudden, unbidden global prominence as a foreclosure epicenter has made it an especially vivid illustration of a new kind of policy milieu in which the political consensus has shifted from ensuring consistent growth to planning for shrinkage: the *shrinkage machine*.

Another matter is how, as a quasigovernmental institution, the CCLRC is held accountable to public oversight and the interests of county residents. The board and staff of the CCLRC are a mix of elected and appointed public officials and representatives from nonprofits. The CCLRC operates as a regional agent for handling REOs, contracting out site maintenance and security, and, upon demolition, returning vacant lots to city inventories. While the CCLRC is required to file audits with the state every year and maintains a strong presence in the media to justify its presence, several issues have been raised by the public that point to the possibility for abuse. First, the primary goal of the CCLRC is to return land to so-called "productive use," with no written guiding principles dictating what kind of productive use this should be or whom it should benefit. While land banking may reduce the "unproductive" effects of land speculation, it does not combat land uses that confer privilege on developers and other powerful private actors. The CCLRC has already been the subject of major local controversies, including allegations of classic city machine patronage in the form of an official targeting demolitions near property he owns with associates—with the idea that the demolitions will boost the value of property nearby. Regardless of the veracity of these allegations, their circulation in local political discourse is significant: whereas a growth machine scandal involves allegations of insider access to information on what is to be *added* to the built environment (e.g., where a new highway is to be built), a shrinkage machine scandal involves accusations of unfair access to *subtraction*—to the destruction of vacant houses.

As this section has indicated, land banks like the CCLRC are private legal entities used to govern populations under the neoliberal view that the "rational" behavior of private actors is the ideal basis for governing. In many ways this represents a break from the urban renewal programs of the mid-20th century, but it also indicates the persistent involvement of the state in both managing and being directed by the logic of private capital. While the managerial land abandonment regime of the CCLRC may be effective in reducing speculation and returning land to productive use, it is worth questioning what price is paid for this flexibility in terms of democratic and participatory planning and oversight. In the regional realignment amidst national economic growth of the 1950s, federal urban renewal had disastrous consequences for the urbanization of racial and class inequality. Now, in an era of variegated but severe national austerity, can there be a socially just outcome when a land bank manages decline in ways that privilege the interests of private capital?

LAND APPROPRIATION BY MUNICIPAL GOVERNMENT

How does the establishment of a quasigovernmental land bank change the methods of land appropriation and redistribution in urban redevelopment, and how does this new scheme compare to those of urban renewal? The influence of private business on urban governments during the 1940s and 1950s led to the use of eminent domain for the appropriation of slum lands that were sold to the private sector at a fraction of the original cost (Hirsch, 1983/1998). Hirsch and others (c.f. Goetz, 2000, 2011) have shown that the results of both postwar and more recent rounds of urban renewal have been highly racialized. In the earlier phase (1940s–1950s), policymakers in cities like Chicago realized that urban redevelopment could not be accomplished without the relocation of existing residents. This set in motion private redevelopment's reliance on public housing facilitated by municipal governments. Chicago's Land Clearance Commission was endowed with the authority to purchase, condemn, clear and resell slum properties to private developers. Commissions such as these had no power to build or operate their own housing and existed solely as a means for reappropriating urban lands (usually occupied by poor people and racial minorities). This residential displacement, facilitated by municipal government and subsequent investment by the private sector, had two primary social effects: it reinscribed and increased racial segregation and did nothing to relieve the city's drastic housing shortage (Hirsch, 1983/1998). These policies set a precedent for economic development as a "public purpose" under which cities could acquire slum and unused lands for the purposes of attracting private reinvestment. Since 1992, HOPE VI urban redevelopment policy has emphasized reuse of cleared public housing sites for urban economic development, not housing for displaced residents (Crump, 2002).

In the contemporary policy environment of permanent-crisis rust belt cities, land bank acquisitions are framed as a public purpose that will lead to renewed economic growth given the proper incentives for private capital. Municipalities across the United States witnessed a new legal framework for eminent domain in the 2005 Supreme Court decision *Kelo v. City of New London*. *Kelo* allowed public takings of private property not only for clearly public facilities like highways, but also for the transfer of land from one private owner to another deemed better able to achieve "highest and best use" activities to generate local economic development. *Kelo* quickly prompted a nationwide backlash, as state legislators tapped into deep property-rights sentiments to prohibit *Kelo*-style takings except for properties clearly designated as "blighted." However, these designations became "extremely malleable, and, in fact, so flexible that most properties could be designated as blighted" (Lovell, 2012, p. 1).

Cleveland and Cuyahoga County have been at the forefront of new developments in the use of public authority to cope with the sedimented legacies of private market catastrophe. While there is considerable variation in how "blight" is defined, in most states the concept is extremely broad, with considerable discretion for implementation at the municipal level. In Cleveland, the definition is at once very specific and all-encompassing, including premises marked by the "occurrence of drug offenses, prostitution, gambling and other criminal acts," or other factors that make them a "place of retreat for immoral or criminal purposes," "or a factor seriously depreciating property values in the neighborhood" (Chapter 324.03 of the Codified Ordinances of the City of Cleveland, cited in City of Cleveland, 2011, p. 9).

In Cuyahoga County, the CCLRC's stated goal is to prevent rampant land speculation by intervening in the reuse of abandoned foreclosed properties. Its mission is to "strategically acquire blighted properties and return them to productive use" (CCLRC, 2009), but its funding mechanisms exemplify the naturalization of "all manner of short-term fixes, band aids, and bromides" (Peck, 2012, p. 630) in the current "austerity moment" of neoliberal urbanism. The CCLRC's operations are funded with penalties and interest on delinquent real estate taxes and assessments, the sale of rehabilitated properties, and the same Wall Street securitization machine that made Cleveland a national and transnational icon of foreclosure: in December 2010, the CCLRC issued $9 million in "special receipts tax exempt" bonds to a single private investor—Capital One Funding, a subsidiary of the well-known credit card issuer that had played its role in the subprime crisis with a specialized Alt-A lender, GreenPoint Mortgage (CCLRC, 2011, p. 29). The CCLRC's land acquisitions come from tax and mortgage foreclosures, but also from a growing number of bank "donations" of abandoned foreclosed properties, which relieve financial institutions of management and maintenance expenses, and also legal liability.

With the CCLRC's increased volume of mortgage foreclosures, its land appropriation and demolitions are necessarily following the spatial contours of the economic crisis in those historically marginalized areas of the city that were targeted by lenders in the boom years. In the last full year before the crash, metropolitan Cleveland posted the worst Black-White racial disparity in subprime mortgage credit of all metropolitan areas nationwide—a more than seven-to-one disparity after controlling for applicant income, loan amount, loan purpose and estimated credit risk (Wyly et al., 2012). Inside the city of Cleveland, the spatial connections between the predatory boom and the policy aftermath persist even after accounting for the characteristics of individual borrowers and different kinds of lenders; these spatial links are remarkably durable.

For a borrower seeking a mortgage loan in the boom years between 2004 and 2006, the likelihood of getting a high-risk subprime loan was driven not only by racism and Wall Street securitization but also by the present *and future* neighborhood effects of foreclosures and demolitions (see Table 1). For a loan originated between 2004 and 2006, increasing the neighborhood's current foreclosure rate by one standard deviation boosts the subprime odds by 1.55; the effect is almost identical (1.53) for foreclosures between 2008 and 2012, and there is also a strong effect (1.49) for demolitions between 2008 and 2012. The persistence of these effects after controlling for a wide variety of supply- and demand-side factors attests to the enduring significance of neighborhood effects in the systemic inequalities of capital investment and disinvestment. The seemingly apolitical "growth first" policy interventions (Cox, 1999) of demolition, therefore, are taking place on a terrain with a deeply racialized history of class inequality. While the CCLRC's acquisition of foreclosed abandoned property is perhaps more progressive than the rampant speculation seen in the "market-only" land abandonment regime of Detroit, the geographies of investment and disinvestment have made the corporation a major force in the institutional dispossession of communities marginalized by class and race.

What does it indicate, then, when former Cuyahoga County Treasurer Jim Rokakis (2013) suggests that these neighborhoods are "primed for a demographic turnaround"? The history of urban renewal efforts in Cleveland indicates that well-intentioned public programs to spur equitable redevelopment have seen any potential redistributive effects defanged by "commonsense" business interests and racism (see Jenkins, 2001). The current demolition plan does little to help those who lost their homes in the crisis and instead looks to the next frontier of development—postcrisis, postforeclosure land cleared for new uses, and if possible, new people—but first and foremost, new profit opportunities. On this new frontier, will the Black neighborhoods that bore the brunt of the crisis see rising home values, more spacious lots and an expansion of green space? If these stated goals of the land bank are accomplished, how quickly will surviving residents be priced out of their neighborhoods?

CLASSIFICATION AND CATASTROPHE: MOBILIZING NEIGHBORHOOD TYPOLOGIES

While there are clear contrasts between Cleveland's historical experiences of redevelopment and the CCLRC in terms of governance structure and methods of land appropriation, in one area the historical legacy remains more profound: the persistence of renewal-style human ecology perspectives in the use of foreclosure demolitions to address urban decline. Much of the history of 20th-century urban planning was defined by the attempt to correlate negative social conditions with spatial patterns (Hall, 1988; Metzger, 2000). This "spatial fetishism" (Sheppard, 2009) is easily recognized in distant historical retrospect—in the meticulous fieldwork and mapping of the human ecology tradition of the Chicago School of sociology (Park & Burgess, 1925) and Hoyt's (1939) map overlays documenting the spatial convergence of each city's worst rental housing submarkets with the highest indicators of poverty and social "deviance" in the metropolis. This approach quickly became part of the practices of the Federal Housing Administration, as well as local housing authorities and private market institutions developing criteria for market potential and risk assessment (Jackson, 1985; Metzger, 2000; Ward, 1976). By the latter third of the 20th century, public and low-cost rental housing were pathologized and spatially joined in the American imagination as "slums" and "ghettos": in popular and policy discourse, social ills were given an ideologically precise location in the poorest, lowest-value neighborhoods.

Two variations on the Chicago School heritage are apparent in Cleveland's foreclosure demolition policies. First, the biological metaphors of *urban growth* that likened poverty to a "cancer" that would expand if left untreated (cf. Wilson, 1987/2012) have been replaced by an even more morbid language of decline: metaphors of disease are used to equate housing dereliction with the bodies that inhabit housing (Weber, 2002) and foreclosures are described as "zombies," as disease vectors that "infect" their communities with social problems and (further) declines in property values (Niquette, 2011). The language is an urban ecology for an age of urban ecological disaster. Given the long-term depopulation of rust belt cities, there is a reduced effectiveness of the old forms of territorial stigmatization that entailed blaming social outcomes on spatial choices—the "where you live affects your life chances" thesis (Slater, 2013, p. 1). Atop this established discourse is something new: the stigma of empty houses and the *potential* of social pathology. Empty houses attract "dangerous" people as well as the *fear* of dangerous people; both are believed to reduce property values in a fragile real estate market.

The second ecological legacy is the tradition of mapping social and market conditions at the neighborhood scale to support policies to reconfigure land use and housing. In the 1930s and 1940s, the newly documented links between deteriorating housing stock and neighborhood social change in the "zone in transition" implied clear policy solutions with targeted interventions in the spatial fabric of the metropolis. This is the recurrent theme from the slum clearance and public housing construction programs of the 1950s and 1960s (Harvey, 1973; Metzger, 2000; Ward, 1976) to the "underclass" poverty de-concentration and project demolition policies of the 1990s and 2000s (Crump, 2002). As John Agnew (1993, p. 258) emphasized, the spatial-ecological nexus "led to an important theoretical slippage: confusion that results from using spatial metaphors as simplified expressions for complex socio-spatial processes" (see also Agnew, 2010). While spatial conditions clearly have some influence on a neighborhood's social character, contemporary critical scholarship demonstrates that simplistic mapping of social process onto spatial form is dangerously reductive—and suppresses the complexity of local, regional, national and transnational processes that intersect at the neighborhood scale (Crump, 2002; Goetz, 2000; Harvey, 1973). Yet the deeply structured constraints of policy and law in American urbanism place severe limits on what can be done at the local scale. Thus the policy solution of razing foreclosed homes—one of the few local interventions that is allowed and enabled by today's policy environment—continues to draw on the cartographic rationality of neighborhood ecology.

Here lies a curious collision of the past and present in ecological perspectives. In the Regional Association of Cleveland's map of residential conditions in Cuyahoga County in 1941 (see Figure 4), we see the clear imprint of concentric-zone expansion of the growing industrial city, with older homes at the core filtering down in a cascade of inner-city "blight" requiring "protective" and "curative" "neighborhood conservation" or "large-scale planned reconstruction." Two-thirds of a century later (see Figure 5), we see a neighborhood taxonomy developed for Cleveland's request for support from the federal Neighborhood Stabilization Program (NSP), the first round of which was passed in the frantic months of economic collapse of late 2008 (see City of Cleveland, 2011). This is the neighborhood classification used by the city to target areas for demolition priority. The map provides a vivid reminder of the persistence of urban ecological thinking in present-day urban policy—despite the fact that it was created amidst a maelstrom of analysis, policy and action, where local neighborhood effects are now understood to be the product of "complex sociospatial processes" (Agnew, 1993, p. 258) at multiple scales. Cleveland's experience in the long economic crisis has been highlighted by repeated public revelations of the ties between local neighborhood conditions and national and transnational processes. A federal judge in Cleveland angrily denounced the "condescending mindset" and "judge, you just don't understand how things work" argument of lawyers for a multinational bank (Deutsche Bank) aggressively trying to foreclose on local families without being able to prove its ownership of the mortgage notes (Boyko, 2007, p. 5). The Federal Reserve Bank of Cleveland (2010) published a meticulous quantitative-spatial analysis of the links between neighborhood foreclosures and the high-risk lending of specific multinational lending institutions. As the foreclosure wave hit hard in 2008, the city sued under public nuisance law to connect the localized neighborhood disaster to the complex webs of securitization spun by more than a dozen financial institutions (City of Cleveland, 2008): the initial complaint, and Mayor Frank Jackson's comparison of the banks' financing for subprime lending to organized crime, could be read as a concise summary of theories of geographical scale and predatory class–monopoly rent. Yet the lawsuit was moved to federal court and dismissed on the grounds that the city lacked authority to regulate mortgage lending, and none of the other research and policy has changed established frameworks of law constraining what can be done at the urban scale. And so the ecological mapping of neighborhoods for demolition priority continues, with more data and advances in spatial and statistical analysis of the fine-grained local contours of disaster and divisive local struggles over who gets their "fair share" of demolitions.

$$\text{Conclusion: } \frac{Dynamo}{Doomsday\ Machine} = Shrinkage\ Machine$$

In a previous era, John Gunther (1947, p. 446) quipped that, with all those industrial smokestacks, Cleveland had "almost as much charm as an automobile cemetery or the inside of a dynamo," but he also praised the tradition of "good government" among the "rich burgher[s]" of the city who returned home each night "with a dutiful prayer, 'have I co-operated well today?'" (p. 444). When the planning technician for the Regional Association of Cleveland (Kayanan, 1943) mapped the residential areas of Cuyahoga County, the urban "density gradient" was clear: new growth at low densities around the edges and old density in a core where the urban fabric was "seriously blighted" and in need of "large-scale planned reconstruction." At this moment, shortly before the 1949 official manifesto of American urban renewal, Cleveland's population was just over 878,000, about 2.3 times the population at the turn of the century (U.S. Bureau of the Census, 1945). A century on, in early 2001, the threat of U.S. budget surpluses at the end of the Clinton Administration led the U.S. Treasury to announce plans to end the world's standard debt benchmark—the 30-year "long bond." Investors cast about for alternatives and began to pour capital into mortgage-backed securities—which had already begun pursuing growth opportunities in "new markets" of low-income people and communities of color. At this point, Cleveland had a population of 478,000 (U.S. Bureau of the Census, 2003), and it had become a textbook case of how the center-to-edge density gradient of models of urban structure continued radiating outward—without enough new investments and immigrants to replace the centrifugal forces of the spinning dynamo. Over the next several years, lenders pushed a flood of high-risk mortgage capital—$1.49 billion in the 2004–2006 period alone—into a shrinking city with a rapidly filtering housing stock. Cleveland was well on its way to losing another 81,000 people, more than a sixth of its already-shrunk population, and ended the decade with barely 1% more people than it had in 1900 (U.S. Bureau of the Census, 2012, p. 34). The industrial "dynamo" (Gunther, 1947) of a local growth machine of "rich burghers" was replaced by a dispersed cast of characters in the transnational "doomsday machine": realtors, brokers, bankers, speculators, hedge fund managers, international investors and all others who earn monopoly rents not from industrial production but from the circulation of "postindustrial widgets" of contemporary financialization (Newman, 2012).

After the damage inflicted by the doomsday machine, most of the industrial dynamo is gone—and so is the logic and rationale of the kind of midcentury urban renewal where "the Federal Bulldozer approach" was similar to a "Corbusian prescription" in "The City of Towers" (Hall, 1988, p. 233). Yet the enduring realities of uneven urban development mean that there is an essential, urgent need for something resembling "renewal" in the strip-mined horizontal single-family housing landscapes of a steady shrinking metropolis. In this context, growth machines of locally dependent interests and institutions confront an array of constraints—from the contradictions of American financial federalism, to the paltry funds available from a paralyzed Washington, DC, to the schemes of "vulture investors" large and small looking for a competitive niche *in* but not *for* Cleveland (Kotlowitz, 2009). As the "preconditions for growth" of classical growth machine theory (Molotch, 1976) have shifted, urban coalitions must adapt while competing against global investors. The CCLRC is one kind of flexible, entrepreneurial institution that mediates the interests of local, land-based elites and public-sector officials working to balance the neighborhood-level conflicts of a depopulating city amidst the decentralized de-industrialization of urban structure and the galactic complexity of neoliberal public-private finance.

In this chapter, we analyzed three features of this "shrinkage machine" and its position in the policy lineage of urban renewal. First, this machine is a new but familiar public authority at the legal and spatial juncture of public/private and local/extralocal domains: its legal and contractual status evades local democracy while drawing from a mixture of local, federal, and Wall Street financing. Second, the machine's land appropriations and demolitions have followed the broad contours of class and racial inequality imposed by the latest round of predatory financialization—even as the long crisis has forced neighborhoods into divisive local struggles to avoid the most severe dysfunctions of disinvestment (Smith, Caris, & Wyly, 2001) signaled by an abandoned, foreclosed house. In Cleveland's bizarre moment of permanent emergency, we see a new form of de-urbanizing White privilege: White urban neighborhoods in a majority-Black city fighting to get their "fair share" of demolitions (Narkin, 2011). Finally, the shrinkage machine is distinguished by a refined tradition of ecological-spatial analysis, mapping parcels and neighborhoods for demolition priority in an era of permanent austerity—even if everyone now understands that those maps of empty houses were rendered not only by sophisticated mapping software but by the competitive evolution of the "genetically modified" political economy (Smith, 2005, p. 891) of global finance and American neoliberalism. In the shrinkage machine, we see the aspirations for investment returns *in* but not *for* cities. Just as industrial profits now surge without corresponding gains in employment, urban entrepreneurialism hopes to renew urban capital without the need for so many urban residents: the latest "back to the city" movement by capital, not people (Smith, 1979).

Table 1. Modeling the Neighborhood Effects of Predatory Capital in Cleveland.

	Model 1: Current Foreclosures		Model 2: Future Foreclosures		Model 3: Future Demolitions	
	Coefficient	Odds Ratio	Coefficient	Odds Ratio	Coefficient	Odds Ratio
Intercept	-1.7476 ***	0.174	-1.705 ***	0.182	-1.2165 ***	0.296
Applicant income ($1,000)	0.0000006008 **	1.033	0.0000004367	1.024	0.0000001655	1.009
Loan to income ratio	-0.1159 ***	0.650	-0.1122 ***	0.660	-0.1041 ***	0.680
Loan to income ratio squared	0.000415 **		0.000401 *		0.000367	
Owner-occupied	-0.7688 ***	0.464	-0.8004 ***	0.449	-0.7576 ***	0.469
Pre-approval requested	-1.9635 ***	0.140	-1.9773 ***	0.138	-1.9429 ***	0.143
Subordinate lien	0.5575 ***	1.746	0.5209 ***	1.684	0.4994 ***	1.648
Home improvement	-0.9962 ***	0.369	-0.9619 ***	0.382	-0.9376 ***	0.392
Refinance	-0.5727 ***	0.564	-0.5529 ***	0.575	-0.5223 ***	0.593
Demographic information missing	0.5391 ***	1.715	0.5918 ***	1.807	0.6222 ***	1.863
Hispanic	0.2345 **	1.264	0.2304 **	1.259	0.2361 ***	1.266
Native American	0.1403	1.151	0.1778	1.195	0.237	1.267
Asian	0.1187	1.126	0.1398	1.150	0.1198	1.127
Non-Hispanic Black, Men	0.5523 ***	1.737	0.6654 ***	1.945	0.7189 ***	2.052
Non-Hispanic Black, Women	0.3698 ***	1.447	0.4787 ***	1.614	0.5662 ***	1.762

Table 1. *Continued*

	Model 1: Current Foreclosures		Model 2: Future Foreclosures		Model 3: Future Demolitions	
	Coefficient	Odds Ratio	Coefficient	Odds Ratio	Coefficient	Odds Ratio
Non-Hispanic Black, Couple	0.2375 ***	1.268	0.3354 ***	1.398	0.4556 ***	1.577
Year 2005	0.8822 ***	2.416	0.8679 ***	2.382	0.8725 ***	2.393
Year 2006	1.4377 ***	4.211	1.4164 ***	4.122	1.4158 ***	4.120
OCC-regulated bank	-0.6976 ***	0.498	-0.7087 ***	0.492	-0.7217 ***	0.486
FDIC-regulated bank	-1.3012 ***	0.272	-1.315 ***	0.268	-1.3094 ***	0.270
OTS-regulated savings & loan	-0.3802 ***	0.684	-0.39 ***	0.677	-0.3933 ***	0.675
NCUA-regulated credit union	-3.3031 ***	0.037	-3.306 ***	0.037	-3.2809 ***	0.038
HUD-reporting independent mortgage company	1.2189 ***	3.383	1.224 ***	3.401	1.2209 ***	3.390
Loan sold to GSE	-1.2971 ***	0.273	-1.3079 ***	0.270	-1.2865 ***	0.276
Loan sold to private purchaser	1.4085 ***	4.090	1.4101 ***	4.096	1.4199 ***	4.137
Loan sold to bank	0.8714 ***	2.390	0.8674 ***	2.381	0.8763 ***	2.402
Loan sold to finance company	0.5174 ***	1.678	0.5261 ***	1.692	0.5439 ***	1.723
Loan sold to affiliate institution	0.4025 ***	1.496	0.4048 ***	1.499	0.4136 ***	1.512
Loan sold to other purchaser (SPV)	0.4182 ***	1.519	0.4259 ***	1.531	0.4325 ***	1.541

Table 1. *Continued*

	Model 1: Current Foreclosures		Model 2: Future Foreclosures		Model 3: Future Demolitions	
	Coefficient	Odds Ratio	Coefficient	Odds Ratio	Coefficient	Odds Ratio
Foreclosure rate, 2004–2006	0.2595 ***	1.557				
Foreclosure rate, 2004–2006 squared	-0.00886 ***					
Foreclosure rate, 2008–2012			0.177 ***	1.526		
Foreclosure rate, 2008–2012 squared			-0.00348 ***			
Demolition rate, 2008–2012					0.3253 ***	1.492
Demolition rate, 2008–2012 squared					-0.0237 ***	
Number of observations (conventional loans originated & collateralized by 1- to 4-family dwellings)	39,756				39,756	
Number of high-cost conventional subprime loans	17,970				17,970	
Max-rescaled R-squared measure of fit	0.51				0.51	
Percent concordant	87.1				86.9	

*Significant at $P<0.05$; **$P<0.01$; ***$P<0.001$.

Fig 1. Initial foreclosure filings, by census tract, as share of occupied housing units.

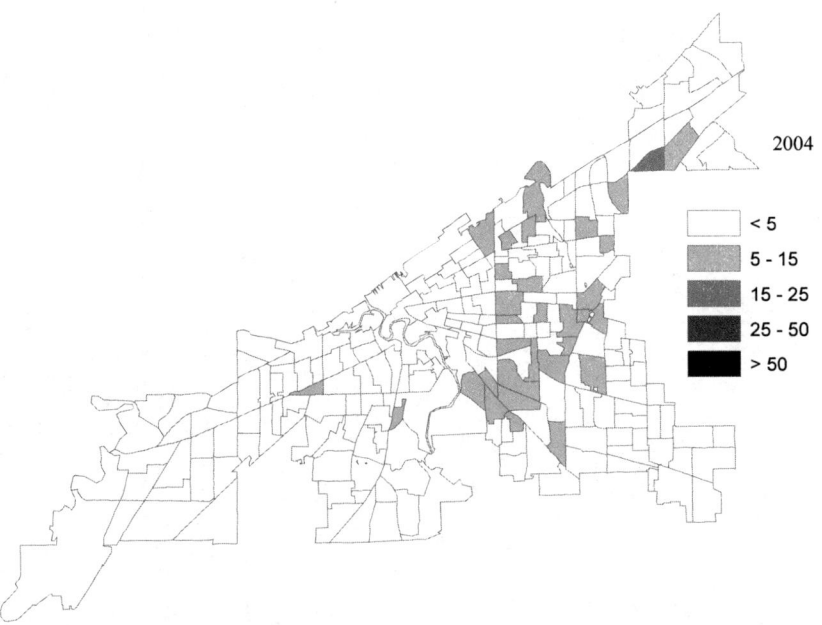

Fig 1. *Continued*

THE SHRINKAGE MACHINE | 61

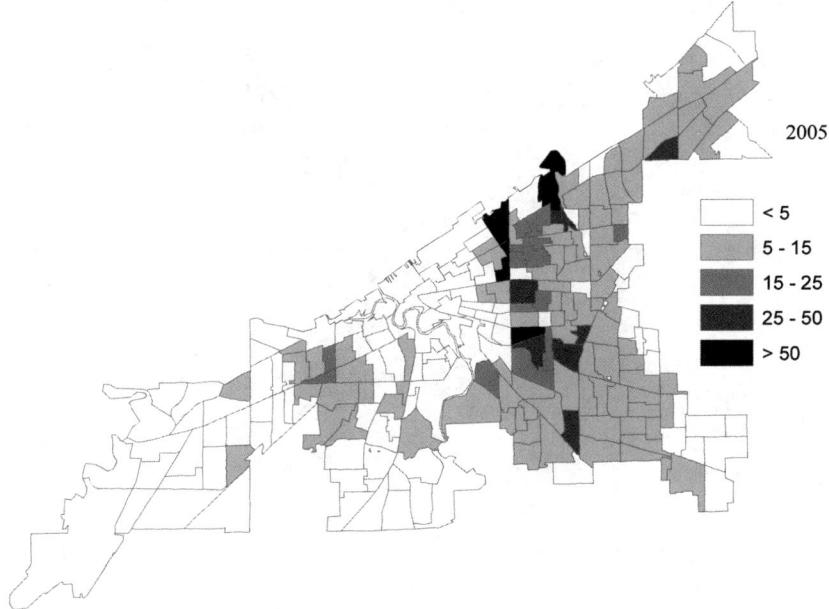

Fig 1. *Continued*

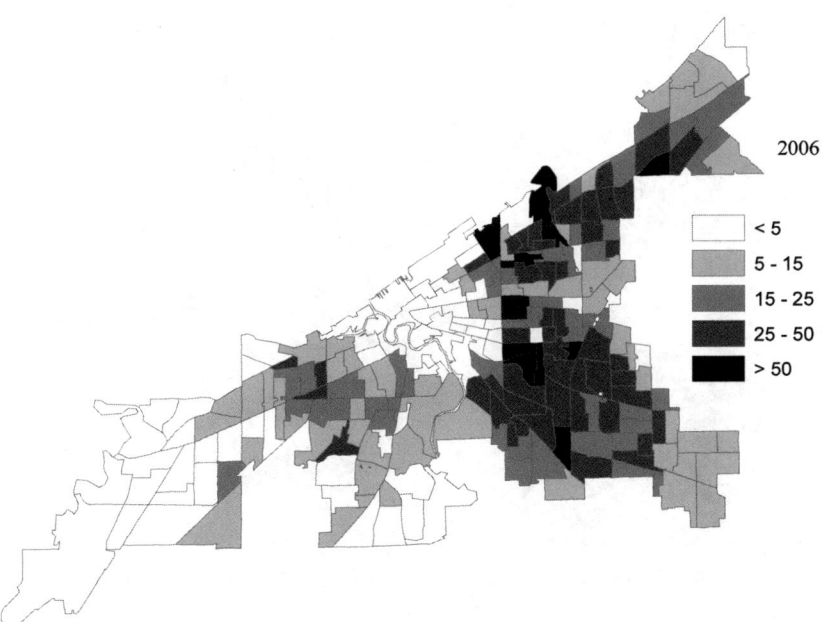

Fig 1. *Continued*

Fig 1. *Continued*

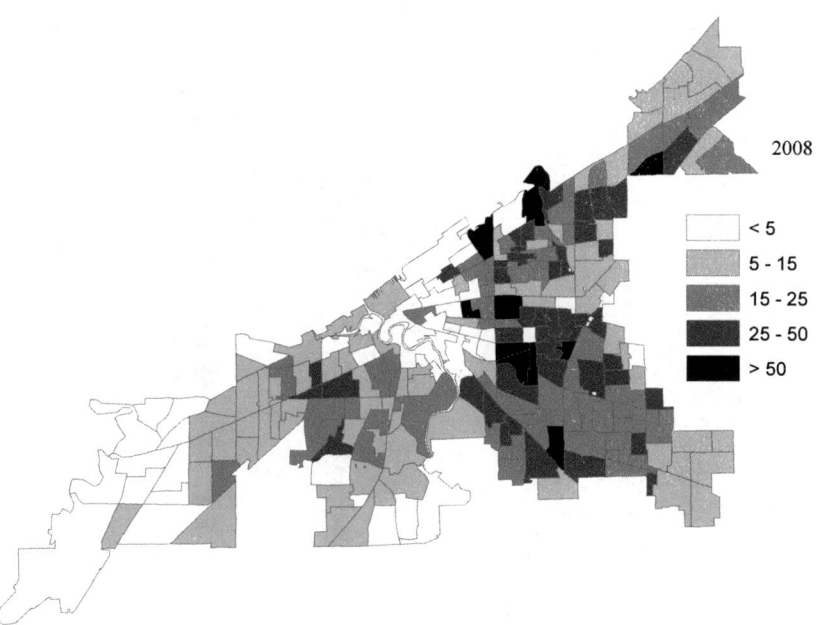

Fig 1. *Continued*

THE SHRINKAGE MACHINE | 63

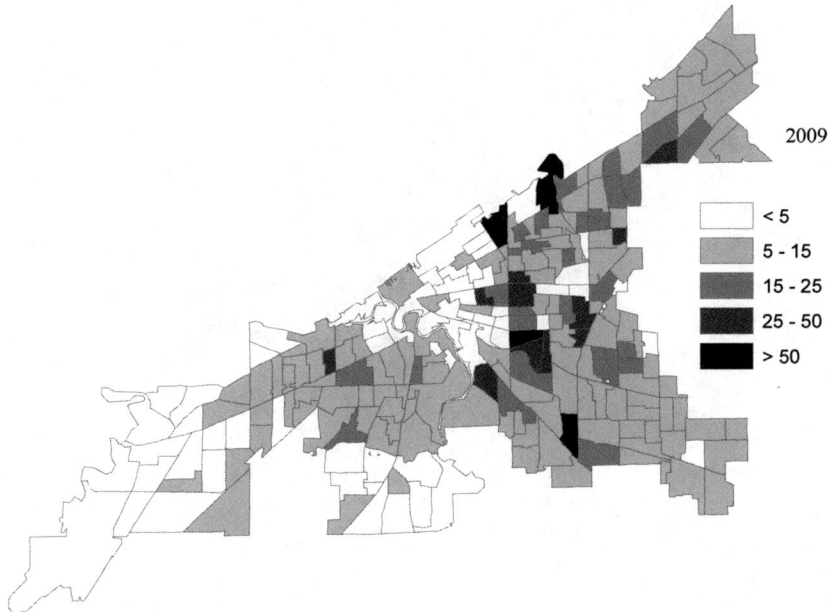

Fig 1. *Continued*

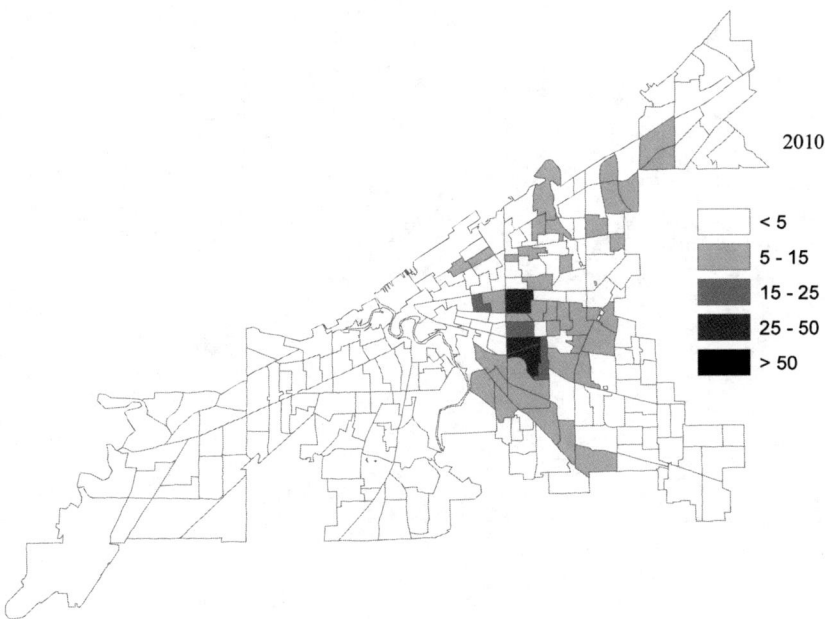

Fig 1. *Continued*

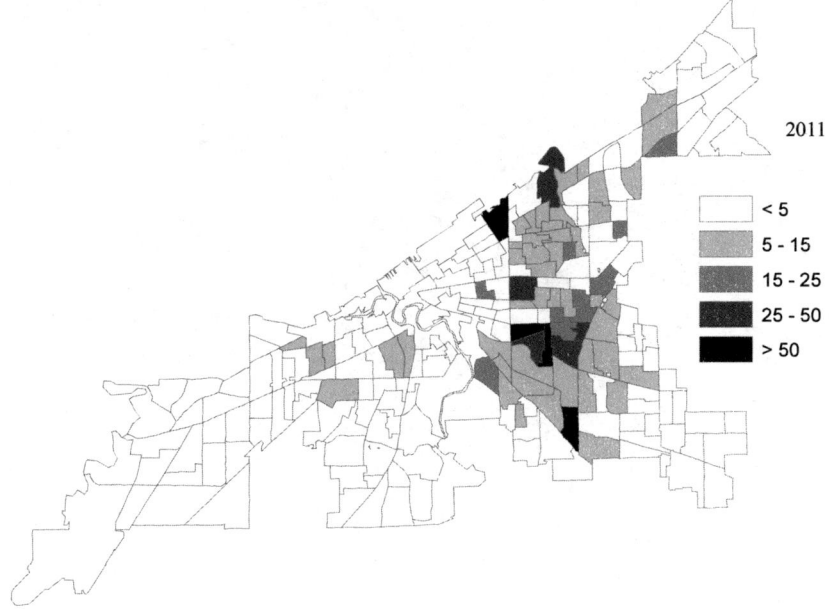

Fig 1. *Continued*

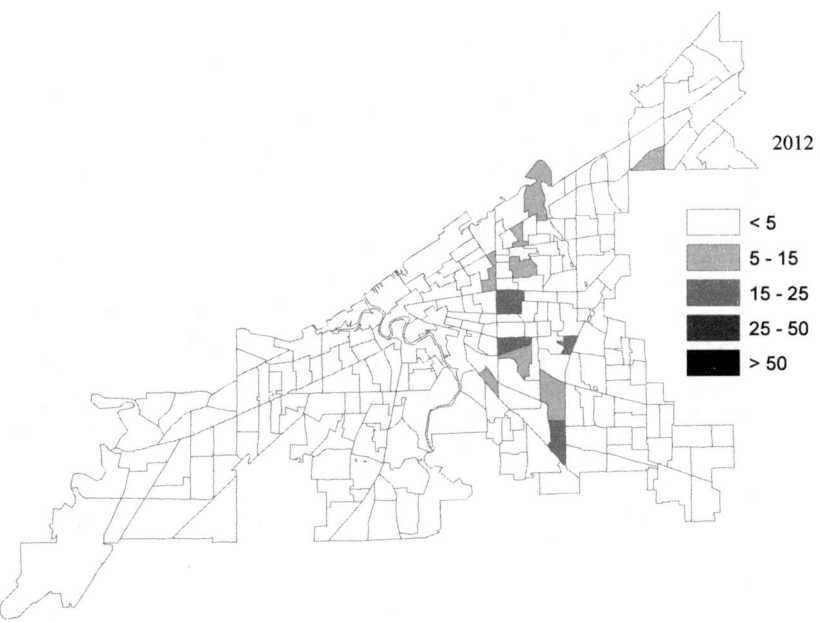

Fig 1. *Continued*

Data sources: City of Cleveland and Cuyahoga County Clerk; U.S. Bureau of the Census, American Community Survey (2005–2009).

THE SHRINKAGE MACHINE | 65

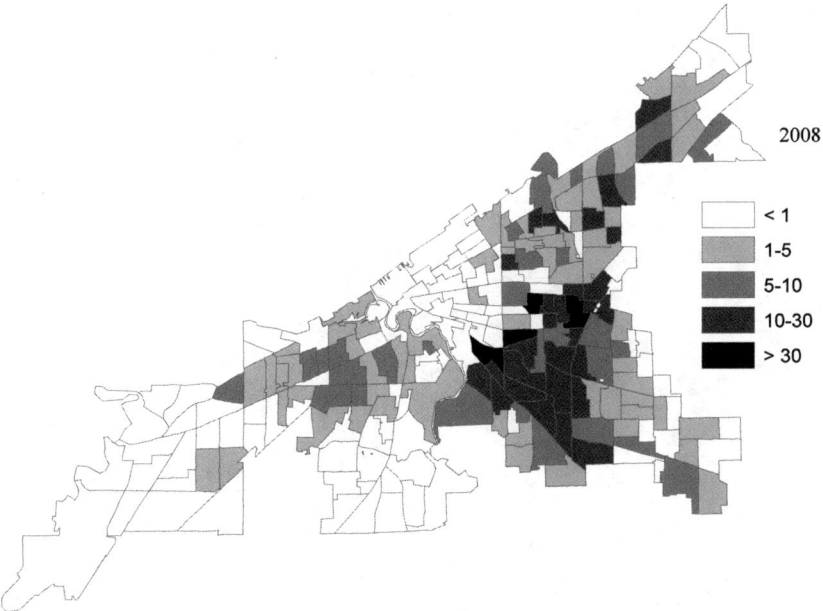

Fig 2. Home demolitions, by census tract, per 1,000 occupied housing units, 2008–2012.

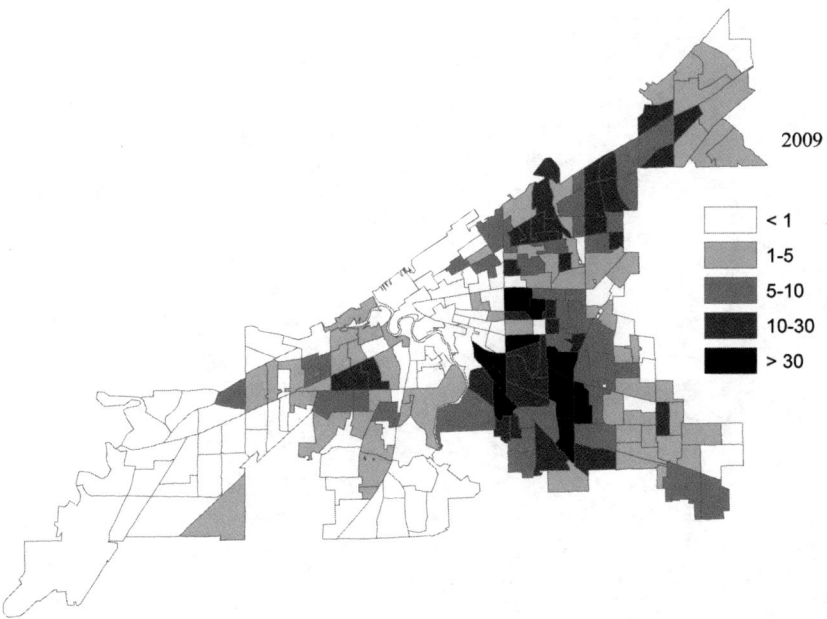

Fig 2. *Continued*

Fig 2. *Continued*

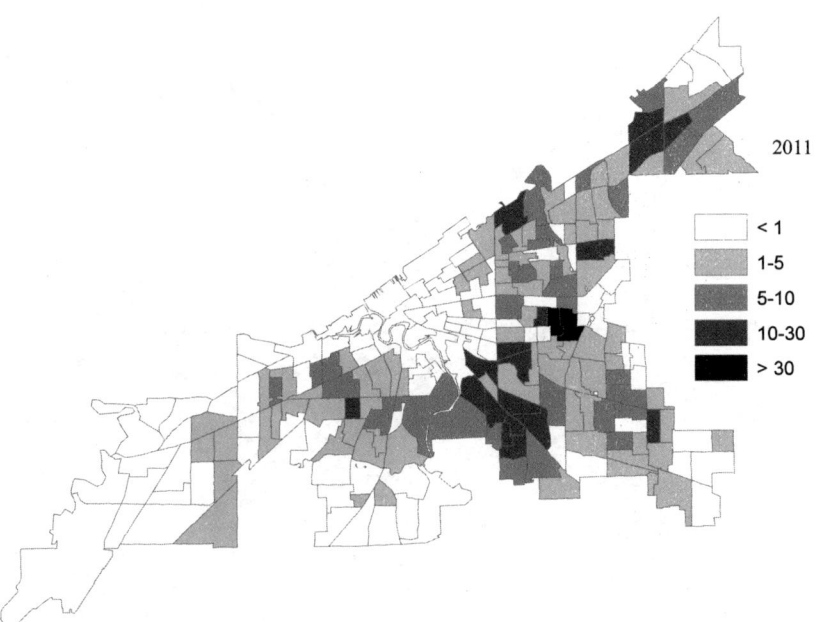

Fig 2. *Continued*

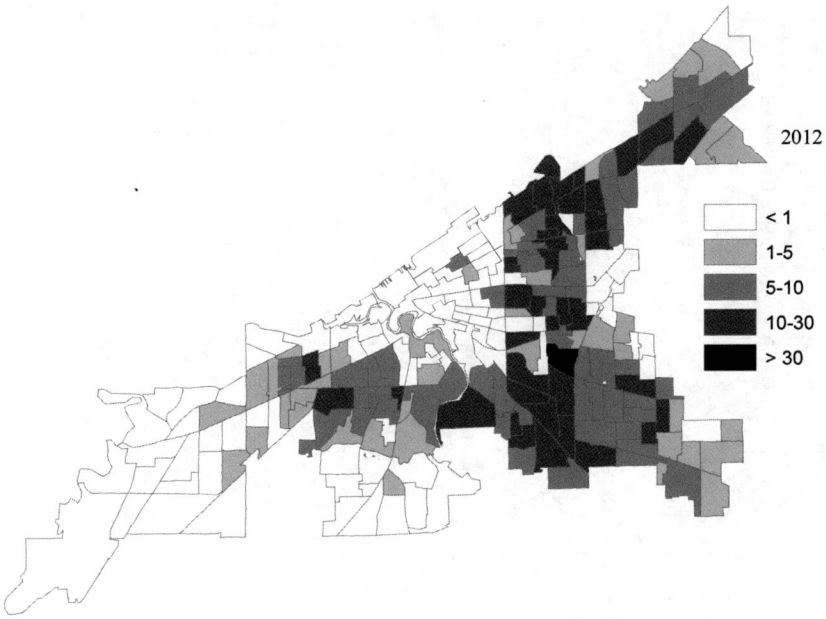

Fig 2. *Continued*

Data Sources: City of Cleveland and Cuyahoga County Clerk; U.S. Bureau of the Census, American Community Survey (2005–2009).

Fig 3. After mortgage foreclosures and evictions, the Cuyahoga County Sheriff's Office sweeps the premises. Deputies enter the houses at gunpoint as a precaution against recalcitrant residents, squatters or drug dealers. Anthony Suau's photograph of such a sweep became a globally circulating image of America's housing catastrophe and global financial crisis. The image was selected from 96,268 photographs submitted by photojournalists for the 2008 World Press Photograph of the Year.

Source: ©Anthony Suau, reproduced with permission.

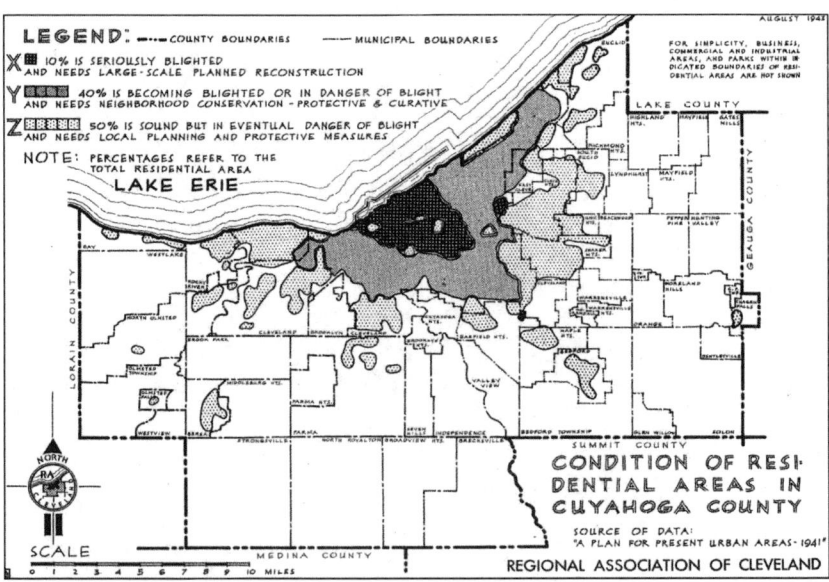

Fig 4. Condition of residential areas in Cuyahoga County, 1941.

Source: Kayanan (1943), public domain image.

THE SHRINKAGE MACHINE | 69

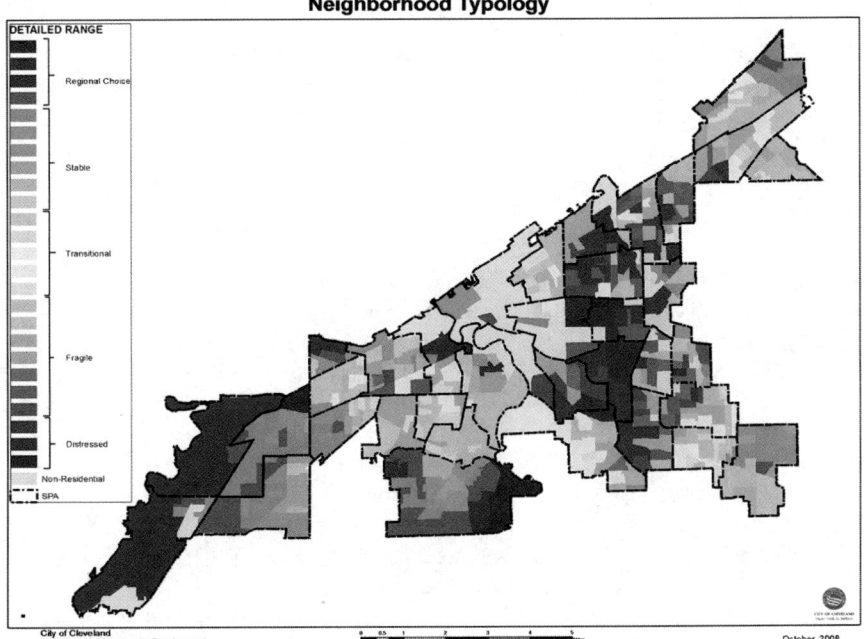

Fig 5. Neighborhood typology and priority areas, Cleveland, 2008. Note: This map was not able to be reproduced in color, and thus much of the detail is lost. However, we include it here to illustrate the use of neighborhood typologies by municipal planners. The vast majority of the tracts labeled "fragile" or "distressed" are in the eastern neighborhoods of Cleveland.

Source: Cummins (2013), public domain image.

Note: odds ratios for continuous variables report the change in odds with a one-standard deviation increase in the predictor.
Data Source: Federal Financial Institutions Examination Council (2005–2007).

NOTES

1. Parts of this chapter are based on a database compiled from public records of mortgage originations, foreclosures and demolitions, built by Sam Walker and Emily Rosenman and edited with assistance from members of the 2013 Urban Research Studio at the University of British Columbia. We thank Victor Douglas Ngo, Joanna Yang, Jenna McNeil, Anna Yuen, Ben Porteous, Andrew Tesarowski, Jennifer Chan and Sven Koberwitz for their excellent research assistance.
2. Hyra (2012) differentiated between the "old" urban renewal of slum clearance and the "new" urban renewal of public housing demolitions that began in 1992 under the "HOPE IV" legislation.
3. Real estate owned properties have failed to sell at a foreclosure auction and thus become the property of the lender, usually a bank or government-sponsored enterprise or agency. By definitions, REOs are properties that cannot be sold except at an extreme loss; lenders therefore

are incentivized to undertake only a minimal level of maintenance until a buyer can be found. It is common for lenders to forego even minimal maintenance expenditures in low-income and minority neighborhoods (see National Fair Housing Alliance, 2012).

4. This program was nearing its end when the CCLRC was incorporated, but some Cleveland demolition funds did go to the CCLRC (Ohio Auditor of State, 2011).

REFERENCES

Aalbers, M. (2009). Geographies of the financial crisis. *Area, 41*(1), 34–42.

Agnew, J. (1993). Representing space, scale, and culture in the social sciences. In J. Duncan & D. Ley (Eds.), *Place/Culture/Representation* (pp. 251–271). New York, NY: Routledge.

Agnew, J. (2010). Slums, ghettos, and urban marginality. *Urban Geography, 31*(2), 144–147.

Akers, J. (2013). Making markets: Think tank legislation and private property in Detroit. *Urban Geography, 34*(8), 1070–1095.

Alexander, F. S. (2011). *Land banks and land banking.* The Center for Community Progress. Retrieved from http://www.communityprogress.net/filebin/pdf/new_resrcs/LB_Book_2011_F.pdf

Atassi, L. (2012, May 6). Cuyahoga County gets nearly $12 million from settlement to eliminate blight. *Cleveland Plain Dealer.* Retrieved from http://www.cleveland.com/metro/index.ssf/2012/05/cuyahoga_county_to_receive_nea.html

Atassi, L. (2013, February 20). Proposed federal legislation could generate millions of dollars for demolition in Cleveland. *Cleveland Plain Dealer.* Retrieved from http://www.cleveland.com/cityhall/index.ssf/2013/02/proposed_federal_legislation_c.html

Boyko, C. (2007). *In re foreclosure cases no. 01:07CV2282 et al., opinion and order.* Cleveland: U.S. District Court, Northern District of Ohio, Eastern Division.

Brenner, N., & Theodore, N. (2002). Cities and the geographies of "actually existing neoliberalism." *Antipode, 34*(3), 349–379.

Bulava, K. (2011, July 9). Wells Fargo, Bank of America donating properties to Cuyahoga Land Bank. *The Collinwood Observer.* Retrieved from http://www.collinwoodobserver.com/read/2011/07/09/wells-fargo-bank-of-america-donating-properties-to-cuyahoga-land

Carmon, N. (1999). Three generations of urban renewal policies: Analysis and policy implications. *Geoforum, 30*(2), 145–158.

Caro, R. (1974). *The power broker: Robert Moses and the fall of New York.* New York, NY: Knopf.

Center on Urban Poverty and Community Development. (2012). Cleveland and Northeast Ohio indicator data: Cleveland area demolitions since 2008, prepared by Justin Fleming, Neighborhood Progress Inc. *Case Western Reserve University, Mandel School of Applied Social Sciences, Center on Urban Poverty and Community Development.* Retrieved from http://neocando.case.edu/cando/cookie.jsp

City of Cleveland. (2008). *City of Cleveland v. Deutsche Bank et al., CV 08646970.* Cleveland, OH: Court of Common Pleas, Cuyahoga County.

City of Cleveland. (2011). *Amendment to FY 2010–2011 Community Development Block Grant Action Plan.* Cleveland, OH: Cleveland Department of Community Development.

Cox, K. (1999). Ideology and the growth coalition. In A. Jonas & D. Wilson (Eds.), *The urban growth machine: Critical perspectives two decades later* (pp. 21–36). Albany: SUNY Press.

Crump, J. (2002). Deconcentration by demolition: Public housing, poverty, and urban policy. *Environment and Planning D, 20*(5), 581–596.

Cummins, B. (2013). *Approaches for improving our neighborhood housing*. Cleveland, OH: Department of Community Development, Cleveland City Council.
Cuyahoga County Land Reutilization Corporation (CCLRC). (2009). *Report to the Ohio General Assembly, November 25, 2009*. Retrieved from http://cuyahogalandbank.org/documents/organizational/cclrc_2009_report_general_assembly.pdf
Cuyahoga County Land Reutilization Corporation (CCLRC). (2011). *Comprehensive ethics policy of the Cuyahoga County Land Reutilization Corporation*. Retrieved from http://cuyahogalandbank.org/documents/organizational/CCLRC_ETHICS_POLICY_20101029.pdf
Fainstein, S. (2005). The return of urban renewal. *Harvard Design Magazine*, Issue 22, Spring/Summer 2005, 1–5.
Federal Financial Institutions Examination Council (Annual). *Home Mortgage Disclosure Act, loan application register raw data*. Washington, DC: Board of Governors of the Federal Reserve System.
Federal Reserve Bank of Cleveland. (2010). *Facing the foreclosure crisis in greater Cleveland: What happened and how communities are responding*. Cleveland, OH: Federal Reserve Bank of Cleveland.
Fitzpatrick, T. J. IV (2009). Ohio's Land Bank Legislation: Modernizing an Aged Model. Journal of Affordable Housing & Community Development Law 19(3), 127–147.
Goetz, E. (2000). The politics of poverty deconcentration and housing demolition. *Journal of Urban Affairs, 22*(2), 157–173.
Goetz, E. (2011). Where have all the towers gone? The dismantling of public housing in U.S. cities. *Journal of Urban Affairs, 33*(3), 267–287.
Gratz, R. (2010, April 19). Shrinking cities: Urban renewal revisited? *Planetizen Blog*. Retrieved from http://www.planetizen.com/node/43826
Gunther, J. (1947). *Inside USA*. New York, NY: Curtis.
Hackworth, J. (2014 forthcoming). The limits to market-based strategies for addressing land abandonment in shrinking American cities. *Progress in Planning*.
Hall, P. (1988). *Cities of tomorrow: An intellectual history of urban planning and design in the twentieth century*. Oxford, UK: Blackwell.
Harvey, D. (1973). *Social justice and the city*. Athens: The University of Georgia Press.
Harvey, D. (1989). From managerialism to entrepreneurialism: The transformation in urban governance in late capitalism. *Geografiska Annaler B*, 71(1), 3–17.
Harvey, D. (2012). *Rebel cities: From the right to the city to the urban revolution*. London, UK: Verso.
Hirsch, A. (1998/1983). *Making the second ghetto: Race and housing in Chicago, 1940–1960*. Chicago: University of Chicago Press.
Hoyt, H. (1939). The structure and growth of residential neighborhoods in American cities. *Federal Housing Administration*.
Hyra, D. (2012). Conceptualizing the new urban renewal: Comparing the past to the present. *Urban Affairs Review, 48*(4), 498–527.
Jackson, K. (1985). *Crabgrass frontier: The suburbanization of the United States*. Oxford, UK: Oxford University Press.
Jenkins, W. D. (2001). Before downtown: Cleveland, Ohio, and urban renewal, 1949–1958. *Journal of Urban History*, 27(4), 471–496.
Kayanan, A. (1943). *Neighborhood conservation*. Cleveland, OH: Regional Association of Cleveland.
Kelo v. City of New London, 545 U.S. 469 (2005). Washington, DC: US Supreme Court.
Kotlowitz, A. (2009, March 8). All boarded up. *The New York Times*. Retrieved from http://www.nytimes.com/2009/03/08/magazine/08Foreclosure-t.html?pagewanted=all&_r=0

Krippner, G. (2012). *Capitalizing on crisis: The political origins of the rise of finance.* Cambridge, MA: Harvard University Press.

Lovell, W. (2012). *The Kelo blowback: How the newly enacted eminent domain statutes and past blight statutes are a Maginot-Line defense for all non-affluent and minority property owners.* Columbus: Ohio State University Moritz College of Law.

MacDonald, C. (2011, February 3). Private landowners complicate reshaping of Detroit. *Detroit News.* Retrieved from http://www.detroitnews.com/article/20110203/METRO01/102030395

Metzger, J. (2000). Planned abandonment: The neighborhood life-cycle theory and national urban policy. *Housing Policy Debate, 11*(1), 7–40.

Molotch, H. (1976). The city as a growth machine: Toward a political economy of place. *The American Journal of Sociology, 82*(2), 309–332.

Narkin, J. (2011, June 1). Grassroots group challenges city of Cleveland officials to better address vacant houses in Ward 14. *Plain Press.* Retrieved from http://plainpress.wordpress.com/2011/06/01/grassroots-group-challenges-city-of-cleveland-officials-to-better-address-vacant-houses-in-ward-14/

National Fair Housing Alliance. (2012). *The banks are back—our neighborhoods are not: Discrimination in the maintenance and marketing of REO properties.* Retrieved from http://www.nationalfairhousing.org/Portals/33/the_banks_are_back_web.pdf

Newman, K. (2012). The new economy and the city: Foreclosures in Essex County, NJ. In M. Aalbers (Ed.), *Subprime cities: The political economy of mortgage markets* (pp. 219–241). Chichester, UK: Wiley-Blackwell.

Niquette, M. (2011, December 21). Cleveland's zombie house scrouge. *Financial Post.* Retrieved from http://business.financialpost.com/2011/12/21/clevelands-zombie-house-scourge/

Ohio Auditor of State. (2011). *Cuyahoga County Land Reutilization Corporation regular audit for the year ended December 31, 2009.* Retrieved from http://cuyahogalandbank.org/documents/organizational/Final_2009_audit_report_issued_060911.pdf

Park, R., & Burgess, E. (1925). *The city.* Chicago, IL: The University of Chicago Press.

Peck, J. (2012). Austerity urbanism: American cities under extreme economy. *City, 16*(6), 626–655.

Peck, J., & Tickell, A. (2002). Neoliberalizing space. *Antipode, 34*(3) 380–404.

Rokakis, J. (2013, April 20). Demolishing vacant homes will help build community: Jim Rokakis. *The Plain Dealer.* Retrieved from http://www.cleveland.com/opinion/index.ssf/2013/04/demolishing_vacant_homes_will.html

Sassen, S. (2012). Expanding the terrain for global capital: When local housing becomes an electronic instrument. In M. Aalbers (Ed.), *Subprime cities: The political economy of mortgage markets* (pp. 74–96). Chichester, UK: Wiley-Blackwell.

Sheppard, E. (2009). Spatial fetishism. In D. Gregory, R. Johnston, G. Pratt, M. J. Watts, & S. Whatmore (Eds.), *The dictionary of human geography, fifth edition* (p. 712). Chichester, UK: Wiley-Blackwell.

Slater, T. (2013). Your life chances affect where you live: A critique of the "cottage industry" of neighborhood effects research. *International Journal of Urban and Regional Research, 37*(2), 367–387.

Smith, N. (1979). Toward a theory of gentrification: A back to the city movement by capital, not people. *Journal of the American Planning Association, 45*(4), 538–548.

Smith, N. (2005). Neo-critical geography, or, the flat pluralist world of business class. *Antipode, 37*(5), 887–899.

Smith, N., Caris, P., & Wyly, E. (2001). The "Camden Syndrome" and the menace of suburban decline: Residential disinvestment and its discontents in Camden County, NJ. *Urban Affairs Review, 36*(4), 497–531.

Stone, S. (1978). Planned shrinkage: Coping with urban decline. *Wayne Law Review, 25,* 987.

The Plain Dealer Editorial Board. (2012, July 9). Costly EPA rule on residential demolitions needs second look: Editorial. *The Plain Dealer.* Retrieved from http://www.cleveland.com/opinion/index.ssf/2012/07/costly_epa_rule_needs_second_1.html

U.S. Bureau of the Census. (1945). *Statistical abstract of the United States, 1944–45.* Washington, DC: U.S. Department of Commerce.

U.S. Bureau of the Census. (2003). *Statistical abstract of the United States, 2003.* Washington, DC: U.S. Department of Commerce.

U.S. Bureau of the Census. (2012). *Statistical abstract of the United States, 2012.* Washington, DC: U.S. Department of Commerce.

Ward, D. (1976). The Victorian slum: An enduring myth? *Annals of the Association of American Geographers, 66*(2), 323–364.

Weber, R. (2002). Extracting value from the city: Neoliberalism and urban redevelopment. *Antipode, 34*(3), 519–540.

Wilson, W. J. (2012 [1987]). *The truly disadvantaged.* Chicago, IL: The University of Chicago Press.

Wyly, E., Atia, M., Foxcroft, H., Hammel, D. J., & Phillips-Watts, K. (2006). American home: Predatory mortgage capital and neighborhood spaces of race and class exploitation in the United States. *Geografiska Annaler B, 88*(1), 105–132.

Wyly, E., Moos, M., & Hammel, D. (2012). Race, class, and rent in America's subprime cities. In M. Aalbers (Ed.), *Subprime cities: The political economy of mortgage markets* (pp. 242–289). Malden, MA: Wiley-Blackwell.

PART TWO

Reorienting Gentrification

Space, Equity and Voice

Englewood, on Chicago's south side, is an African American community rich in cultural history but challenged by decades of residential poverty and economic disinvestment. Intense capital withdrawal in the 1970s catapulted much of the vicinity into financial distress, transforming once active industries, lavish parks, three-story gray stones and a busy shopping concourse into its current landscape of foreclosed homes, boarded-up buildings and vacant lots. As of 2011, the area's 30,000-plus inhabitants represented less than one-third of its 1960 population. In spring 2012, Mayor Rahm Emanuel targeted Englewood as an "opportunity area" for public and private investment. His plan is to first address the community's food desert crisis by erecting a Whole Foods market and turning empty lots into vegetable gardens. An additional fiscal boost is coming from the railway company, Norfolk Southern, which, over the past several years, has been procuring neighborhood land and demolishing homes for business expansion purposes.

On the surface, urban agriculture and an end to food deserts appears favorable for Englewood. Without question, residents strongly desire a rebirth of the existing infrastructure. The problem is that many of the decisions being made, with respect to the area's future, have been top-down and with very little civic input. How, for example, will Englewood citizens (42% being low income) afford the upscale commodities of Whole Foods? What is more, community groups have shown that Norfolk Southern's development plans are environmentally hazardous. The increased use of freight cars and trucks has health impacts related to noise pollution, as well as toxic diesel air contamination that can exacerbate chronic lung diseases such as asthma. In September 2013, however, after months of disputes, Englewood constituents and other community organizations successfully negotiated with Norfolk Southern and the city of Chicago for new green space and toxic waste reductions.

Push back from Englewood citizens has less to do with averting the process of revitalization and more with voicing their needs and concerns around their quality of life and well-being. Residents, both young and old, most definitely wish to be included in renewal policies that intimately impact the space around them. They are indeed the experts of their own sociocultural contexts and have fundamentally different notions on how to equitably improve their lived conditions. The authors in this section assert that while urban renewal has its advantages for improving struggling communities, when it is done in ways that are irresponsible to the environment and that neglect democratic participation, then we must expose, critically examine and contest those profit-driven, top-down arrangements that, on the surface, appear as opportunities.

HRH

CHAPTER FOUR

Sustainable Urban Development AND Environmental Gentrification

The Paradox Confronting the U.S. Environmental Justice Movement

DANIEL FABER AND SHELLEY McDONOUGH KIMELBERG[1]

INTRODUCTION

Sustainable urban redevelopment projects are becoming increasingly prevalent across the United States and Canada. In fact, sustainability principles are now commonly integrated into both the vernacular and content of urban policy, planning and development processes of most major cities (Bunce, 2009, pp. 652–653). These include energy- and resource-efficient buildings; bike lanes, walking paths and better mass transportation systems; alternative energy and technology; improved parks and green spaces; antisprawl initiatives; climate mitigation and adaptation measures; environmental cleanup of the land, air and water; urban gardening and sustainable agriculture; and recycling programs, to name but a few.

At first glance, such urban sustainability initiatives seem to offer an unassailable public good. However, a number of scholars have raised questions about the distributional impacts of these efforts. As Curran and Hamilton (2012) noted, "Many visions of the green city seem to have room only for park space, waterfront cafes, and luxury LEED-certified buildings, prompting concern that there is no place in the 'sustainable' city for industrial uses and the working class" (p. 1027). In fact, the urban sustainability initiatives often launched to stimulate private investment in sectors such as high technology, real estate, energy or finance are explicitly designed to restore or build the environmental amenities and reputational status thought to attract a highly educated and skilled workforce (Florida, 2002; Tretter, 2013). In this context, "greening the city" can be seen as part of a broader neoliberal strategy to remake the urban landscape to satisfy the lifestyle demands of the middle and professional classes (Bunce, 2009, pp. 652–654). North America is not unique in this respect. According to Tretter (2013), "The compatibility between an agenda for sustainable urban development and the neoliberal economic restructuring of urban space has been observed within cities in developed countries across the globe" (p. 297).

This treatment of environmental quality as yet another consumer "amenity" with the potential to fuel economic growth (see Clark, Lloyd, Wong, & Jain, 2002) has important implications for the affordability and accessibility of the city. Environmental quality has long been known to exercise a significant influence over the price of housing (Boyle & Kiel, 2001). "In short, land values tend to be relatively higher near [environmental] amenities and relatively lower near [environmental] disamenities and tend to adjust as those amenities and disamenities change" (Eckerd, 2011, p. 34). Therefore, green projects that create open space or otherwise aim to improve the environmental profile of a neighborhood can trigger increases in real estate prices, rents and property taxes, leading to the economic displacement of the existing residents who had endured the deleterious effects of pollution and ecological degradation (Banzhaf & McCormick, 2012). At the same time, environmental upgrading can result in what Marcuse (1986) termed "exclusionary displacement," rendering the sustainable city inaccessible for future residents of limited economic means. Housing price increases hit the poor—the vast majority of whom are renters—especially hard. It is typically only landlords and homeowners who stand to capture the property value gains associated with enhanced environmental amenities (Banzhaf & McCormick, 2012).

Thus, the risk of *environmental gentrification*[2] (Sieg, Smith, Banzhaf, & Walsh, 2004) has emerged as a critical concern in the discourse surrounding the proliferation of urban sustainability programs. Specifically, the extent to which the elimination of environmental disamenities (such as toxic waste sites) and the creation of environmental amenities (such as parks) can exact a toll on vulnerable residents, especially poorer people of color, is paramount (Curran & Hamilton, 2012; Gamper-Rabindran & Timmins, 2011). In fact, Essoka (2010) offered empirical evidence to highlight the racial dimension of environmental gentrification, documenting decreases in the local Black and Latino populations after the revitalization of land contaminated with toxic chemicals (i.e., "brownfield" redevelopment). Similarly, as Gould and Lewis (2012) revealed, the restoration of Brooklyn's Prospect Park in the 1990s led to a significant increase in new construction around the park and a corresponding decrease in the racial and socioeconomic diversity of those areas. They concluded that rhetoric around the "greening" of New York is actually code for the "whitening" of the city's revitalized urban areas.

As such, urban sustainability initiatives, while providing and/or rehabilitating parks, playgrounds, bicycle paths and waterfront promenades, can also often serve as a "discursive guise" by which middle- to upper-income people can reinhabit (or gentrify) existing city spaces (Bunce, 2009). As Checker (2011) argued, "environmental gentrification operates through a discourse of sustainability which simultaneously describes a vision of ecological and socially responsible urban planning, a 'green' lifestyle which appeals to affluent, eco-conscious residents, and a technocratic, politically neutral approach to solving environmental problems" (p. 212).

It is clear that a failure to adequately address the social justice dimensions of neoliberal sustainability initiatives can contribute to environmental gentrification (Dooling, 2009; Gibbs & Krueger, 2007; Pearsall, 2012; Quastel, 2009). Such a failure is typically grounded in "asymmetrical power relations…[that] continually influence how and what kinds of 'environmental' issues are addressed" (Tretter, 2013, p. 308). Resolving this tension requires balancing the demands of environmental activists for such ecological benefits as less automobile-dependent cities and the reduction of urban sprawl, and the concerns of social justice advocates about the impacts of neoliberal development approaches that magnify income disparities and social dislocation (Tretter, 2013).

While a focus on the inequitable outcomes associated with top-down, elite-sponsored environmental improvement efforts is crucial, it does not obviate the need to also examine the consequences of bottom-up, community-driven struggles by people of color and working-class Whites to advocate for cleaner, healthier neighborhoods. In poor African American, Latino and Native American communities all across the country, people who have traditionally been relegated to the periphery of the environmental movement are challenging the ruination of their land, water, air and health by corporate polluters and indifferent governmental agencies. Combining elements of civil rights, social justice, the struggle for land rights and respect for the environment, oppressed peoples of color have formed environmental justice (EJ) movements to contest the disparate ecological and economic burdens placed upon their communities. Working-class Whites are also part of this fight (Faber, 2008). As a result, the federal government and municipalities are responding to demands by the EJ movement for environmental restoration and sustainable development. Is it possible that successful efforts by the EJ movement to force the cleanup of their communities can likewise result in environmental gentrification?

If the answer is "yes," then the EJ movement could be facing what Checker (2011, p. 211) called a "pernicious paradox." Her concerns echo the reports of the federal government's National Environmental Justice Advisory Council (NEJAC, 2006):

> Gentrification has placed populations in urban areas in direct competition for inner city space with relatively powerful and privileged groups. Environmental cleanup of these formerly industrialized, now residential, communities can be a powerfully displacing force.... Citizens living in urban, poor, and people-of-color communities are currently threatened by gentrification, displacement and equity loss on a scale unprecedented since the Urban renewal movement of the 1960s. (p. 2)

Are successful EJ struggles by the working class to make their urban environments "greener" likely to yield unintended consequences in the form of the eventual displacement and relocation of these same residents into other (potentially more heavily) polluted communities, where rents and housing prices are cheaper? Indeed, Checker (2011) underscored the vexing nature of this dilemma: Must lower income residents "reject environmental amenities in their neighborhood in order to resist gentrification that tends to follow..." (p. 211)? This is the essence of the paradox confronting the EJ movement.

ENVIRONMENTAL INJUSTICE AND GENTRIFICATION

While gentrification remains a contested term, most scholars identify it as a predominantly urban process marked by an infusion of capital in particular neighborhood(s); visible changes to the physical, social, cultural and economic landscape; and the displacement of lower-income households by an increasingly affluent resident base (Atkinson, 2003; Brown-Saracino, 2010; Kennedy & Leonard, 2001; Smith, 1996). Key disagreements in the literature center on whether gentrification can best be explained by market forces (Harvey, 1985; Smith, 1996) or the preferences and tastes of consumers (Ley, 1996; Lloyd, 2005), the extent to which the motivations of gentrifiers vary in meaningful ways (Brown-Saracino, 2009; Lloyd, 2005; Spain, 1993; Zukin, 1987), and how to identify, measure and characterize the consequences of gentrification for residents, neighborhoods and cities (Brown-Saracino, 2010; Freeman, 2006; Newman & Wyly, 2006).

The recent attention devoted to the potential for environmental gentrification is based, at least in part, on the recognition that the very environmental injustices that have earned the condemnation of scholars and activists may also make an affected area ripe for future gentrification. Consistent with the argument advanced by some gentrification scholars that decades of economic and political disinvestment created a "rent gap" between current and potential rents that developers and landlords could readily exploit (Smith, 1979), the environmental neglect, pollution and/or willful destruction that plague many neighborhoods likewise depress local housing values and associated rents.

For example, studies reveal that all things being equal, property values tend to be lower near Superfund sites (Eckerd, 2011, 2013; Gayer, Hamilton, & Viscusi,

2000). Likewise, housing prices have been shown to fall coincident with increased pollution levels from larger industrial facilities (Sanders, 2011), water pollution (Legget & Bockstael, 2000) and the opening of new power plants (Davis, 2011; Smith & Desvousges, 1986). In one Minnesota community, a landfill was found to adversely affect home values in the range of 12% at the landfill boundary and 6% at about one mile (Nelson, Genereux, & Genereux, 1992). Housing prices in one Nevada county declined over 15% in response to the discovery of a local cluster of childhood leukemia (Davis, 2004). In fact, the recognized economic impacts are so great that some local policymakers have taken steps to mitigate the potential consequences of environmental hazards. Tompkins County, New York, for example, adopted a program to insure existing residents for any loss of income, decline in property values or other damages due to environmental pollution stemming from a landfill in their neighborhood (Kiel & McClain, 1995).

While the relationship between lower housing values and environmental disamenities is perhaps not surprising—indeed, the classic Tiebout (1956) model would predict such an outcome—the stark reality that in the United States, communities of color and lower-income households overwhelmingly shoulder the burden of these environmental hazards has been a source of continuing concern among environmental justice advocates. Despite the declaration of President Clinton's Executive Order on Environmental Justice in 1994, which mandated federal agencies to incorporate environmental justice into their work and programs, the displacement of ecological problems onto minority neighborhoods persists. In fact, for the first time in history, people of color now comprise the majority of the population living near the nation's commercial hazardous waste facilities. These neighborhoods are 56% people of color (compared to 30% in nonhost areas) and are often economically distressed (Bullard, Mohai, Saha, & Wright, 2007). In Massachusetts, for instance, communities of color average over 48 hazardous waste sites per square mile (psm), a rate that is *more than 23 times greater* than the average of two sites (psm) in predominantly White communities (Faber & Krieg, 2005).

In general, America's Asian, African American, Hispanic and working-poor populations bear the greatest health risks from pollution traced to various sources, including industrial facilities and transportation infrastructure. For example, Pastor, Morello-Frosch, and Sadd (2005) found significant racial and income disparities in air toxic exposures in California. In December 2005, the Associated Press (AP) released an analysis of a little-known U.S. Environmental Protection Agency (EPA) research project revealing that Black Americans are 79% more likely than Whites to live in neighborhoods where industrial pollution is suspected of posing the greatest health danger. Residents of neighborhoods with the highest pollution scores are also poorer, less educated and suffer unemployment rates 20% higher than the national average.

The AP also found that in many states, Blacks, Hispanics and Asians are more than twice as likely as Whites to reside in neighborhoods where air pollution poses the greatest health dangers (Pace, 2005). According to the EPA, 57% of all Whites nationwide live in areas with poor air quality, compared with 80% of all Latinos (Wernette & Nieves, 1992). In Massachusetts, communities of color receive 10 times as many pounds of industrial pollutants per square mile as do predominantly White communities (Faber & Krieg, 2005). Nationwide, 68% of African Americans live within 30 miles of a coal-fired power plant, among the worst air polluters and a major contributor to the formation of smog (Keating & Davis, 2002).

Explanations for the racialized and class-based spatial patterning of environmental hazards are complex and not always clear cut. Some accounts point to the increased likelihood that people of color and poorer households are more likely to move to areas that are already polluted because their limited economic resources restrict their housing choices and force them to privilege low-cost housing over environmental amenities. Banzhaf and McCormick (2012), for example, concluded that "minorities are more likely than whites to 'come to the nuisance'" (pp. 39–41). Likewise, Been and Gupta (1997) demonstrated that census tracts become poorer and also witness an increase in the proportion of minority residents after a hazardous waste facility is built. A related explanation centers on the mobility patterns of higher income households, suggesting that lower income households are more likely to be "left behind" in areas with hazardous facilities due to the out-migration of more advantaged residents (Banzhaf & Walsh, 2008). Alternative theories focus on the deliberate siting of industrial hazards and other environmental disamenities in poorer communities and communities of color (Pastor, Sadd, & Hipp, 2001), based in part on the belief that such communities lack the social or political capital to mount a successful opposition. Finally, some scholars point to the tendency of firms to zero in on places where land is inexpensive when selecting a site for their facilities, rather than targeting a specific demographic group (Wolverton, 2009, 2012).[3]

As Banzhaf (2012) noted, developing a clear understanding of the root causes of environmental injustice is critical to crafting appropriate policy solutions that will disrupt wide-scale patterns of inequity in the distribution of environmental amenities. Indeed, if the chief culprits of environmental justice lie in local housing markets, this calls for strategies that are quite different than those that would be necessary if the practices of the private business sector are primarily to blame. In the meantime, there have been numerous efforts across the United States to address immediately the proximate health hazards that plague many disadvantaged communities. Billions of pounds of highly toxic chemicals, including mercury, dioxin, PCBs, arsenic, lead and heavy metals such as chromium, have been dumped or left behind in thousands of unsuspecting neighborhoods. These "brownfield" sites poison the land, contaminate drinking water and potentially cause cancer, birth defects, nerve and liver damage and other illnesses (Becerra, 2013). There are an estimated 450,000 brownfields in the United States, a disparate proportion of which are located in or near poorer working-class neighborhoods and communities of color (EPA, 2013). For example, since its inception, the EPA's Brownfields program has successfully leveraged more than $14 billion in cleanup and redevelopment funding from the private and public sectors to clean up contaminated or potentially contaminated properties (Pearsall, 2010). Local municipalities, including New York City, which has plans to remediate over 7,000 acres of contaminated former industrial sites by 2030, are also increasingly likely to pursue brownfield redevelopment (Pearsall, 2010).

These and other efforts to clean up dirty, unhealthy neighborhoods are often (and understandably) celebrated for rectifying the wrongs suffered by existing residents for far too long and striving to dramatically improve the overall quality of life in the affected area. By doing so, cleanup campaigns hold the potential to increase what Logan and Molotch (1987), in their urban sociology classic *Urban Fortunes*, termed the "use value" of residents' homes. Unfortunately, however, many residents are unable to enjoy the benefits of their rehabilitated neighborhoods, as "cleaning up the environment may increase housing costs for the poor by more than their willingness (or ability) to pay" (Sieg et al., 2004, p. 77).

In an extensive review of the literature, Banzhaf and McCormick (2012) concludes that although there are exceptions, "the evidence seems clear that in most cases improvements in local environmental conditions do trigger increases in property prices" (pp. 39–41). For example, numerous studies have found that the cleanup of Superfund and other brownfield sites—and in some cases, even the *anticipation* of future remediation—results in rising land values, housing values and/or rents (see, e.g., Bryson, 2012; Dale & Newman, 2009; Gamper-Rabindran & Timmins, 2013; Gayer et al., 2000; McCluskey & Rausser, 2003; Pearsall, 2010).

In this way, the redevelopment of environmentally blighted areas, and related initiatives undertaken in the name of urban sustainability, increase not only the "use value" of affected homes by rendering them more livable, but also (and more notably), their "exchange value," or financial worth in the market (Logan & Molotch, 1987). While these two values always exist in some degree of tension, the pursuit of exchange value by landlords, developers and planners eager to capitalize on revitalized "green" urban spaces can price vulnerable residents—that is, low-income tenants—out of the market, effectively limiting their ability to enjoy the long-term benefits of the now healthier environment. Thus, as Dillon (2013, p. 3) argued, "Brownfield redevelopment, rather than representing a clean break with an industrial past, often reproduces the social relations of an older, industrial economy—particularly those related to racial and health injustices."

Pearsall's (2010) analysis of brownfield redevelopment and vulnerability to gentrification in New York City provided a pointed illustration. Drawing on interviews with residents in affected neighborhoods, Pearsall demonstrated that "elderly, low-income, and minority populations were particularly sensitive to rapid increases in the cost of living" (p. 879). Across neighborhoods in several boroughs,

> [r]enters were less fortunate in that they had fewer options stemming from the inflated housing market than the homeowners. Long-time renters found the informal landlord-renter agreement of a stable rent that varies little from year to year to be forgotten with increasing frequency. More and more landlords sought to realize the full potential of their rental units—during the housing boom—and increased the rents to the same level as other offers in the neighborhood. (Pearsall, 2010, p. 880)

Indeed, while there are important exceptions (see, e.g., Eckerd, 2011), the literature suggests that, left unchecked, it would be reasonable to expect some degree of residential turnover as a result of local environmental improvements. However, even in those cases where existing households are not physically displaced, potential changes to the economic, social and cultural landscape in the wake of environmental remediation may be cause for concern. In particular, when such efforts are undertaken as part of broader urban sustainability initiatives, the "greening" of urban spaces is often accompanied by new retail stores, restaurants, and amenities explicitly targeted toward middle- and upper-class residents (Bryson, 2012; Dale & Newman, 2009). This remaking of the commercial and social aspects of neighborhoods can serve to alienate or marginalize lower-income residents and especially the homeless (Dooling, 2009, 2012), not only because participation is cost-prohibitive to those with limited means but also because it serves to reinforce class-based symbolic and social boundaries within the community (Lamont & Molnar, 2002).

ACHIEVING ENVIRONMENTAL JUSTICE WITHOUT GENTRIFICATION: POLICY APPROACHES AND STRATEGIES

While debates about neighborhood gentrification occasionally feature philosophical arguments about whether "urban development" is inherently good or bad, most scholars and activists maintain that the more germane question is whether development initiatives are *equitable* in their approaches and outcomes (see PolicyLink. org for more on the concept of equitable development). Similarly, the pressing issue concerning the remediation of environmental injustices is not whether cleaning up hazardous sites or building healthy urban spaces represents, on balance, a positive or negative thing, but rather the extent to which the decisions undertaken to accomplish those efforts, and the consequences that stem from them, are fair and equitable. This necessitates ensuring that certain groups do not reap most or all of the rewards of environmental cleanup initiatives and urban sustainability programs, while others disproportionately suffer the costs (or are excluded from sharing in the benefits altogether). Again, as with broader discussions of gentrification, opinions regarding what must be done in order to prevent environmental gentrification or to mitigate its damaging consequences vary considerably, reflecting a wide range of beliefs about the root causes of the problem, the feasibility or appropriateness of different tactics and the specific urban institutions implicated. In the following section, we briefly review a few of the general policy approaches featured in the literature. While this sampling is by no means exhaustive, it is intended to illustrate a range of strategies embraced by those concerned about environmental gentrification in the United States.

MARKET-BASED APPROACHES: FOCUSING ON THE CONSUMER

As economists such as Banzhaf (2008) and others argue, environmental gentrification is fundamentally about markets. Classic economic models (Coase, 1960; Tiebout, 1956) would predict adverse consequences for lower-income households as a result of environmental cleanup, either because the improvements rendered translate into higher-priced housing that lower-income people can no longer afford or because they reduce the compensation that firms pay to neighborhoods in order to pollute, thus depleting public funds that can be utilized for other urban institutions and infrastructures (Banzhaf, 2008). From this vantage point, the root problem is poverty and how it constrains the choices and bargaining power of individuals and households; gentrification is merely a manifestation or symptom of that poverty. Thus, solutions focus on recognizing poor people as consumers in the urban marketplace and providing them with the resources needed to expand their choice set and "compete" in a changing residential market. Banzhaf (2008) maintained the following:

> If public policy intends to address inequality, it would do better to attack poverty directly through transfer policies or by targeting its causes. This would give more people the ability to "purchase" environmental quality through markets (or to use their resources in other ways they see fit). In addition, policies could facilitate markets in pollution by reducing transactions costs, thereby allowing them to work for all groups. (p. 21)

In other words, market-based approaches typically do not hinge on challenging or altering the expected outcomes (e.g., by capping housing-price increases or preventing a change in the neighborhood landscape) but rather on increasing the negotiating power and financial capability of individual consumers within a given market.

INSTITUTIONAL APPROACHES: HOUSING-MARKET INTERVENTIONS

Other experts emphasize that urban sustainability initiatives cannot and should not be viewed in isolation from the pursuit of other forms of social justice, including the right to affordable housing (Dooling, 2009; Foy, 2012). Thus, some scholars advocate for direct interventions in the local housing market to prevent or offset the consequences of the shifts in housing stock and rents that are likely to stem from neighborhood environmental improvements. These can include efforts to build new low-income housing in areas slated to be cleaned up, as well as mechanisms to protect existing housing from price increases after environmental remediation has occurred. For example, Pearsall (2012) noted that rent stabilization was a key factor increasing residents' resilience in the face of gentrifying forces in their New York City neighborhood. A related approach would work to increase affordable housing in places that *already* enjoy healthy environmental conditions, to expand the number of good housing options available to lower-income individuals and families.

The importance of affordable housing as a component of urban sustainability efforts is underscored in the research of Pollack, Bluestone, and Billingham (2010), which explored the effects of transportation investments on proximate communities. Consistent with the findings discussed in this chapter, the researchers found that "transit investment frequently changes the surrounding neighborhood…the most predominant pattern is one in which housing becomes more expensive, neighborhood residents become wealthier and vehicle ownership becomes more common" (p. 1). Despite the fact that the expansion of public transportation systems is often heralded as a means to increase the environmental health and sustainability of urban areas, Pollack et al. discovered that "…a new transit station can set in motion a cycle of unintended consequences in which core transit users—such as renters and low income households—are priced out in favor of higher-income, car-owning residents who are less likely to use public transit for commuting" (p. 1). In light of the potential for these undesirable outcomes, Pollack et al. stressed the need for direct housing-market strategies and programs—for example, the proactive acquisition of impacted land, the preservation and production of affordable housing and so forth—to be included in transit-oriented development plans.

ACTIVIST APPROACHES: CHANGING THE NARRATIVE

Alternative approaches focus on harnessing the power of residents to challenge the inevitability of gentrification as a by-product of environmental change.

Here, the work of Winifred Curran and Trina Hamilton (2012) is particularly instructive. There is an important distinction to be drawn, these scholars argued, between "gentrifying" a community and "stabilizing" a community or forestalling further commercial or physical deterioration. In other words, in the narrative of urban change, neighborhood improvement is often conflated with large-scale reinvestment, suggesting that there is predictability and a presumed antagonism around the outcomes of neighborhood change. Yet, when low-income people are involved in the process, and their needs are represented at the table, the "expected" path can be interrupted and redefined. Applied to the case of Greenpoint, Brooklyn, this thinking underpinned a local strategy to push for environmental remediation that was "just green enough" or "just clean enough," wherein "as much of the environmental hazard as possible is removed in order to assure community health while still allowing for industrial uses on the waterfront for the explicit purpose of maintaining the area's working-class population" (p. 1039). Thus, the false dilemma that typically dictates the discourse on environmental remediation was contested, demonstrating a possible world where "…cleanup does not automatically or exclusively lead to the 'parks, cafes, and a riverwalk' model of a green city" (p. 1028).

Such a change in narrative is often best accomplished via the efforts of community alliances, particularly those that cut across class and residential tenure lines (Hamilton & Curran, 2013). Environmental activists and longtime residents can shape the course of neighborhood change by teaching gentrifiers about the needs and worries of the local community and identifying ways that each group can complement the strengths and supplement the weaknesses of the other. Activist approaches thus draw heavily from the social movements literature, focusing in part on the ways in which framing the issues in "collective" terms—that is, emphasizing the risks, opportunities and responsibilities that different groups share—can lead to equitable outcomes (Hamilton & Curran, 2013; see also, Gin & Taylor, 2010, for more on the deployment of different frames in response to gentrification).

CONTRACTUAL APPROACHES: COMMUNITY BENEFITS AGREEMENTS (CBAs)

Finally, any of the aforementioned approaches may be buttressed by formal contracts detailing the specific community benefits or mitigations that will accompany urban sustainability initiatives. One example is a Community Benefits Agreement (CBA; Janis, 2007). Since the early 2000s, CBAs have been deployed numerous times throughout the United States, typically in the context of large municipal economic development projects that stand to have a significant impact on low-income communities (the "L.A.

Live" CBA, negotiated as part of the development of a multibillion dollar entertainment complex in downtown Los Angeles, is a prime example; Saito, 2012). While CBAs may have different purposes and applications depending on the project and market in question, their primary goal is to ensure that vulnerable residents will both share in the anticipated gains of a planned development, as well as be spared undue burdens associated with it (Larsen, 2009).

Thus, in the case of economic development projects, emphasis may be placed on the creation of new jobs for local residents, the enforcement of certain hiring and wage requirements and plans to build or reserve space for needed community organizations, services and amenities. For urban sustainability or environmental remediation initiatives, CBAs may focus on these common ideas but also include explicit conditions concerning the protection or creation of affordable housing units, the provision of opportunities for community input on the development plans and/or mechanisms to capture for public use a certain percentage of the expected price increases stemming from the cleanup.

CONCLUSION

To avoid the pernicious paradox of environmental gentrification, we agree with Checker (2011) that the EJ movement must combat "postpolitical" discursive practices and technocratic governance systems that shun politics and de-link sustainability from social justice considerations. The danger is that postpolitical planning and policymaking approaches appear to be politically neutral and consensus-based, as well as ecologically and socially sensitive. In actual practice, however, they too often disallow spaces for a politics of resistance, and they do not allow for more democratic modes of governance (Swyngedouw, 2007). As such, justice and equity considerations are subordinated in favor of profit-minded development. In such a circumstance, the proponents of neoliberal forms of sustainable development and environmental gentrification can deflate resistance by appropriating (or even co-opting) the material and discursive successes of the EJ movement to serve high-end redevelopment (Checker, 2011, p. 212). To combat this possibility, the EJ movement must claim the high ground in the public debates around sustainable urban development and insure that the "greening" of America's cities serves all residents—both present and future generations.

NOTES

1. The authors would like to thank Hannah Bergam and Hannah Gartner for their research assistance on this project.

2. Related terms in the literature include ecological or "eco" gentrification, green gentrification and the political ecologies of gentrification. While conceptual and practical distinctions among these terms exist, space considerations prevent us from addressing them here.
3. See Banzhaf (2012) for a helpful summary of these and other explanations for environmental injustice.

REFERENCES

Atkinson, R. (2003). Introduction: Misunderstood saviour or vengeful wrecker? The many meanings and problems of gentrification. *Urban Studies, 40*, 2343–2350.

Banzhaf, H. S. (2008). Environmental justice: Opportunities through markets. *Property and Environment Research Center Policy Series, 42*, 1–25.

Banzhaf, H. S. (2012). The political economy of environmental justice: An introduction. In H. S. Banzhaf (Ed.), *The political economy of environmental justice* (pp. 1–22). Stanford, CA: Stanford University Press.

Banzhaf, H. S., & McCormick, E. (2012). Moving beyond cleanup: Identifying the crucibles of environmental gentrification. In H. S. Banzhaf (Ed.), *The political economy of environmental justice* (pp. 23–51). Stanford, CA: Stanford University Press.

Banzhaf, H. S., & Walsh, R. R. (2008). Do people vote with their feet?: An empirical test of Tiebout's mechanism. *American Economic Review, 98*(3), 843–863.

Becerra, M. (2013, April). Environmental justice for whom? Brownfield redevelopment and gentrification in the city of Chicago (Master's thesis). University of Michigan, Ann Arbor.

Been, V., & Gupta, F. (1997). Coming to the nuisance or going to the barrios? A longitudinal analysis of environmental justice claims. *Ecology Law Quarterly, 24*(1), 1–56.

Boyle, M. A., & Kiel, K. A. (2001). A survey of house price hedonic studies of the impact of environmental externalities. *Journal of Real Estate Literature, 9*(2), 117–144.

Brown-Saracino, J. (2009). *A neighborhood that never changes: Gentrification, social preservation, and the search for authenticity*. Chicago, IL: The University of Chicago Press.

Brown-Saracino, J. (2010). *The gentrification debates*. New York, NY: Routledge.

Bryson, J. (2012). Brownfields gentrification: Redevelopment planning and environmental justice in Spokane, Washington. *Environmental Justice, 5*(1), 26–31.

Bullard, R. D., Mohai, P., Saha, R., & Wright, B. (2007). *Toxic wastes and race at twenty: 1987–2007—Grassroots struggles to dismantle environmental racism in the United States* (Report for the United Church of Christ Justice and Witness Ministries, pp. 1–68).

Bunce, S. (2009). Developing sustainability: Sustainability policy and gentrification on Toronto's waterfront. *Local Environment, 14*(7), 651–667.

Checker, M. (2011). Wiped out by the "greenwave": Environmental gentrification and the paradoxical politics of urban sustainability. *City & Society, 23*(2), 210–229.

Clark, T. N., Lloyd, R., Wong, K., & Jain, P. (2002). Amenities drive urban growth. *Journal of Urban Affairs, 24*(5), 493–515.

Coase, R. H. (1960). The problem of social cost. *Journal of Law and Economics, 3*(1), 1–44.

Curran, W., & Hamilton, T. (2012). Just green enough: Contesting environmental gentrification in Greenpoint, Brooklyn. *Local Environment, 17*(9), 1027–1042.

Dale, A., & Newman, L. L. (2009). Sustainable development for some: Green urban development and affordability. *Local Environment, 14*(7), 669–681.

Davis, L. W. (2004). The effect of health risk on housing values: Evidence from a cancer cluster. *American Economic Review, 94*(5), 1693–1704.

Davis, L. W. (2011). The effect of power plants on local housing values and rents. *Review of Economics and Statistics, 93*(4), 1391–1402.

Dillon, L. (2013). Race, waste, and space: Brownfield redevelopment and environmental justice at the Hunters Point shipyard. *Antipode.* doi:10.1111/anti.12009, 1–17. Retrieved from http://www.academia.edu/2426509/_Waste_Race_and_Space_Brownfield_Redevelopment_and_Environmental_Justice_at_the_Hunters_Point_Shipyard._Antipode_A_Journal_of_Radical_Geography_online_early_view_

Dooling, S. (2009). Ecological gentrification: A research agenda exploring justice in the city. *International Journal of Urban and Regional Research, 33*(3), 621–639.

Dooling, S. (2012). Sustainability planning, ecological gentrification and the production of urban vulnerabilities, In S. Dooling & G. Simon (Eds.), *Cities, nature and development: The politics and production of urban vulnerabilities* (pp. 101–119). Farnham, UK: Ashgate.

Eckerd, A. (2011). Cleaning up without clearing out? A spatial assessment of environmental gentrification. *Urban Affairs Review, 47*(1), 31–59.

Eckerd, A. (2013). Policy alternatives in adaptive communities: Simulating the environmental justice consequences of hazardous site remediation strategies. *Review of Policy Research, 30,* 281–301.

Essoka, J. D. (2010). The gentrifying effects of brownfields redevelopment. *Western Journal of Black Studies, 34*(3), 299–315.

Faber, D. (2008). *Capitalizing on environmental injustice: The polluter–industrial complex in the age of globalization.* Lanham, MD: Rowman & Littlefield.

Faber, D., & Krieg, E. J. (2005). *Unequal exposure to ecological hazards 2005: Environmental injustices in the commonwealth of Massachusetts* (Report by the Philanthropy and Environmental Justice Research Project, Northeastern University). Retrieved from http://nuweb9.neu.edu/nejrc/wp-content/uploads/final_unequal_exposure_report_2005_1012051.pdf

Florida, R. (2002). *The rise of the creative class.* New York, NY: Basic Books.

Foy, K.C. (2012). Home is where the health is: The convergence of environmental justice, affordable housing, and green building. *Pace Environmental Law Review, 30*(1), 1–57.

Freeman, L. (2006). *There goes the 'hood: Views of gentrification from the ground up.* Philadelphia, PA: Temple University Press.

Gamper-Rabindran, S., & Timmins, C. (2011). Hazardous waste cleanup, neighborhood gentrification, and environmental justice: Evidence from restricted access census block data. *American Economic Review Papers and Proceedings, 101*(3), 620–624.

Gamper-Rabindran, S., & Timmins, C. (2013). Does cleanup of hazardous waste sites raise housing values? Evidence of spatially localized benefits. *Journal of Environmental Economics and Management, 65*(3), 345–360.

Gayer, T., Hamilton, J. T., & Viscusi, W. K. (2000). Private values of risk trade-offs at Superfund sites: Housing marking evidence on learning about risk. *Review of Economics and Statistics, 82*(3), 439–451.

Gibbs, D. C., & Krueger, R. (2007). Containing the contradictions of rapid development? New economy spaces and sustainable urban development. In R. Krueger & D. C. Gibbs (Eds.), *The sustainable development paradox: Urban political economy in the United States and Europe* (pp. 95–122). New York, NY: Guilford Press.

Gin, J., & Taylor, D. E. (2010). Movements, neighborhood change, and the media—Newspaper coverage of anti-gentrification activity in the San Francisco Bay Area: 1995–2005. In D. E. Taylor (Ed.), *Environment and social justice: An international perspective* (Research in Social Problems and Public Policy, Vol. 18) (pp. 75–114). Bingley, UK: Emerald Group.

Gould, K., & Lewis, T. L. (2012). The environmental injustice of green gentrification: The case of Brooklyn's Prospect Park. In J. DeSena & T. Shortell (Eds.), *The world in Brooklyn: Gentrification, immigration, and ethnic politics in a global city* (pp. 113–146). Lanham, MD: Lexington Books.

Hamilton, T., & Curran, W. (2013). From "five angry women" to "kick-ass community": Gentrification and environmental activism in Brooklyn and beyond. *Urban Studies, 50*(8), 1557–1574.

Harvey, D. (1985). *The urbanization of capital.* Baltimore, MD: The Johns Hopkins University Press.

Janis, M. (2007). Background on community benefits agreements: The process, the projects, and the prospects for the future. In Annie E. Casey Foundation (Ed.), *Community benefits agreements: The power, practice, and promise of a responsible redevelopment tool* (pp. 10–20). Baltimore, MD: Annie E. Casey Foundation.

Keating, M. H., & Davis, F. (2002). *Air of injustice: African Americans and power plant pollution.* Retrieved from http://www.catf.us/publications/view/20

Kennedy, M., & Leonard, P. (2001). Dealing with neighborhood change: A primer on gentrification and policy choices (Discussion paper). Washington, DC: Brookings Institution.

Kiel, K. A., & McClain, K. T. (1995). House prices during siting decision stages: The case of an incinerator from rumor through operation. *Journal of Environmental Economics and Management, 28,* 241–255.

Lamont, M., & Molnar, V. (2002). The study of boundaries in the social sciences. *Annual Review of Sociology, 28,* 167–195.

Larsen, L. (2009). *The pursuit of responsible development: Addressing anticipated benefits and unwanted burdens through community benefit agreements.* Ann Arbor: Center for Local, State, and Urban Policy, University of Michigan.

Legget, C. G., & Bockstael, N. E. (2000). Evidence of the effects of water quality on residential land prices. *Journal of Environmental Economics and Management, 39,* 121–144.

Ley, D. (1996). *The new middle class and the remaking of the central city.* London, UK: Oxford University Press.

Lloyd, R. (2005). *Neo-bohemia: Art and commerce in the post-industrial city.* New York, NY: Routledge.

Logan, J., & Molotch, H. (1987). *Urban fortunes: The political economy of place.* Berkeley: University of California Press.

Marcuse, P. (1986). Abandonment, gentrification, and displacement: The linkages in New York City. In N. Smith & P. Williams (Eds.), *Gentrification of the city* (pp. 153–177). London, UK: Unwin Hyman.

McCluskey, J. J., & Rausser, G. (2003). Hazardous waste sites and housing appreciation rates. *Journal of Environmental Economics and Management, 45,* 166–176.

National Environmental Justice Advisory Council (NEJAC). (2006). *Unintended impacts of redevelopment and revitalization efforts in five environmental justice communities* (pp. 1–26). Washington, DC: Author.

Nelson, A. C., Genereux, J., & Genereux, M. (1992). Price effects of landfills on house values. *Land Economics, 68*(4), 359–365.

Newman, K., & Wyly, E. (2006). The right to stay put, revisited: Gentrification and resistance to displacement in New York City. *Urban Studies, 43*(1), 23–57.

Pace, D. (2005, December 13). More blacks live with pollution. *Associated Press.* pp. 1–3. Retrieved from http://www.familiesagainstcancer.org/?id=287

Pastor, M., Morello-Frosch, R., & Sadd, J. L. (2005). The air is always cleaner on the other side: Race, space, and ambient air toxics exposures in California. *Journal of Urban Affairs, 27*(2), 127–148.

Pastor, M., Sadd, J., & Hipp, J. (2001). Which came first? Toxic facilities, minority move-in, and environmental justice. *Journal of Urban Affairs, 23*(1), 1–21.

Pearsall, H. (2010). From brown to green? Assessing social vulnerability to environmental gentrification in New York City. *Environment and Planning C: Government and Policy, 28,* 872–886.

Pearsall, H. (2012). Moving out or moving in? Resilience to environmental gentrification in New York City. *Local Environment, 17*(9), 1013–1026.

Pollack, S., Bluestone, B., & Billingham, C. (2010). *Maintaining diversity in America's transit-rich neighborhoods: Tools for equitable neighborhood change.* Dukakis Center Publications. Retrieved from http://iris.lib.neu.edu/cgi/viewcontent.cgi?article=1003&context=dukakis_pubs

Quastel, N. (2009). Political ecologies of gentrification. *Urban Geography, 30*(7), 694–725.

Saito, L. T. (2012). How low-income residents can benefit from urban development: The LA Live community benefits agreement. *City & Community, 11*(2), 129–150.

Sanders, N. J. (2011). *Toxic assets: How the housing market responds to environmental information shocks.* Stanford, CA: Stanford Institute for Theoretical Economics Summer Workshop, Stanford University.

Sieg, H. V., Smith, K. Banzhaf, H. S., & Walsh, R. (2004). Estimating the general equilibrium benefits of large changes in spatially delineated public goods. *International Economic Review, 45*(4), 1047–1077.

Smith, K. V., & Desvousges, W. H. (1986). The value of avoiding a LULU: Hazardous waste disposal sites. *Review of Economics and Statistics, 68*(2), 293–299.

Smith, N. (1979). Toward a theory of gentrification: A back to the city movement by capital not people. *Journal of the American Planning Association, 45*(4), 538–547.

Smith, N. (1996). *The new urban frontier: Gentrification and the revanchist city.* London, UK: Routledge.

Spain, D. (1993). Been-heres versus come-heres: Negotiating conflicting community identities. *Journal of the American Planning Association, 59*(2), 156–171.

Swyngedouw, E. (2007). Impossible "sustainability" and the postpolitical condition. In R. Krueger & D. C. Gibbs (Eds.), *The sustainable development paradox: Urban political economy in the United States and Europe* (pp. 13–40). New York, NY: Guilford Press.

Tiebout, C. (1956). A pure theory of local expenditures. *Journal of Political Economy, 64*(5), 416–424.

Tretter, E. M. (2013). Contesting sustainability: "SMART growth" and the redevelopment of Austin's eastside. *International Journal of Urban and Regional Research, 37*(1), 297–310.

Wernette, D. R., & Nieves, L. A. (1992). Breathing polluted air: Minorities are disproportionately exposed. *EPA Journal, 18,* 16–17.

Wolverton, A. (2009). Effects of socio-economic and input-related factors on polluting plants' location decisions. *The B. E. Journal of Economic Analysis & Policy, 9*(1), Article 14.

Wolverton, A. (2012). The role of demographic and cost-related factors in determining where plants locate: A tale of two Texas cities. In H. S. Banzhaf (Ed.), *The political economy of environmental justice* (pp. 199–222). Stanford, CA: Stanford University Press.

Zukin, S. (1987). Gentrification and capital in the urban core. *Annual Review of Sociology, 13,* 129–147.

CHAPTER FIVE

Visualizing Change

Using Technology and Participatory Research to Engage Youth in Urban Planning and Health Promotion

ANTWI AKOM, AEKTA SHAH AND AARON NAKAI

INTRODUCTION

What technological tools are youth using to transform their communities? What role can young people play in urban planning and health promotion? Do youth of color enjoy the same environmental protections as adults? This chapter describes a model (Streetwize) for using technology and Youth Participatory Action Research (YPAR) to engage youth in health promotion and urban planning, paying special attention to intersections of race, space, place and waste. In particular, we address an issue that is often overlooked by geographic information systems practitioners: the planning needs of the local community and youth-driven innovations in geographic information system (GIS) mapping that can help meet those needs. Our chapter describes how a planning team from San Francisco State University and the University of California, Berkeley, in partnership with an innovative, community-based, non-profit organization—Institute for Sustainable Economic Educational and Environmental Design (I-SEEED)—joined with 150 youth to integrate technology with community planning.

Our research builds on works by several scholars on race, space and community mapping, including Powell (1998), Al-Kodmany (2000), Flicker (2008), Corburn (2002), Corburn, Osleeb and Porter (2006), Bonam (2010), London, Zimmerman, and Erbstein (2003), Zimmerman, Phelps, and Lerner (2008) and Parikh, Javid, Ghosh, and Toyame (2006). In particular, it builds on Flicker's work on augmenting GIS with multimedia in planning information support systems. The research presented in this chapter is an attempt to describe how low-income communities and communities of color are using technology in grassroots planning processes. A youth-led team focused its efforts on refining tools that are most effective in conducting community-based planning. Finding ways to digitally democratize access to data—from a community-driven perspective—is a critical innovation, particularly with growing access to smart phones, the Internet and wireless technology in low-income communities and communities of color. This chapter argues that the integration of traditional tools, such as paper maps with cutting-edge and breakthrough technology, is the most powerful way to bridge the digital divide and increase health promotion in low-income communities and communities of color.

Drawing on original research from 150 youth and four community partners, this chapter seeks to accomplish two goals: The first is to provide a theoretical basis of the Streetwize model and provide an overview of how the model was developed, along with implications for practice and research. The second is to demonstrate how our team enhanced the utility of GIS mapping by incorporating several other public participation methods. This chapter has four sections. The first section frames recent demographic shifts and the ways in which the fields of urban planning, civic engagement and participatory GIS can be strengthened by better integrating issues of youth voice and technology into development strategies. The second section outlines the theoretical basis of the Streetwize model and provides an overview of how the model was developed. The third section details the process of data collection, and the compilation and creation of GIS layers, and describes the integration of images into GIS based on various urban design models. The fourth section describes the key innovations of youth-driven community GIS and how young people responded to issues of race, space, place and waste.

VISUALIZING CHANGE

Profound demographic inflows and outflows are fundamentally shifting the U.S. population and culture in unprecedented ways. People of color and indigenous people currently comprise one-third of the U.S. population, and by 2042, they will be the majority. These dramatic population shifts have been accompanied by an explosion in new technology. The emergence of new technologies is changing society, the way we live, they way we work, the way we play, the way we communicate and do business—and perhaps most importantly, the way we learn. New technology has many potential benefits for youth. All one has to do is observe the prolific use of cell phones, iPads, iPods, blogs, Facebook, Twitter, Instagram and Snapchat, to see that today's youth are born digital—seamlessly integrating technology into their everyday lives (Lenhart, Madden, & Hitlin, 2005). This normalization of technology presents a unique opportunity to appeal to youth culture and engage youth in health promotion, urban planning and social justice.

In the last decade, there has been a growing recognition by planners and community development practitioners that community members and youth are important populations to involve in the planning of their communities. Researchers have shown that engaging youth through technology in planning raises the self-esteem of youth, fosters their sense of environmental and community responsibility and encourages them to become more civically minded (Santo, Ferguson, & Trippel, 2010). Community and youth engagement also leads to better planned communities, capable of responding to the unique needs of their resident populations. In particular, youth can be key stakeholders; since they have the largest stake in the future of their community (given that they will be around the longest), their input often strengthens the sustainability of community planning. Furthermore, their enthusiasm, creativity and idealism bring a fresh perspective to planning initiatives. Despite the acknowledged positive outcomes that stem from community and youth engagement in planning, there are still many barriers to their participation. Municipal staff, community organizers, policymakers and planners may have limited time and resources, competing interests and few tools to effectively engage youth in planning processes. Because of these barriers, many youth and adult allies do not venture into the field.

This chapter aims to address these concerns by providing practical ideas that are easy to use, effective and have been tested in the field by community organizations, planners and cities. Our chapter describes a model for using technology and YPAR to engage youth in community planning and health promotion. The Streetwize model was developed in collaboration with 150 youth and four community partners through a project funded by the Robert Woods Johnson Foundation. The model is designed to be used with youth working with a facilitator within a school, youth or community-based organization. In addition to outlining the theoretical basis of the Streetwize model, this chapter provides an overview of how the model was developed, along with implications for practice and research.

STREETWIZE: YPAR AND TECHNOLOGY

The knowledge youth have about their experiences of living with multiple environmental health hazards and chronic disease is one of the fundamental assets they can contribute to schools and community-based organizations and is often a key resource the community organizes to avoid being exploited and exposed to health risks and social toxins (Corburn, 2002, Corburn et al. 2006). When youth engage in environmental health research, their primary goal is often to help themselves and their communities by generating usable or actionable knowledge—information that goes beyond description and analysis and suggests proactive or precautionary intervention strategies. By taking action to transform the social and material conditions in their communities, youth build self-respect, self-confidence and self-determination while fostering positive relationships with caring adults (Camino, 2005).

Because YPAR utilizes theories of empowerment, resistance, problem-posing education and popular education, its implementation with youth in educational and community settings expands the goals of positive youth development to include personal and community transformation (Freire, 1973, 1997; Ginwright, Noguera, & Cammarota, 2006; Rappaport, 1987). By providing participants with opportunities for meaningful engagement in problem identification, analysis, planning, civic engagement and youth-led evaluation, YPAR teaches young people to "read the world" and develop skills, which can contribute to a sense of mastery, power and control over their environment (Freire, 1973). Further, today's youth often utilize technology as a means to engage with, understand and literally navigate their environments.

We are said to be living in a digital age with the vast majority of American (87%) youth accessing the Internet regularly, either on computers or on their phones/mobile devices (Lenhart et al., 2005; Media Awareness Network, 2005). Increasingly, youth are using technology to create and maintain social networks (Flicker, 2008; Montgomery, Gottlieb-Robles, & Larson, 2004) and to promote activism (Lombardo, Zakus, & Skinner, 2002, Sheffield & Landrigan, 2011, Strack, Magill, & McDonagh, 2004). Just looking at the use of smart phones, iTouches, blogs and text messaging in the average Black or Brown classroom and community, it is clear that for many youth, technology is a seamless part of how they conduct their lives (Lenhart et al., 2005; Lombardo et al., 2002; Montgomery et al., 2004). This technological fluidity presents a unique opportunity to appeal to youth culture, connect to GIS technology and engage youth in neighborhood and school planning (Skinner et al., 1997). In this vein, many researchers have begun to utilize an e-PAR (Electronic Participatory Action Research) approach, which encourages young people to use technology and social media to identify, understand and transform structural issues in the community (Flicker, 2008). These researchers have focused on integrating multimedia and hyper-media components (Van Wart & Parikh 2013, Parikh et al., 2006); into GIS mapping (i.e., pictures, video, audio, text messages), developing low-tech GIS interfaces for youth to collect, analyze and upload both quantitative and qualitative data (Parikh et al., 2006; Van Wart & Parikh, 2013) and incorporating local knowledge, such as feelings about the safety of their community, in the building of stock GIS databases (Bosworth & Donovan, 1998; Parikh et al., 2006; Van Wart & Parikh, 2013).

Streetwize draws from the rich traditions of e-PAR, YPAR, and participatory GIS mapping and extends the questions posed by critical youth studies and participatory scholars (Bourgois, 1995; Fine & Weis, 1998; Giroux, 1983; Kelley, 1994; MacLeod, 1987; McRobbie, 1991; Oakes & Rogers, 2006; Rasmussen, Rofes, & Talburt, 2004; Sullivan, 1989; Willis, 1977). How do youth learn the tools of inquiry and GIS skills, and take transformative action within formal and informal youth development and/or educational settings? How is it possible for youth inquiries to grow into challenges to institutional practices of systematic oppression? Under what conditions can GIS be a tool of youth development? Streetwize offers a new tool for community development, neighborhood advocacy and citizen planning that can support youth and community-driven action, specifically for urban design and public planning projects.

STREETWIZE IN PRACTICE

The Streetwize model, while grounded in the theory of e-PAR, is rooted in practice. Between 2011 and 2013, with funding from the Robert Woods Johnson Foundation, the model was implemented over six times in collaboration with four youth-serving community organizations. The participating organizations included two in school, one after school and one summer program. Each project was completed by a team of young people in collaboration with I-SEEED. The youth met weekly for between 6 and 12 weeks (see Table 1). Generally, I-SEEED staff approached youth-serving organizations (YSO) that worked with low-income youth and youth of color and asked if they would be interested in partnering to test our model. In some instances, I-SEEED staff facilitated the process; in others, a cofacilitating approach with YSO staff was used.

Table 1. Overview of the Streetwize Model in Practice for Participating Projects.

Project	Technology/Mapping	PAR
National City, San Diego, CA: 10 youth (14–18 years old); 6 weeks	*Group Technology:* Internet and video, issue identification focused on health-related issues, in particular clean drinking water and empty lots. *Group GIS Mapping:* GIS/field mapping using Streetwize platform, water testing kits and pictures to collect data on: toxins in drinking water, location of drinking water facilities, location of empty lots, current state of empty lots.	*Action:* Youth present a student health plan including GIS visualizations to their school principal, advocating for clean drinking water stations as well as a longer-term project to revitalize an abandoned lot near their school for a school garden and a place to be physically active outdoors.

Table 1. *Continued*

Project	Technology/Mapping	PAR
Merced/Planada, CA: 6 youth (14–18 years old); 6 weeks	*Group Technology:* Internet and video, issue identification focused on health-related issues, in particular public transportation availability and reliability and parks and open space. *Group GIS Mapping:* GIS/field mapping using Streetwize platform, data tables and pictures to collect data on: location of public transportation, reliability of public transportation, availability of bike lanes, availability and current state of public parks.	*Action:* Youth present their community health assessment including GIS visualizations and physical health plan to the county transportation and planning board.
Leadership High School, Bayview Hunter's Point, CA: 15 youth (14–16 years old); 12 weeks	*Group Technology:* Internet and video, issue identification focused on health-related issues, in particular healthy food vs. fast food in low-income and communities of color vs. higher-income communities *Group GIS Mapping:* GIS/field mapping using Streetwize platform, data tables, audio recordings and pictures to collect data on: location of food outlets, availability of healthy, fresh food in food outlets, ingredients used by food outlets, frequency of unhealthy food outlets.	*Action:* Students use GIS mapping visualizations to present to their school leaders to ban food or drinks that contain "RED" chemicals—commonly found in popular youth snacks including "Flaming Hot Cheetos" and "Red Bull"—from their school. Additionally, students developed an innovative solution to promote health in their school by creating, branding and marketing a homemade healthy juice drink to students at their high school. Additionally, students presented their findings and mapping visualizations to experts at the American Education and Research Association and the Community Based Participatory Action Research Conference.

Table 1. *Continued*

Project	Technology/Mapping	PAR
Oakland High School, East Oakland, CA: 40 youth (14–18 years old); 10 weeks	*Group Technology:* Internet and video, issue identification focused on health-related issues, in particular healthy food vs. fast food in low-income and communities of color and specific health issues facing East Oakland neighborhoods (including transportation, food access, park access, air quality, soil quality). *Group GIS Mapping:* GIS/field mapping using Streetwize platform, data tables, audio recordings and pictures to collect data on: location of food outlets, availability of healthy, fresh food in food outlets, ingredients used by food outlets, frequency of unhealthy food outlets.	*Action:* Youth present a student health plan including GIS visualizations to their school principal and school board, advocating for healthy snack machines in schools. Students also conducted a health impact assessment in individual East Oakland neighborhoods, producing place-based field reports and GIS visualizations identifying specific health issues for each neighborhood and proposed solutions for each.

Note. PAR = Participatory Action Research; GIS = Geographic Information System.

While many scholars and community planners have advocated for the use of GIS in community planning, few have described the role the youth can play in building a community-driven GIS database. In the YSOs our team worked with, few individuals at the beginning of the project understood the data collection process that is necessary to even begin using GIS. The primary goal of our projects in these communities was to help create a vision that would reduce childhood obesity, increase health promotion, develop GIS capacity and skills in youth and adult partners and guide climate action plans and future community development. A secondary goal was to create a mutually respectful partnership among I-SEEED, universities and neighborhood residents.

The foundational tool of the project was a community-based GIS tool that the team named Streetwize consisting of local data, parcel-level maps and photographic images of neighborhoods at the local level. As the team began compiling data for GIS, we found the following: (1) Most of the available GIS data is not available at a resolution that is useful for neighborhood planning, and (2) the youth themselves were often the individuals who have the most current and accurate understanding of race, space, place and waste in the neighborhood.

To address these needs, we developed Streetwize, a GIS mapping and data visualization tool that integrates the Local Ground GIS data collection platform developed by researchers Sarah Van Wart and Tapan Parikh at the University of California, Berkeley. The Streetwize tool consists of local and national data sets, paper maps and photographic and video images of neighborhoods. It allows local residents to document and rank community health and well-being by combining a list of indicators of well-being and the corresponding data for each community, neighborhood, city and school. The datasets are grouped into 10 categories, including Demographics, Education, Economy, Health, Safety, Transportation and Environment. Users are able to select up to 5 indicators in which to rank their neighborhoods, and the platform generates a thematic map showing how strongly the set of indicators is represented in those neighborhoods. The selected indicators can be further weighted so that the values of certain indicators are more heavily represented in the map. In addition to the map analysis, the application also produces tabular data showing the exact values of each selected indicator for each neighborhood. Charts and graphs showing the distribution of an indicator set and statistical information on each indicator are available as well. For further analysis, a variety of geographic reference points, such as what are the spaces in a community that bring youth a sense of peace and joy, or happiness, or access to tutorial services, child care or public transportation, are also recorded and can be overlaid on the thematic map. Overall Streetwize makes it easy for youth to move from surviving to thriving by providing the best information available and empowering them to share digital stories and critical feedback on their social and structural environments. By democratizing data, Streetwize increases youth's ability to transform the structures that affect their daily lives.

All geographic data are accessed through ERSI's ArcGIS server and is operable on personal computers and smartphones through an online interface. Streetwize contains many layers of mapped, digital, tabular, qualitative and quantitative information to support the work of YSOs to increase community involvement in planning. Layers that our four YSOs used included the following: digitized paper maps, vacant land, grocery stores, liquor stores, land use, aerial photography, public schools, transport, major industry and employers.

YOUTH DEVELOPING, CONSTRUCTING, AND EXTENDING GIS LAYERS

In the first stage of GIS layer development, our team followed three urban design models to create the Streetwize GIS community database and web knowledge tool: (1) Lynch's concept of Imageability, (2) Nasar's concept of Likeability/Dislikeability, and (3) Al-Kodmany's model of thematic image maps. Lynch (1960) describes the built environment by organizing it into five "imageable" elements: nodes, paths, districts, edges and landmarks. The youth design teams applied this model to neighborhoods, took digital pictures of these significant elements and then uploaded them to GIS maps. This model was useful in visualizing the major components of the overall community. We used Nasar's (1998) Likeability/Dislikeabilty model to document young people's own evaluation of the walkability, bikeability, and transitability of the neighborhood One of the criticisms of national walkability services is the limits of their accuracy. Specifically, walk scores often do not calculate whether there are sidewalks, how many lanes of traffic one must cross or how much crime occurs in the neighborhood. They also do not differentiate among types of amenities, for example a supermarket grocery store versus a small food mart selling mostly liquor and fatty foods. Following Jane Jacobs's work, *The Death and Life of Great American Cities* (1961), our point of departure was that the most important evaluation of a city's built environment is the opinion of those who live there and use it on a daily basis. For example, one group of young people were able to assign walk scores, bike scores and transit scores to points on maps and then ask residents to point out areas that were safe, walkable, bikeable and transitable and used cameras to take shots of those areas. Another youth team went to sites with local residents to visually and orally recorded residents' racial, spatial and environmental health experience in terms of access to healthy food and liquor stores. This group, supported by the local school districts and several community organizations, used Streetwize and the Local Ground GIS data collection platform to assess the accuracy of an existing "official grocery store" dataset from the local county public health department. Students then submitted photos, audio clips, comments and ratings of the stores they visited in their neighborhoods and presented their data sets to school officials. Their findings showed that only about a third of the stores in the "official grocery store" data set provided even moderately healthy food options (see Figure 1).

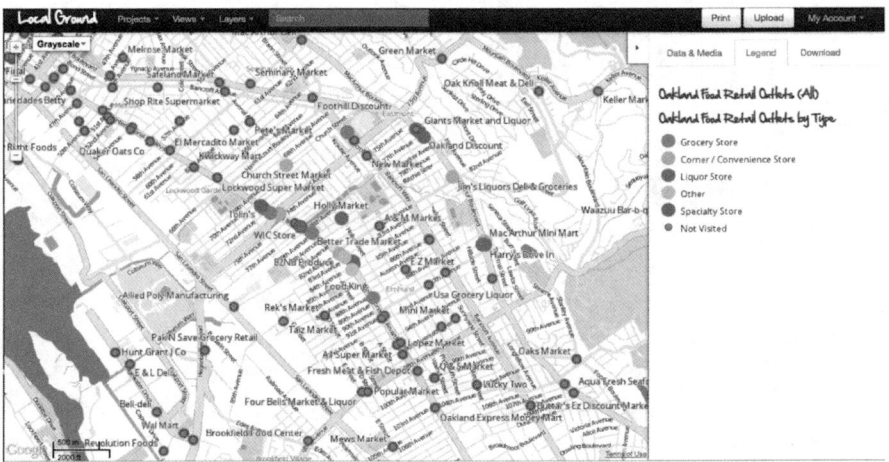

Fig 1. Youth Mapping/Rating of "Grocery Stores" (screenshot taken from Local Ground.org).

In our work, Lynch's model was useful for visualizing the overall appearance of the neighborhood and capturing some of the more salient characteristics, while Nasar's model was useful for visualizing the feelings of students and residents about specific areas in their community. Additionally, using paper maps allowed for the incorporation of comments that enriched the image database and made the photos more personal and meaningful. The implementation of Lynch's model was not a top-down approach, because the youth were the planning "experts" that decided on the pictures to be taken, thus the pictures reflect their unique view of the community. To further underscore the importance of a participatory approach, our implementation of Nasar's model could be considered "bottom-up," because the youth pictures reflect their evaluative feelings about their neighborhood.

Most recently, we have been testing Al-Kodmany's model of creating thematic image maps. In this model, images are collected according to themes that reflect the different aspects of a neighborhood such as economic, educational, cultural and environmental assets and challenges in relation to racial and ethnic residential patterns. Once mapped, these images reveal opportunity structures and can reduce spatial mismatch. By generating maps showing the relationship between Black and Brown communities and opportunity, we can begin to frame opportunity structures in a geographical perspective and create more targeted and transformative initiatives.

Not merely a simple indicator of well-being, "opportunity" is defined here as the structures and environmental conditions that contribute to community stability and individual advancement. Spanning a range of economic and social factors such as sustainable employment, high-quality educational institutions and experiences, healthy and safe communities, stable and safe housing and access to healthcare, "opportunity" is an entire environmental context consisting of structure, or the macrolevel, and culture, or the microlevel. Access to structures of opportunity is critical to an individual's success. Structures may mediate personal choice in implicit and explicit ways.

YOUTH INNOVATION: INTEGRATING EMERGING TECHNOLOGY INTO GIS

In our two years of extensive experience in implementing a community GIS, young people identified various ways to augment existing capabilities of the GIS system (such as adding social media features and networks like Facebook, Twitter, Instagram and Snapchat) as well as additional tools to be used alongside the GIS (such as 3-D modeling software, animation, video streaming, photography, music and participant drawing on maps). These features paved the way for the creation of a "youth yelp" for social services, where young people conceptualized, developed and designed a way to instantly collect crowd-sourced data and tag it to relevant GIS layers that meet the needs of specific communities and YSOs. Additionally, youth integrated social media to instantly disseminate maps, charts digital stories, and geo-referenced reports for actionable community change to a diverse group of stakeholders, including their peers, parents, teachers, urban planners and decision makers. These features will substantially enhance overlay analysis as they enable the ability to focus on both the "systems level" and the "micro-interactions" of community development—indeed, harnessing the power of GIS mapping and social media ensures that communities of color and poor people have greater participation in the research and decision making that affect their everyday lives (Corburn, 2002).

CONCLUSION

Streetwize uses an innovative approach to measure accessibility to health and educational related resources by embedding grassroots constituencies in GIS mapping, data visualization and sustainable urban design. Our approach combines youth participatory action research with GIS data collection in order to develop tools for neighborhood advocacy and citizen planning that make schools and communities healthier places to live, learn, work and play. Our unique approach places students at the center of learning, teaches them how to make evidence-based decisions, to participate in public discourse about educational and environmental design strategies and demonstrates how they can influence important community, neighborhood and policy decisions.

For those interested in transforming the current conditions of Black and Brown communities, it is imperative that young people play a central role. Streetwize offers a powerful model of youth engagement, one that affirms the legitimate role that young people can and should play in the process of collecting and analyzing data that reveals the contemporary conditions of urban schools and communities. Our approach of "democratizing data" is of particular significance because it builds collaborations among teachers, students, urban planners, and public health and scientific communities—while also training the next generation of climate scientists, health advocates, urban planners and social justice educators.

Streetwize offers an innovative approach to translate the on-the-ground experiences of communities working on an array of health, education or environment-related issues. Youth can learn and use GIS to (a) communicate their neighborhood/community perceptions of health disparities, workforce development or environmental health; (b) describe their neighborhood/community in their own words using both quantitative and qualitative forms of data; (c) evaluate the social, structural and environmental assets and challenges in their neighborhood/

community; and (d) develop recommendations and action plans for community development. Importantly, Streetwize both is a culturally relevant technology *tool* to conduct research and utilizes YPAR and e-PAR *methodologies* of community engagement—adding to existing planning and survey methods by offering a new approach to integrating youth's local knowledge that expands beyond more traditional approaches.

Finally, it is important to consider the significance of youth-initiated research to our systems of planning and development. When the young people most implicated by the inequitable conditions of neighborhoods develop the tools and capacities to conduct research, they ask fundamentally different questions and contribute local, grounded knowledge in important and powerful ways. Planners, developers, policymakers, and district and community leaders would benefit greatly from increasing the opportunities for young people to develop Streetwize projects in their schools and communities. The theory of change conveyed by Streetwize is grounded in the belief that local knowledge is key to creating the power shift that will make significant community change a reality. When communities on the margins move to the center, when they can develop and implement new strategies for social change and have the tools and resources to demand how institutions big and small should act—a more equitable society will emerge. Streetwize is a technological innovation that may allow this dream to become a reality.

REFERENCES

Al-Kodmany, K. (2000). Extending geographic information systems to meet neighborhood planning needs: The case of three Chicago communities. *URISA Journal, 12*(3), 19–37.

Bonam, C. M. (2010). *Devaluing black space: Black locations as targets of housing and environmental discrimination.* Stanford, CA: Stanford University.

Bosworth, M., & Donovan, J. (1998). *A mapmaker's dream. Public involvement applications utilization of GIS.* Position paper presented at Project Varenius Specialist Meeting: Empowerment, Marginalization, and Public Participation GIS, Santa Barbara, CA.

Bourgois, P. (1995). *In search of respect: Selling crack in el barrio.* New York, NY: Cambridge University Press.

Camino, L. (2005). Pitfalls and promising practices of youth–adult partnerships: An evaluator's reflections. *Journal of Community Psychology, 33*(1), 75–85.

Corburn, J. (2002). Combining community-based research and local knowledge to confront asthma and subsistence-fishing hazards in Greenpoint/Williamsburg, Brooklyn, New York. *Environmental Health Perspectives, 110*(Suppl 2), 241.

Corburn, J., Osleeb, J., & Porter, M. (2006). Urban asthma and the neighbourhood environment in New York City. *Health & Place, 12*(2), 167–179.

Flicker, S. (2008). Who benefits from community based participatory research? *Health Education & Behavior, 35*(1), 70–86.

Fine, M., & Weis, L. (1998). *The unknown city: Lives of poor and working class young adults.* Boston, MA: Beacon Press.
Freire, P. (1973). *Education for critical consciousness* (Vol. 1). New York, NY: Continuum.
Freire, P. (1997). Paulo Freire. *Education Review, 67,* 2.
Ginwright, S. A., Noguera, P., & Cammarota, J. (Eds.). (2006). *Beyond resistance! Youth activism and community change: New democratic possibilities for practice and policy for America's youth.* New York, NY: Routledge.
Giroux, H. A. (1983). *Theory and resistance in education: A pedagogy for the opposition.* South Hadley, MA: Bergin & Garvey.
Jacobs, J. (1961). *The death and life of great American cities.* New York, NY Random House.
Kelley, R. (1994). *Race rebels: Culture, politics, and the black working class.* New York, NY: Simon & Schuster.
Lenhart, A., Madden, M., & Hitlin, P. (2005). *Teens and technology: Youth are leading the transition to a fully wired and mobile nation.* Washington, DC: Pew Internet & American Life Project.
Lombardo, C., Zakus, D., & Skinner, H. (2002). Youth social action: Building a global latticework through information and communication technologies. *Health Promotion International, 17*(4), 363–371.
London, J. K., Zimmerman, K., & Erbstein, N. (2003). Youth-led research and evaluation: Tools for youth, organizational, and community development. *New Directions for Evaluation, 2003*(98), 33–45.
Lynch, K. (1960). *The image of the city* (Vol. 11). Cambridge, MA: The MIT Press.
MacLeod, J. (1987). *Ain't no makin' It: Leveled aspirations in a low-income neighborhood.* Boulder: CO: Westview Press.
McRobbie, A. (1991). *Feminism and youth culture: From "Jackie" to "Just seventeen."* London, UK: Macmillan.
Media Awareness Network. (2005). *Young Canadians in a wired world. Phase II: Student survey.* Ottawa, Ontario, Canada: Media Awareness Network.
Montgomery, K., Gottlieb-Robles, B., & Larson, G. (2004). *Youth as e-citizens: Engaging the digital generation.* Washington, DC: Center for Social Media, American University.
Nasar, J. L. (1998). *The evaluative image of the city.* Thousand Oaks, CA: Sage.
Oakes, J., & Rogers, J. (2006). *Learning power: Organizing for education and justice.* New York, NY: Teachers College Press.
Parikh, T. S., Javid, P., Ghosh, K., & Toyama, K. (2006). Mobile phones and paper documents: Evaluating a new approach for capturing microfinance data in rural India. In *Proceedings of the SIGCHI conference on human factors in computing systems* (pp. 551–560). New York, NY: ACM.
Powell, J. A. (1998). Race and space: What really drives metropolitan growth. *The Brookings Review, 16*(4), 20–22.
Rappaport, J. (1987). Terms of empowerment/exemplars of prevention: Toward a theory for community psychology. *American Journal of Community Psychology, 15*(2), 121–148.
Rasmussen, M. L., Rofes, E., & Talburt, S. (2004). *Youth and sexualities: Pleasure, subversion and insubordination in and out of schools.* New York, NY: Palgrave Macmillan.
Santo, C. A., Ferguson, N., & Trippel, A. (2010). Engaging urban youth through technology: The youth neighborhood mapping initiative. *Journal of Planning Education and Research, 30*(1), 52–65.
Sheffield, P. E., & Landrigan, P. J. (2011). Global climate change and children's health: Threats and strategies for prevention. *Environmental Health Perspectives, 119*(3), 291.

Skinner, H., Morrison, M., Bercovitz, K., Haans, D., Jennings, M., & Megenko, L. (1997). Using the Internet to engage youth in health promotion. *International Journal of Health Promotion & Education, 4,* 23–25.

Strack, R. W., Magill, C., & McDonagh, K. (2004). Engaging youth through photovoice. *Health Promotion Practice, 5*(1), 49–58.

Sullivan, M. L. (1989). *"Getting paid": Youth crime and work in the inner city.* Ithaca, NY: Cornell University Press.

Van Wart, S., & Parikh, T. S. (2013). Increasing youth and community agency in GIS. In Proceedings of the SIGEHI conference on human factors in computing systems. New York, NY: ACM.

Van Wart, S., Tsai, K. J., & Parikh, T. (2010, December). Local ground: A paper-based toolkit for documenting local geo-spatial knowledge. In *Proceedings of the First ACM Symposium on Computing for Development* (p. 11). New York, NY: ACM.

Wallerstein, N. (1992). Powerlessness, empowerment, and health: Implications for health promotion programs. *American Journal of Health Promotion, 6*(3), 197–205.

Willis, P. E. (1977). *Learning to labour: How working class kids get working class jobs.* New York, NY: Columbia University Press.

Zimmerman, S. M., Phelps, E., & Lerner, R. M. (2008). Positive and negative developmental trajectories in US adolescents: Where the positive youth development perspective meets the deficit model. *Research in Human Development, 5*(3), 153–165.

CHAPTER SIX

Reframing Spatial Inequality

Youth, Photography and a Changing Urban Landscape

STUART GREENE, KEVIN BURKE AND MARIA McKENNA

> Places are never finished but always the result of processes and practices. As such, places need to be studied in terms of the "dominant institutional projects," the individual biographies of people negotiating a place, and the way in which a sense of place is developed through the interaction of structure and agency. (Cresswell, 2004, p. 37)
>
> What I do or take pictures of inspires others to change the way people think about things. (Leah, fifth grade)

Kinloch (2012) has called attention to the "silenced dialogue" that excludes the perspectives and voices of youth. Unfortunately, studies of low-income minority students' underachievement emphasize pathology in a "culture of poverty," the lack of parental support and stability, high rates of students dropping out and the extent to which schools support a school-to-prison pipeline (e.g., Wright Edelman, 2012). The statistics can indeed be alarming. However, reports can mask larger inequalities in school and in economic development that create spaces of inequality and distort our perceptions of youth such as Leah, who grow up in unsafe neighborhoods amid abandoned houses and a dearth of material resources. (All names throughout are pseudonyms.) As we show, embedded within these spaces are stories of "community…resilience, and identity formation" that make visible the challenges that youth and their families face in their day-to-day lives and their hopeful responses to these challenges (Schein, 2006, p. 14). Material spaces can limit or enable youth and their families when they seek opportunities for growth, development and economic well-being. Just as important are the ways youth, in particular, reimagine those spaces in order to take ownership of their lives and envision possible futures that transcend the material and racialized contexts of abandonment and conflict that leave families of color behind.

The aim of this chapter on children's images of space is to emphasize the extent to which place, a literal and metaphorical space, is a fluid concept that emerges through children's sense of self, their interactions with friends, family, and community, and often through the ways others perceive those spaces (e.g., urban, inner city, threatening, dangerous). Aitken (2001) underscored the importance of place for youth as "contexts that play a large role in constructing and constraining dreams and practices" (p. 20). All too often assumptions about place slip into judgments about the race of the children and families who occupy those spaces (Ferguson, 2001). How children identify, interpret and respond to their surroundings, however, is overlooked in descriptions of place; as Leah explained earlier, the pictures she takes as a participant in a PhotoVoice project provide her with an opportunity to tell a significant story that seeks "to change the way people think about things," particularly deficit perspectives of youth in urban neighborhoods. Thus the images she provides have the potential to transform the ways others see the spaces she inhabits and to attach significance to seemingly abstract spaces with her own gaze drenched in history, relationships and familiarity. Unfortunately, deficit perspectives about "urban" and "inner-city" youth limit adults' understanding and appreciation of youths' capacities for understanding and contributing to meaningful change in the places they inhabit.

Amid the changing landscape of gentrified spaces, children can help us reorient our understanding of place: what it means to have a sense of belonging, the effects of displacing children and their families, children's need for safe spaces where they can flourish and a sense of what is a just and equitable use of space that can serve the common good. Hearing children's voices through the images and texts they produce is especially important at a time when researchers (Miller, 2011; Neuman & Celano, 2001) have begun to map spatial inequality and demonstrate the extent to which low-income minority children do not have access to important material resources that can help them flourish in and out of school. Children help us see both the strengths and value of real spaces and the imagined spaces that nurture their hopes and life paths.

We are reminded in our work with children of how Cisneros (1989) wrote so eloquently about Esperanza and the child's imaginative constructions about places that shape identity. *The House on Mango Street* deserves revisiting for the value the text holds in creating a model narrative of geospatial inequality, child voice and the child imaginary (Greene, Burke, & McKenna, 2013). Esperanza uses her writing to galvanize herself against a nun's observation that equates Esperanza's sense of who she is with the neglected and deteriorating house where she shared the upstairs apartment with her family. "You live *there*," the nun repeated three times. Esperanza's sense of place is bound up with people and has no real limits, so she resists any identification with the house on Mango Street. "I am going to tell you a story about a girl who didn't want to belong...I am too strong for her to keep me here forever." Indeed, Esperanza resists her friend Alicia who tells her, "Like it or not you are Mango Street." Esperanza reacts sharply: "Not me. Not until someone makes it better" (Cisneros, 1989, p. 107). As Kinloch (2010) pointed out, youth are border crossers: from school to home, from impoverished neighborhoods devastated by loss to hopes and dreams of futures still undefined; relationships with friends, family, teachers, tutors that help them maintain a stable sense of self; and their faith. But there are also barriers—psychological, racial and geographical—that a child like Esperanza expresses in trying to just "be and be...against so many bricks" (p. 75). These border crossings and barriers are especially important to see from children's perspectives.

PHOTOVOICE AS A TOOL FOR PARTICIPATORY ACTION RESEARCH

PhotoVoice provides children with a space to represent their social worlds and order their place within particular environments. In the study we report on in this chapter, photography gave youth the opportunity to express how they felt about a changing neighborhood landscape that was at once familiar and strange. In turn, listening to children tell their stories enabled us to think differently about how economic policy had begun to alter the lived and perceived realities of children.

Photography afforded the children with the opportunity to use different tools to communicate nuanced thinking about their surroundings. The New Literacy Studies have described the multiple "sign systems" that children use to make meaning and author an identity connected to family and community. As Urciuoli (1995) has observed, each action system has a "different potential for enacting a self" or "allows a different way to 'be your-self'" (p. 191). Children's use of multiple sign systems enabled them to develop a sense of agency as authors of their space within the structural forces of racial segregation and economic development that would naturally marginalize and silence their voices. PhotoVoice helped youth see familiar spaces in new ways (Hull & James, 2006) by reimagining possibilities in what can be conceived of as youth's performance of literacy (i.e., through reading, writing, speaking and taking photographs). It seems appropriate to characterize children's enactments of self through image and text as performance, because the ways in which children construct meaning are always in concert with others (Hull & Katz, 2006). They use cultural tools of language, image and text in new contexts and reappropriate these tools in ways that enable them to develop a sense of agency.

As a method of inquiry, PhotoVoice draws upon Freire's (1970) argument that critical reflection not only leads to discovery and action but ideally to the social transformation of inequality through counternarratives. The value of counternarratives for research using visual images as a method is perhaps best asserted by Delgado (1989) as the need for "naming one's own reality" (p. 2073). Indeed, counterstories provide a window into the experiences of marginalized groups and create spaces of possibility by sharing strategies of transformation (Solórzano & Yosso, 2002, p. 32). Importantly, Delgado Bernal (2002) explained that counterstories not only affirm youth as both "holders and creators of knowledge" (p. 113) but also serve as tools that enable youth to navigate obstacles in their day-to-day lives in school and in the community. In the case of PhotoVoice, the counternarratives are strengthened by the visceral nature of visual images.

CRITICAL GEOGRAPHY AND GENTRIFIED SPACES

The rhetoric of gentrification calls attention to the blight of low-income neighborhoods when, in reality, lending institutions have long rendered these neighborhoods unsafe for investment. As Simon (2013) pointed out, "long-term disinvestment" in many urban neighborhoods "led to the deterioration of buildings and the seeming decline of community life." Gentrification seeks to refashion a cultural identity that exists within a framework of race, class and ethnicity, particularly recent efforts to attract new sources of wealth to a given city. This is true with the construction of privately owned single-family homes and condominiums in newly gentrified urban spaces, historically associated with individuality and the White middle class in suburbs after World War II. Unfortunately, newly constructed homes have displaced low-income minority families who rented apartments and homes often owned by absentee landlords. This sense of privacy brings with it a sense of exclusion that once manifested itself through racial covenants, real estate steering and government lending programs. Although fences may not serve as physical markers of privacy, the wealth and privilege associated with new housing serve as markers of "class distinction," "racial identity" and White privilege (Harris, 2006, p. 148). As Kevin Fox Gotham has noted, "privacy starts at the property boundary and is related to the deed of ownership and an entire ideology that is encapsulated within it, such that the discussions of private property rights must be viewed as a language of exclusion" (cited in Harris, 2006, p. 130). Excluded are those perceived as "other," including youth, and especially youth of color, who threaten the safety of White families of relative wealth.

One of the outcomes that Hall identifies in his introduction to this volume is "the 'pushing out' and 'keeping out' of urban poor and minority occupants in order to add to the attractiveness of city living." The result may be to remove poverty from the urban core by removing minority, low-wage earners and their respective living spaces, but little is done to distribute resources more equitably. We encounter this theme of inequity in children's photographs and their perceptions that new housing should go hand-in-hand with making provisions for the poor and the homeless.

The "aesthetic appeal of gentrification" has simply given way to developers who provide abstract and decontextualized descriptions of urban spaces that rarely match the day-to-day experiences of the people who occupy those spaces. Instead, developers proffer what Haymes (1995) has described as a "landscape of consumption" (p. 97), where people with relative wealth can purchase a lifestyle that is urban but that also lacks the ethnic, racial and socioeconomic diversity often associated with urban areas. This we find to be increasingly true in the erosion of public spaces.

In what follows, we hear 12 African American students tell about their lives, their city and the attendant structural power inequities latent in policies that affect them. In the process of the research we report on here, the students have created viable counternarratives informed (outlined and inscribed) by racial and socioeconomic realities often ignored both in policy and in research. Throughout the text, we alternate referring to the children participating in this work as co-researchers, youth and children. The term co-researcher stems from a methodological lens that pays the same deference to the lived realities of individuals participating in a given study as the researchers' own interpretations (Christensen & James, 2000; Mazzoni & Harcourt, 2013; Norton, 2006) This is deliberate on our part because we privilege children's voice in the adult-centered space of academic writing. Their counterstories are what we attempt to make sense of in this chapter.

THE CONTEXT FOR A UNIVERSITY-COMMUNITY PARTNERSHIP

We met twice a week for seven weeks with a group of 12 low-income African American youth, three boys and nine girls in the fourth and fifth grades, at a community center that developed out of a collaboration between the city and a local private university. For the past 12 years, the center has fashioned a gathering space for learning and building relationships through cultural and academic enrichment, arts education and youth development. As one of the children put it, "The center is…about inspiring other people to do great things…the people who work at the center really care for the kids, they're not here because it's just a job that they have to do."

The center is set back just about 100 feet from a busy thoroughfare that connects the northeast neighborhood to downtown. Just to the south of the center is an angled road, a state highway, that is one of the area's most frequently travelled intersections, where high-speed traffic and congestion create "hazardous" conditions for pedestrians (City plan, n.d.). The entrance to the center faces another busy thoroughfare, and the rear touches on a neighborhood where there is a park and rows of houses, whose occupants live at or below the poverty line. However, according to a recent development plan, the area surrounding the center has been vulnerable to negative economic and social forces, including White flight and slow economic growth. Moreover, families near the center have limited access to goods and services, such as a pharmacy, a full-service grocery store, healthcare providers, a library, employment opportunities or reliable public transportation. As it turns out, the grocery store and a Goodwill store occupied the building where the community center is now located.

To the north and slightly west of the center is a neighborhood that city planners, the neighborhood association and others have sought to rehabilitate. Large single-family homes have replaced rental properties and vacant, deteriorating houses across a 60-acre section of land in the neighborhood. Older, cottage-style homes have been refurbished. Within sight of the center directly to the north is a private university that is part of the community collaboration and the transformation of a once-impoverished neighborhood that now caters to students, alumni and faculty from the university, as well as White, affluent professionals with clothing shops, slow-fast food chains, bars, a salon and a high-end restaurant. As Anyon (2005) has suggested, it is hard not to translate spatial demographics like these into racial disparities between those who live in newly gentrified neighborhoods and those who live just to the south without access to affordable goods and services.

The 12 fourth- and fifth-grade students with whom we worked were participating in a 12-week course that combined an oral history project of local residents with photography which the city's Center for History now includes in its archives. The children's photographs are also displayed on the walls of the community center. As one child put it so poignantly, their photographs "tell a story with a beginning, middle and end—without words." We collaborated with the instructor, a long-time resident of the neighborhood surrounding the center, and a local not-for-profit neighborhood organization, which has sponsored a PhotoVoice project to engage youth in community involvement and social change. The children in the photography class agreed to participate in the PhotoVoice project, and the two projects overlapped a great deal.

The children used their artwork, including collage, drawing, map making and whiteboard art, to help adults understand what the children saw as assets in the neighborhood surrounding the center, as well as some of the problems they felt city planners could help solve. To document children's stories, we audiotaped and then transcribed focus group discussions in which the children explained the thinking behind their art, writing and photography. We transcribed children's oral presentations to the larger group about children's priorities. In addition, we collected and analyzed children's writing, mapping and artwork samples. (See Greene et al., 2013, for a fuller discussion of the PhotoVoice methodology).

CHILDREN'S "I AM" POEMS: WRITING ABOUT IDENTITY AND PLACE

We wanted to position the children to tell stories about themselves, their families and their sense of place. This sense of place is where children sought to carve a space for themselves. Indeed, the stories in *The House on Mango Street* (Cisneros, 1989), which we refer to at the beginning of this chapter, resonated with the children and helped them reflect upon their connection to the neighborhood surrounding the community center. More specifically, the stories we shared with the children helped them think about their relationships to people and places and provided a space in which they could imagine alternative futures to the kind of spatial inequality and displacement they experienced. In fact, children often punctuated their observations of the older houses where many of their friends still lived with metaphors they recalled Esperanza using to describe the houses in her neighborhood. For example, one of the children mentioned the "tight steps" in front of one house we walked past and that the windows in another were "so small you'd think they were holding their breath" (Cisneros, 1989, p. 4). They enjoyed the play of language that gave them the opportunity to visualize a place they may have taken for granted. Moreover, the children's writing from their "I Am" poems (Christensen, 2000) gave voice to their own personal histories ("I am from pictures of my grandpa that passed"), their memories ("I am from Grannie's sweet potato pie"), their sense of self ("I'm from where chocolate is sweet and smooth") and family ("Where your mom just read you a book in bed and your dad tucks you in").

Importantly, children's written work gave them additional opportunities to think about the complex relationship between place and identity. For instance, Leah tried to make sense of the memories of where she came from and where she is now in both space and time.

I am a bed with the soft pillow and warm covers.

I am a Playstation 2 moving from place to place on Rise to Honor and Street Basketball.

I am the tree with dead leaves during troubling times that has hoped to keep swaying on.

I am a little girl day dreaming about my brothers and sisters and closest friends—Louis, Chantelle, Jamal, Anthony, Althea, and Corwin. With my closest friends Isaiah, Raini, Ivontre, and last but not least, the best dancer ever, Carmen.

I am an 11-year old girl enjoying macaroni and cheese, chicken, and greens.

I am emotion feeling overboard and fading.

I am the USA Skating Rink enjoying the vibration of the hip-hop songs.

I am that 5-year old girl staring at my old street and looking at the abandoned houses and cracked up sidewalks.

We learn from our conversations with Leah that she has moved from the urban decay of a large city to live with her aunt who works at the community center and who offers stability, warmth and a future unimaginable just a few years ago. Thus Leah writes with a sense of hope and renewal ("I am the tree with dead leaves during troubling times that has hoped to keep swaying on"), the friends who help her feel connected and define who she is (and is becoming), the comfort foods that bring her closer to her family and the music that binds her with a generation. Leah is also that little girl who "stares at her old street and look[s] at the abandoned houses and cracked up sidewalks." The deteriorating houses with bricks crumbling near the center connect her past and present.

Finally, we provided different spaces to help foster children's sense of agency and collective power, particularly when they wrote together about ways they could change the neighborhood around the community center for the better and when they took on the identities as researchers. Together, we filled sheets of paper and a whiteboard with lists of assets and problems in the community and then the children began to "code" what they wrote. Naming problems gave the children a sense of authority, and they developed their ideas in a systematic fashion by labeling and categorizing things that need to be "preserved, fixed or kept," such as houses and buildings, and "actions" they felt they could take. Such an approach gave legitimacy and credibility to the proposal they submitted to the local neighborhood association and presented to the mayor.

CHILDREN'S IMAGES OF A CHANGING URBAN LANDSCAPE

Four themes emerged from the children's writing and photographs: the circumstances of attachment and belonging, the lack of safe spaces in their environment, the unequal distribution of material resources and displacement stemming from the gentrification that was beginning to transform the neighborhood surrounding the community center. In large part, these are artificial distinctions, because we never really discussed these themes in isolation of one another: The kids always wove them together in their art, in our discussions with them and even in the ways they began to "code" the assets and problems they identified. The children's photographs and narratives provided a context for thinking critically about their neighborhood and ways to address these problems in order to create meaningful change. They also served as entry points for defining shared problems outside of traditionally rigid adult-oriented viewpoints.

CIRCUMSTANCES OF ATTACHMENT AND BELONGING TO THE NEIGHBORHOOD

The children's photographs brought into focus the significance of the community center as a place where the children felt cared for, respected, listened to. They could be themselves. This was particularly true of the photographs most of the children took of the mural on the side of the community center. The mural juxtaposes images of people living together and the language of "devotion" that children in past years have used to describe the center's significance in shaping their lives. For example, Leah explained that she took a picture of the mural because the images reflect the children's perspectives. In a focus group discussion with other children (October 9, 2012), she explained that "The pictures tell about the kids, what they were thinking in their mind, their imagination." Leah wanted to emphasize that children drew upon their own inspiration and did not rely on adults for their ideas: "People randomly drew what they were thinking in their mind and just didn't like take, well, the teachers didn't tell them what to draw and it's their imagination of what they think" (Focus Group, October 9, 2012). The images and words in the mural capture what children value and identify with.

In turn, we learn through their narratives that the children's sense of place embeds memories of ongoing relationships with tutors who have helped them with their schoolwork, the other children with whom they shared meals, and programs, too, like Legos, photography and Shakespeare. The children spoke very openly about their attachment to the center and tutors because recent construction and development have threatened the location of the center. Developers and the local university have discussed and debated proposals to move the center, and the children clearly resisted such an idea.

Similarly, several children took pictures of an elementary school they attended before moving on to middle school. The school, just a quarter mile south of the community center, is across a busy state highway. More than anything, the children seemed to understand the important role that a school plays in a neighborhood by helping build relationships among children. In this case, they recalled the ways the teachers and the principal there helped engage the children in the community. As Jana informed us, each year the children participated in Earth Day by picking up litter in the neighborhood surrounding their school. For Jana, this was the children's way of "helping the community" (Focus Group, October 9, 2012). In this way, too, the photographs of the school and community center not only reflected a sense of place and belonging, but children continually reminded us of their hope-filled and profound belief that people need to be committed to helping one another.

Finally, children also captured their sense of history in their photographs of a street sign and a small patch of green space named after a well-known resident of the neighborhood. In each instance, the children identified with these spaces in ways that went beyond their own well-being and encompassed family members, neighbors and community members. However, the children do not take this sense of community for granted. They know the real threat that gentrification poses to homes, schools and even their community center in the area. Therefore, they appreciate the value of those spaces in the here and now. Leah's photograph of a tall oak with red and yellow leaves reflects the beauty, stability and strength that children seem to be drawn to. We like to think that this is the tree that Leah identifies herself with in the "I Am" poem she wrote: "I am the tree…that has hoped to keep swaying on."

THE LACK OF SAFE SPACES

The children's photographs also made clear the limits on the their ability to get together with friends and develop social relationships in safe spaces. That few safe spaces exist for these children may explain why so much of their identity and sense of belonging are generated by the community center. When we had asked the children to take pictures of things they thought could be changed, they photographed the crumbling sidewalks in older, less well-kept parts of the neighborhood, while also commenting on the clean sidewalks that surrounded the newer, privately owned homes and storefronts. For the children, these new houses served as a boundary between the university and the older neighborhood where many of their friends lived. The children took pictures of an abandoned park with little or no grass that was fenced in, a rotting bench and a rusty newspaper stand near a building that some speculate might have been a bookstore at one time. The children explained that no one really knows what goes on in the building. They do know that people come and go at night and in the early morning.

The children appeared to have a fairly consistent and permanent mental map of which parts of the neighborhood are safe and which aren't. The further they walked into the neighborhood southwest of the community center, the less safe they felt. When asked what she felt, Shondra observed that, "It's very dark...even when the sun's out, like all those trees surrounding it makes it look all dark and scary to be at, especially at night" (Focus Group, October 9, 2012). Shondra's sense of fear stands in stark contrast to the strength and beauty all of the children experienced closer to the community center. Shondra went on to explain that, "It's like those types of neighborhoods where your parents say 'make sure you have the door locked' and like they say it all like stern because it's important." Here she distances herself from the neighborhood just blocks from the community center. "There's so much trash around those areas but sometimes the trash is like blended in like in there you can see all those leaves and everything surrounding it, yeah. It's been like that for a long time" (Focus Group, October 9, 2012). Another student, Kaden, recounted a similar story of the lack of safe space around his home: "Yeah I play outside a lot but my house is in a dangerous neighborhood so I need my mom to be out there and I don't like to play with my mom out there. It makes me feel like a little kid" (Focus Group, October 9, 2012).

We cannot ignore the sad irony when we listen to Kaden. Unable to play freely, he feels infantilized in an environment where he does not get to be a kid. He is, in essence, unable to become a subject (Lefebvre, 1992) in the ways that other children in safer environments might be, particularly because his chance to create space is circumscribed first by a concerned adult, but mostly because of the "power geometries" that "refract and reflect global economic processes" (Aitken, 2001, p. 136) that result in his living in an unsafe urban space. And while "adults almost always see the importance of creating spaces for young people," they remain "loath to let them do so themselves" (Aitken, 2001, p. 169). Relying on adults to create safe spaces for them, children begin to lose a sense of self-reliance and agency. However, our efforts in the PhotoVoice project have sought to provide children with a sense of ownership and control by asking them what they value and what they want to change in the spaces they inhabit. Shifting the perspective to what children think and feel opens up a world of possibility often foreclosed by adults' decisions about economic development and change.

SPATIAL INEQUALITY AND THE UNEQUAL DISTRIBUTION OF MATERIAL RESOURCES

Gentrification brings into focus a set of contradictions around hope and loss that the children, our co-researchers, grappled with in the photographs they took and in their analyses of how changes in the urban landscape could provide advantage for more people. That is, the children could not always reconcile the seemingly inefficient use of space to build large single-family homes that often existed in close proximity to abandoned houses that could have been "rehabilitated," to use the language of the city's report on economic development. In this conversation between Jana and Shondra, Jana points out a house that many of the other children also noticed with a puzzled, questioning tone. For Jana, this is a house that developers simply left behind amid the newly gentrified spaces just blocks from the community center. "Part of the porch, and it's like broken off of the roof and it looks pretty dangerous for like anybody who goes up to the door and they could get seriously hurt" (Focus Group, October 9, 2012). Jana assumes that the "city" should take responsibility for doing something, because to her mind, the city, with the university, has bought up all of the property near the community center.

Jana asks Shondra in the same focus group discussion what she would do and Shondra is very direct: "We'd like fix that little porch part and we'd re-install the ramp and then maybe their house isn't, maybe the inside of their house isn't that great so we'd like to remodel the house." In turn, given the opportunity to create change, Shondra would explain to city planners "all the facts that I know." Moreover, Shondra speculates that someone might still live in this house. "There's a ramp, which means that they're handicapped probably and that means that they really need help. And then I'd also just show [city planners] the picture, which would probably convince them enough that they really need to help the people living in the house." Leah contributes to the conversation and provides an ethic of care in another conversation with one of the authors that she believes should guide decisions about economic development:

Leah:	'Cause it's not caring about the houses.
Author:	What do you mean it's not caring?
Leah:	Well they don't care about those houses that [people] live in.
Author:	Who's "they"?
Leah:	Um, the city, just put it that way.
Author:	Who in the city?
Leah:	Um, the people who buy the property here. (Focus Group, October 11, 2012)

The children's conceptions of economic development focus on equity—what is fair and unfair—given their perception that some people are hurt when distress in blighted neighborhoods is ignored. Aaron pointed out that, "what would be useful for the neighborhood [is], like, the houses, like, for the homeless and the people… and we talked about how it will be better if all the bad things were changed like the houses" (Focus Group, October 9, 2012). Jana readily agreed, although it's not always clear who should take responsibility: "I don't get why people buy big houses when there's like homeless people out there. They won't help them out at all, and like give them space or like something to eat; and, yeah they are selfish cause they're not helpin' out and everything. They could fit two families [in the house] but they just wanna have one big family" (Focus Group, October 9, 2012). The children are aware of the differences that are as racialized as they are socioeconomic and the reason why some houses are abandoned amid new construction: "Probably because of the area maybe, like where they're located." In the end, the children point to the moral consequences of inequality and a commitment to the common good. For many of our co-researchers, gentrification is more about a lack of attention on the part of city planners to the needs of those children and families in the neighborhood than it is about accumulating wealth and purchasing a given lifestyle.

A SENSE OF DISPLACEMENT

The children's sense of equity and fairness served as a powerful lens through which they perceived changes in the neighborhood, particularly how space is used and the displacement that has fragmented the neighborhood. Jana, Leah and Kaden expressed their concern that people are being pushed out of the neighborhood and complained that developers were "buying up everything" (Focus Group, October 9, 2011). Gentrified spaces may provide hope for some, but Kaden expressed a mature sense of loss when he shared the photograph he took of the empty lot where the house his mother rented once stood. The now-empty lot sits across from stores and next to a house that inexplicably has not been torn down. Kaden took a picture of the classic car that sat in the driveway of this house. Kaden recalled seeing the car there when he was growing up and the memories of a place where he felt loved and nurtured, where he could play outside and feel safe, and where he shared a meal that "tasted like heaven."

> My house got torn down 'cause of the university. There was only like five other houses there that had to move. They wanted to extend their land and so we had to move cause we couldn't pay rent.... There was like two across the street and three of us and we used to have a huge backyard and we used to have a big tree and we could walk, like, I was sayin' the other day, a couple days ago, we could walk back home from [the community center], so it wasn't that very far. There used to be a big tree with a tire swing on it, we used to play on. I remember one night we had spaghetti and it was delicious, we got garlic bread and it tasted like heaven. (Focus Group, October 9, 2012)

When he describes having to move, Kaden points to the economics that have contributed to change—that owners of rental properties raise rents that low-income families cannot afford and then sell their properties for a profit given the changing market. Thus families like Kaden's are expendable, despite the history and sense of community that was so vital for so many families that have been pushed out. The irony is that economic development has worked against the kind of community and social capital that strengthened low-income families and helped them flourish in the area around the community center and the school just south of the center (Greene, 2013). Fortunately, Kaden and his family were able to move in with his grandparents, but it is unclear what happens to families who are forced to leave their homes.

Kaden's memories of growing up in this house also help explain the way he sees what other children might perceive as the inefficient use of space in the large, newly constructed one-family houses that had begun to swallow up the neighborhood around the community center. Not at all complaining, Kaden recalled with some joy the extent to which the house his mother rented was a bit cramped:

> There were three bedrooms, my room, my sister's room, and my mom's room. And there was a living room and a bathroom, and then there was a kitchen. It was a really fun house. I remember that my little sister had to sleep in my mom's room for a couple of years because of her age. (Focus Group, October 9, 2012)

However, when he visited his teacher's large home around the corner from the community center, he was puzzled that any one family needed so much space.

> I've been in that house before and there's like seven rooms, including the master bedroom, and the master bedroom is really big and I'd say in each room you could have, in each bedroom you could have ten people in there. In the lower level you can have thirty people and in the basement it's huge! I mean huge, real big; as big as the center actually. That's how huge it is and it can hold up to at least fifty people. So that would be 150 people because ten people in seven bedrooms, that's seventy, thirty people on the lower level, fifty people on the basement, that equals up to 150. (Focus Group, October 9, 2012)

Kaden's world is colored by his sense of closeness and belonging in a modest house where a meal of spaghetti and garlic bread "tasted like heaven." Thus it seemed almost inconceivable that a home this large could not accommodate multiple families in need. Children's conversations about their lived spaces were not marked with jealousy or contempt. Instead, they seemed disappointed that more sharing of existing resources isn't the norm.

Writing of racial containment, as a White strategy for maintaining property values and the segregation of the city of Chicago, Philpott (1991) [noted that "restrictive covenants" were "contractual agreement[s] among property owners that none of them would permit a 'colored person' to occupy, lease, or buy…property" (p. 189). Their options severely limited, Blacks in the post-Great Migration city paid exorbitant rents for the limited housing stock available to them because "that was just it. There was no place to go" (p. 198). While we might note that restrictive covenants as overt public policy rarely surface, and there are now federal protections in place explicitly abjuring racial protectionism in housing, more subtle forms of steering still exist, often justified along the lines of social class (which, of course, intersects explicitly with race) and under the cleansing terms of urban renewal and gentrification. But tearing down houses to build new, more expensive ones continues to force upon kids the dilemma of having no place to go, to go home to.

CONCLUSION

As we suggest at the outset, places are always in process, reflecting individuals' practices and developing through the interaction between structure and agency. "Places are constructed by people doing things and in this sense are never 'finished' but are constantly being performed" (Cresswell, 2004, p. 38). This is what could be construed as Soja's (2010) *thirdspace*, "the practiced and the lived, rather than simply the material, and provides the ground for understanding place as a lived, practiced, and inhabited space" (Cresswell, 2004, p. 38). Place then provides the conditions of possibility for creative practice that challenge normative conceptions of institutional spaces that reproduce inequality and the constructions of "us" and "them." Place does not need to be exclusionary, marked by permanence and boundedness, but openness and change, constructed by people in social interaction with one another, marked by reciprocity and an equal distribution of power. Thus reimagining place in this way allows us to see place as the confluence of dynamic, constantly evolving networks that are social and cultural—where paths lead in and out.

Unfortunately, groups already at the edge are further marginalized by economic development and gentrification, and children especially are seemingly invisible on the landscape of economic policy. Thus, we see a need to investigate urban landscapes as children "see them" from the perspective of their own multiple lifeworlds shaped by race, class, gender and location. It is important to recognize that children have very different values about place and space than do adults. Rather than assuming children know less than adults, we suggest that they may know "something else." This is evident in the ways that children in this study understand the depths of spatial inequality and the disconnect that many children experience, particularly Kaden, whose house was demolished to make room for upscale stores and expensive condominiums. The children's stories, poems and narratives also speak to the importance of distributing resources more equitably to mitigate the effects of poverty and homelessness. The children imagine safe spaces where children and families could develop a sense of community, identification, stability and belonging. In some sense, we are arguing that for a poverty of opportunities and a uniquely adult inability to listen, children are not being allowed to engage in civic discussions. We argue that their voices, always in formation, provide ready lenses for understanding gentrification and cityscapes differently.

In moving forward, we argue that it is important that adults root their understanding of place in the experiences of children's diverse and multifaceted everyday life. After all, children's experiences embody their own richness and intrinsic value, which is often not obvious to adults who ignore what we have termed the Adult-Child Imaginary (Greene et al., 2013). The Adult-Child Imaginary reminds the adult involved with children to take heed of youth perspectives and visions because of the clarity they possess. By using this construct, we distinguish between the privileged discourses that we used to describe youth empowerment, spatial inequality and transformative change and youth's own visions for this project rooted in their experiences and their own analyses.

The children in this study demonstrate the extent to which art empowers children to have a voice that will help them gradually begin to take greater responsibility for creating communities that fulfill their needs. To be fully engaged, youth need opportunities to tell their stories in ways that enable them to express their ideas across time and space through different modes of inquiry and presentation (e.g., photographs, narrative, maps, digital storytelling, video, performance and the like). Furthermore, youth must have access to these multimodal ways of telling their stories as vital components of personal agency. In the case of the children in our study, the project culminating in sharing the photos and the corresponding narratives in public places became their first point of public action and active resistance to the attempts by those outside of their space to define and limit their lived realities. This is an important step in children's sense that they can begin to take ownership of the spaces they inhabit, to speak for themselves and break with what Schutz (cited in Greene, 1973) called "the social world as taken for granted" (p. 183). Our work with children also shows us that there is much to understand about the ways children can more fully participate as partners to affect social change and the decisions that shape the urban landscape as places of inclusion.

In the end, youth's stories prompt us to reimagine afterschool programs as places that can serve as bridges between the struggles students experience in the neighborhoods they inhabit and the vision for community they may offer with their stories. Programs like the one we describe here provide a logical starting place for reinvesting time and energy in opportunities for relationship building with children. Relationships are a necessary precondition, we learned, for creating space for counternarratives to be told and heard safely, for fostering connections to community and for cultivating an ethic of civic engagement. For all too many children in the United States, schools are no longer places of relationship building or community connection. Afterschool programs can fulfill this purpose by creating spaces for youth to develop a sense of agency in solidarity with others, that give them the tools to remake their communities as places of hope and possibility and resist the conditions that silence the voices of low-income youth.

CODA

We have discussed the multiple affordances that art, reading, writing and speaking can provide youth in developing a sense of agency, particularly in a context where adults take their ideas seriously. In turn, their various uses of literacy to represent their worlds complicates the idea of counternarrative and offers us a multitextured understanding of the ways children make sense of the urban landscape. Children come to voice and a belief in their capacities that we wouldn't see without these multiple affordances. We see changes in the children's belief in themselves and the role they believe they can play in creating meaningful change in a world where little space is given to children to make decisions that affect them.

We are optimistic that the children we describe in this chapter will see the extent to which their research has the potential to create safe spaces where they can be themselves. Some six months after we first met with the children to write "I Am From Poems" and take photographs, they decided that that they wanted to redesign a space in the neighborhood that used to be a park but that has now deteriorated; few children play there. They presented their proposal to the members of the Northeast Neighborhood Council where they were giddy but also subdued. They were halfway between the roles they know—being kids at the community center—and roles they were now trying on. Leah explained to council members, "We learned a lot about becoming a researcher...viewing things around the neighborhood and deciding which is the good part or which is the bad part. So we started reviewing things around the neighborhood, doing observations, and brainstorming." Joseph urged the adults to think about the value of a park in the neighborhood. "If you see kids playing at the park, it will attract more people to move into the area. The community can use this as a place to exercise and hang around and meet other people. That's what I want the park to be like." Kaden, too, moved us all when he described what used to be his home. "I used to have a house...in this neighborhood and it got torn down a couple of years ago and now I don't want to see that happen to the park where people have a lot of good memories, and I want it rebuilt and nourished with energy and young life. If we can make that happen, I will be really happy."

In a way, the children's proposal harkens back to the 19th-century idea of parks as "democratic spaces" (Rutkow, 2012), and we can't help but be reminded of the wisdom of the 1971 Dr. Seuss classic, *The Lorax*, where all of the trees are cut down for a profit-making venture. Eventually only a few individuals even remember what real trees look like. In the end, it is a child who plants the last remaining seed to regrow the forest.

The members of the neighborhood association treated the children's ideas with dignity and respect. The chair offered to write a letter on the children's behalf and perhaps apply for a grant, so that they could garner support for creating a park. She also conveyed to the children that they were sitting among a powerful group of people and that they knew the mayor. Indeed, she invited the mayor to meet with the children.

A week later the children greeted the mayor at the community center where he told them, "I think that grown-ups could probably do a better job of listening to young people who are, after all, experts of our neighborhoods." He listened intently to their proposal and said that he was struck by how often they mentioned the word "we." He asked the children and all of the others in the room—tutors, children and parents—if they would help refurbish the park. Hands shot up, and he told the children that everyone's willingness to help was "a good sign." After all, he explained, "not everyone can make changes alone, not even the mayor," and he agreed to support their proposal, even if it meant completing their design in stages.

Kids are rarely asked what they think, and with increased disinvestment in the places where they live, they often express doubt that they have a voice. The mayor's affirmation may have prompted the children to see that change is possible, but they will wait to see tangible results of their research.

Fig 1. Mural art at the community center that youth participants talked about.

REFRAMING SPATIAL INEQUALITY | 125

Fig 2. A tree that has significance for the youth participants.

Fig 3. The park that the youth participants want to refurbish.

Fig 4. A construction site near the community center.

REFERENCES

Aitken, (2001). *Geographies of young people. The morally contested spaces of identity.* London, UK: Routledge.
Anyon, J. (2005). *Radical possibilities: Public policy, urban education, and a new social movement.* London, UK: Routledge.
Christensen, L. (2000). *Reading, writing, and rising up: Teaching about social justice and the power of the written word.* Milwaukee, WI: Rethinking Schools.
Christensen, P., & James, A. (Eds.). (2000). *Research with children: Perspectives and practices.* London, UK: Falmer Press.
Cisneros, S. (1989). *The house on Mango Street.* New York, NY: Vintage Books.
City plan (n.d.). Housing. Office of the Mayor.
Cresswell, T. (2004). *Place: A short introduction.* Malden, MA: Blackwell.
Delgado Bernal, D. (2002). Critical race theory, Latino critical theory, and critical raced-gendered epistemologies: Recognizing students of color as holders and creators of knowledge. *Qualitative Inquiry 8*, (1), 105–126.
Delgado, R. (1989). Storytelling for oppositionists and others: A plea for narrative. *Michigan Law Review, 87*, 2411–2441.
Ferguson, A. (2001). *Bad boys: Public schools in the making of black masculinity.* Ann Arbor: University of Michigan Press.
Freire, P. (1970). *Pedagogy of the oppressed.* New York, NY: Continuum.
Greene, M. (1973). The matter of justice. *Teachers College Record, 75*(2), 181–191.
Greene, S. (2013). *Race, community, and urban schools: Partnering with African American families.* New York, NY: Teachers College Press.
Greene, S., Burke, K., & McKenna, M. (2013). Forms of voice: Exploring the empowerment of youth at the intersection of art and action. *The Urban Review, 45*, 311–334.

Harris, D. (2006). Race, class, and privacy in the ordinary postwar house, 1945–1950. In R. Schein (Ed.), *Landscape and race in the United States* (pp. 127–155). New York, NY: Routledge Taylor & Francis Group.
Haymes, S. (1995). *Race, culture, and the city: A pedagogy for Black urban struggle.* Albany, New York: SUNY Press.
Hull, G., & James, M. (2006). Geographies of hope: A study of urban landscapes, digital media, and children's representations of place. In P. O'Neill (Ed.), *Blurring boundaries: Research and teaching beyond a discipline* (pp. 255–289). Cresskill, NJ: Hampton Press.
Hull, G., & Katz, M. (2006). Crafting an agentive self: Case studies of digital storytelling. *Research in the Teaching of English, 41*(1), 43–81.
Kinloch, V. (2010). *Harlem on our minds: Place, race, and the literacies of urban youth.* New York, NY: Teachers College Press.
Kinloch, V. (2012). *Crossing boundaries: Teaching and learning with urban youth.* New York, NY: Teachers College Press.
Lefebvre, H. (1992). *The production of space.* London: Wiley-Blackwell.
Mazzoni, V., & Harcourt, D. (2013). An international experience of research with children: Moving forward on the idea of children's participation. *Qualitative Research.* doi:10.1177/1468794112468470.
Miller, P. (2011). Mapping educational opportunity zones: A geospatial analysis of neighborhood block groups. *Urban Review, 44*(2), 189–218.
Neuman, S. B., & Celano, D. (2001). Access to print in low-income and middle-income communities: An ecological study of four neighbornhoods. *Reading Research Quarterly, 36*(1), 8–26.
Norton, N. (2006). Talking spirituality with family members: Black and Latina/o children co-researcher methodologies. *The Urban Review, 38*(4), 313–333.
Philpott, T. L. (1991). *The slum and the ghetto: Neighborhood deterioration and middle class reform, Chicago 1880–1930.* New York, NY: Wadsworth.
Rutkow, E. (2012). *American canopy.* New York, NY: Scribner.
Schein, R. H. (Ed.). (2006). *Landscape and race in the United States.* New York, NY: Routledge Taylor & Francis Group.
Seuss, D. (1971). *The lorax.* New York, NY: Random House.
Simon, E. (2013). *How closing schools hurts neighborhoods.* Retrieved from http://www.washingtonpost.com/blogs/answer-sheet/wp/2013/03/06/how-closing-schools-hurts-neighborhoods/
Soja, E. W. (2010). *Seeking spatial justice.* Minneapolis: University of Minnesota Press.
Solórzano, D., & T. Yosso. (2002). Critical race methodology: Counter-storytelling as an analytical framework for educational research. *Qualitative Inquiry, 8,* 23–44.
Urciuoli, B. (1995). The indexical structure of visibility. In B. Farnell (Ed.), *Human action signs in cultural context: The visible and the invisible in movement and dance* (pp. 189–215). Metuchen, NJ: The Scarecrow Press.
Wright Edelman, M. (2012). *America's public schools: Still unequal and unjust.* Retrieved from http://www.huffingtonpost.com/marian-wright-edelman/public-schools-minority-students_b_1408878.html

PART THREE

Is There a Class-Conscious, Racially Sensitive Doctor in the House?

The Affordable Care Act is the cornerstone legislation for which President Obama will be remembered. The controversial "Obamacare" has been a topic of debate around kitchen tables, water coolers and news talk shows because sickness doesn't discriminate. It welcomes the young and the old, the rich and the poor—regardless of race, ethnicity, class or gender. Sickness is the one issue everyone understands because we all experience it or know someone who has in some form. If we have the misfortune of becoming ill and desire medical treatment, we want expeditious, quality care regardless of our socioeconomic status. Ironically, likelihood of illness, access to medical treatment and timeliness and quality of the treatment are greatly affected by these very demographics: race, ethnicity, class, socioeconomic status (SES)—specifically where we live—as the authors in this section demonstrate.

Lopez explores the three phases of gentrification, aligning them with health issues existing residents experience as a result of living in communities in transition. He calls for preventative measures prior to the start of gentrification to reduce the negative impact on residents, which range from the socioemotional to the physical.

Akintobi, Braithwaite and Dodds begin with the premise that African Americans living in segregated communities suffer from higher levels of stress and poorer overall health. Like Lopez, they find that this remains true even during gentrification (which brings changes to the demographic composition of the communities).

Barr's chapter on physician preparation emphasizes the importance of training medical students to be culturally sensitive to the needs of the patient. This is increasingly important at a time when racially and ethnically diverse communities are experiencing shortages of primary care physicians and at the same time increasing the service population as Obamacare now extends healthcare to this population. The chapter presents a strong and compelling argument for changes in the gatekeeping mechanisms (which negatively impacts students of color entrance to and retention in medical school) used for predicative analytics of medical school performance and future career success. The sociodemographic data and perspectives of patients of color with lower SES make evident the huge disconnect between the values of medical schools and those receiving treatment, which begs the question, "Is There a Class-Conscious, Racially Sensitive Doctor in the House?"

CCR

CHAPTER SEVEN

Training Physicians FOR THE Demographics OF THE 21st Century

The Importance of Diversity and Cultural Competency

DONALD A. BARR

As we move into the 21st century, the United States has been experiencing two important trends: rising economic inequality and rising demographic diversity. Between 2000 and 2011, real median household income in the United States fell from $54,841 to $50,054, a drop of 8.7%. (U.S. Census Bureau, 2012a). While the median household was experiencing this drop in income, the top 1% of households experienced an increase in their income over this same time period. Especially in the recovery that followed the recession of 2008, those at the 99th percentile of household incomes ($367,000 in 2011) experienced an 11.2% increase in income, while those in the lower 99% of households experienced a *drop* of 0.4%. Between 1993 and 2011, the top 1% of households accounted for more than half of the overall growth in real incomes (Saez, 2013).

Income inequality, however, is not only an issue of numbers. Consistent with the historical pattern in the United States, median income of different racial and ethnic groups continues to vary substantially. The median income in 2011 of Asian households was $64,995, and of non-Hispanic White households it was $55,412, while that of Hispanic households was $38,624, and of non-Hispanic Black households it was $32,229 (U.S. Census Bureau, 2012a).

This inequality of income becomes even more important when we look also at the demographic changes expected to take place in the United States. In 2011, the U.S. population was 63.4% non-Hispanic White, 12.3% non-Hispanic Black,

4.8% Asian, 2.0% American Indian and 16.7% Hispanic. By 2050, those percentages are expected to change to 46.6% non-Hispanic White, 13.0% non-Hispanic Black, 7.7% Asian, 1.5% American Indian and 28.0% Hispanic (U.S. Census Bureau, 2012b). We will become a majority-minority society, albeit one with a long history of income inequality.

INEQUALITY IN ACCESS TO HEALTH-CARE AND THE STEPS TAKEN BY THE AFFORDABLE CARE ACT

While income inequality has been growing, inequality in access to healthcare has also been growing. Between 2001 and 2011, the portion of the U.S. population who had no health insurance grew from 14.7% to 16.9% (U.S. Census Bureau, 2011). A central cause of this growing number of uninsured Americans has been the rapidly rising cost of health insurance. Over this same period, the average cost of providing health insurance for an employee and his or her family rose from $7,061 to $15,073—an increase of 113% (Kaiser Family Foundation, 2011). This was the total cost of the policy, which is typically shared by the employer and the employee. With the rapidly rising cost of health insurance, employers had little choice but to shift more of that cost onto the employee. Over this 10-year period, the employee's share of the premium went from $1,787 to $4,129, an increase of 131%, while the employer's share went from $5,269 to $10,944, an increase of 108%.

The rising cost of healthcare coupled with the growing percentage of Americans who were uninsured were major stimulants to the passage in 2010 of the Affordable Care Act (ACA), fondly (and not-so-fondly) referred to by Democrats (and Republicans) as "Obamacare." While ACA makes some efforts at constraining the rising cost of healthcare nationally, its principal thrust is the expansion of access to affordable health insurance. It does this through two principal mechanisms: a major expansion of the Medicaid program and the establishment of an individual mandate for health insurance coupled with creation of health benefit exchanges (HBE) for each state.

While the Supreme Court's 2012 decision removing the mandatory nature of the Medicaid expansion will have some impact on the eventual outcomes from the expansion, it seems apparent that most (if not all) states will eventually adopt the expansion. Failure to do so would place burdensome costs onto the providers within a state, while leaving large numbers of low-income residents out of the expanded coverage. The abhorrence some state governments have for Obamacare aside, sooner or later these incentives will likely persuade most states to participate in the expansion. This certainly has been the experience with Medicaid historically. When federal funding for state Medicaid programs first became available in 1967, only 26 states chose to participate. Within two years, however, 41 states had elected to participate; within five years, all but one state had Medicaid programs (Kaiser Family Foundation, 2012). It seems reasonable to expect the same pattern for the Medicaid expansion under ACA.

Federal estimates indicate that between 27 million and 30 million people who currently are without health insurance will gain coverage through ACA. Of these, 12 million are expected to enroll in the expanded Medicaid program, with the others gaining coverage through the health benefit exchanges (Congressional Budget Office, 2013). This expansion of coverage is expected to have a powerful impact on the access to care for those gaining coverage. An estimated 52% of currently uninsured adults have no usual source of healthcare, as compared to 10% of those with Medicaid or private insurance. Among uninsured adults, 42% do not obtain routine preventive care. Among uninsured adults with chronic medical conditions, 39% have no usual source of care, and 43% have not had a visit to a doctor within the previous year (Kaiser Family Foundation, 2010).

THE NEED TO EXPAND THE PRIMARY CARE WORKFORCE

For the nearly 30 million gaining coverage under ACA, many of whom have either delayed or forgone needed care, the availability of health insurance will open the doors to the healthcare system. This raises a crucially important question. Once they enter those doors, will there be a doctor there to see them? And for the many non-White patients who now disproportionately make up the uninsured population and who gain coverage under ACA, will there be a doctor there with whom they can connect on a cultural and linguistic basis?

National attention has recently been focused on two core issues: (1) a growing shortage of primary care physicians, and (2) continued problems in attaining the goal of having the diversity of the medical profession reflect the diversity of the population it serves.

In 2010, there were about 210,000 primary care physicians practicing in the United States. ("Primary care" refers to physicians trained in family medicine, general internal medicine, or general pediatrics. Some include those trained in obstetrics/gynecology as primary care physicians.) A combination of three factors will increase the need for primary care physicians: the aging of the population, population growth, and the expansion of health insurance under ACA. These factors will increase the number of primary care physicians needed in the year 2025 by more than 50,000 physicians (Petterson et al., 2012). The need for this level of expansion in the primary care workforce is complicated by waning student interest in primary care, as compared to subspecialty medical practice. Of the nearly 23,000 medical school graduates in 2012 in the United States, only 20% are likely to end up in primary care practice following their residency training (Schwartz, 2012).

The potential shortage of primary care physicians is even more acute for patients covered by Medicaid. Because state Medicaid programs typically pay physicians substantially less than other private or public insurers, fewer and fewer physicians are willing to treat those covered by Medicaid. One study from 2006 found a "striking trend...that care of Medicaid patients is becoming increasingly concentrated among a smaller proportion of physicians..." (Cunningham & May, 2006, p. 1). Nationally, 58% of primary care physicians will not accept new Medicaid patients, while nearly half of specialists will not accept new Medicaid patients (Cunningham, 2011). The shortage of primary care physicians is even more acute for African Americans living in segregated residential areas. Compared to majority White zip codes, the odds of a majority African American zip code being within a primary care shortage area were increased by 67% (Gaskin, Dinwiddie, Chan, & McCleary, 2012).

For many Medicaid patients, as well as those without insurance, a federally qualified health center (FQHC) is the "safety net" on which they can rely to obtain needed medical care. Organized as community-based, nonprofit organizations, FQHCs receive extra funding to provide care to these populations, without regard for a patient's ability to pay. In 2010, "about 93% of health center patients are below 200% of the federal poverty level, 62% are racial or ethnic minorities, and about 38% are uninsured" (Shi et al., 2013, p. 57). Under the reforms included in ACA, FQHCs will receive substantial new funding to expand the number of physicians and other providers to meet the expected increase in demand for care placed on them as a result of the Medicaid expansion.

Expanding the Racial and Ethnic Diversity of the Medical Profession

In addition to the question of whether there will be sufficient numbers of primary care physicians and other types of physicians available to treat the 30 million or more individuals who will gain health insurance under ACA, there is a second, equally important question. Given the high proportion of racial and ethnic minority patients who will be among the newly insured, will the physicians and other practitioners be sufficiently diverse to provide care that is culturally appropriate and linguistically accessible?

The Office of Minority Health of the U.S. Department of Health and Human Services publishes its National Standards on Culturally and Linguistically Appropriate Services (CLAS; U.S. Department of Health and Human Services Office of Minority Health 2013a). The first of these standards states, "Health care organizations should ensure that patients/consumers receive from all staff members [sic] effective, understandable, and respectful care that is provided in a manner compatible with their cultural health beliefs and practices and preferred language." In order to provide this type of culturally appropriate care, physicians must "hav[e] the capacity to function effectively as an individual…within the context of the cultural beliefs, behaviors, and needs presented by consumers and their communities" (U.S. Department of Health and Human Services Office of Minority Health 2013b).

Unfortunately, research has shown that for many patients, especially those from racial or ethnic minority groups, the experience of care does not always meet these standards. LaVeist and colleagues surveyed both Black and White patients who had chronic heart disease requiring frequent visits to a physician about their perception of the care they received. Overall, Black patients were less satisfied with their care. With further questioning, the authors were able to identify the possible source of this reduced level of satisfaction. Black patients reported lower levels of trust in their physician, associated with perceptions of race bias on the part of the physician. When these two factors—trust, and the perception of bias—were taken into account in their statistical analysis, the level of satisfaction between Black and White patients did not differ (LaVeist, Nickerson, & Bowie, 2000).

A second study by Cooper-Patrick and colleagues explored patients' perceptions of the extent to which their physician involved them in the process of making decisions about care. They studied both Black and White patients, as well as Black and White physicians. Interestingly, while the average "Participatory Decision Making" score was the same for Black and White physicians, both Black and White patients perceived more involvement in their care if their physician was of the same race (Cooper-Patrick et al., 1999).

The Association of American Medical Colleges (AAMC) has focused considerable attention on the need to enhance the racial and ethnic diversity of the medical profession in the United States. In a campaign that targets students from underrepresented minority (URM) groups who may be interested in a medical career, the AAMC (2012a) states that, "One of the most pressing health care challenges facing the nation is the critical need for more minority physicians. In the next 15 years, the nation is projected to confront an overall shortage of physicians, but the need is, and will continue to be, particularly great for minority physicians" (p. 1).

Prior to the early affirmative action policies of the 1960s and 1970s, there were few URM students enrolled in U.S. medical schools. In 1950, 55 black students, 8 Latino students and 1 Native American student graduated from a U.S. medical school, representing about 1% of all graduates (Association of American Medical Colleges, 2006). This number rose somewhat, reaching 3.6% of graduating medical students in 1968, although more than half of these students were enrolled at either Howard University College of Medicine or Meharry Medical College, both historically Black universities.

The 1960s saw the beginnings of efforts to increase the number of URMs in medicine through race-based affirmative action programs. By 1978, the graduating class of U.S. medical schools included 724 Black students, 313 Hispanic students and 34 Native American students. Combined, these students from URM groups represented 7.4% of all graduates (Association of American Medical Colleges, 2010).

Then the U.S. Supreme Court decided what has come to be referred to as "the Bakke case" (*Regents of the University of California v. Bakke*, 1978). Alan Bakke was a White man who in both 1973 and 1974 applied to the University of California, Davis (UC Davis), medical school, both times failing to gain admission. As part of its affirmative action program, the UC Davis medical school had accepted Black applicants who had somewhat lower grades and/or test scores than Bakke. Bakke filed suit against the University of California for race bias. In 1978, the Supreme Court ruled that giving Black applicants an explicit preference violated the U.S. Constitution. From that point on, race could not be used by medical schools to establish explicit quotas for admission, although race could still be used as one among many factors in deciding admission. In 2003, the use of race as one of many factors was again found to be permitted under the Constitution in a case involving the University of Michigan law school (*Grutter v. Bollinger*, 2003; although the use of race in this context was limited somewhat in a case [*Fisher v. University of Texas*] decided by the Supreme Court in 2013).

Between 1978, when the Bakke case was decided, and 2011, the number of URM medical school graduates gradually increased. In 2011, about 15% of the more than 17,000 students graduating from medical school were Black, Hispanic, or American Indian (Association of American Medical Colleges, 2012b). Most of the increase from 1978 had taken place by 2000. The percentage of URM medical school graduates increased from 7.4% in 1978 to 11.3% in 1990 and to 14.6% in 2000. Between 2000 and 2011, the percentage of URM medical graduates increased only slightly.

While this represents real and meaningful progress in the inclusion of URM students among those trained to be physicians, we must recall that in 2011, these racial and ethnic groups made up 31% of the overall U.S. population. Despite decades of efforts at making medical education more available to students from groups that had historically been largely excluded, the percentage of URM physicians being trained represents less than half of the URM population nationally.

This brings us back to the potential problem facing the U.S. healthcare system following full enactment of ACA, as described by Marc Nivet (2011), Chief Diversity Officer at the AAMC: "The combined forces of health care reform, demographic shifts, continued economic woes, and the projected worsening of physician shortages portend major challenges for the health care enterprise in the near future" (p. 1487).

Thirty million or more Americans will gain access to health insurance under ACA. Since nearly half of those who are currently uninsured are either Black or Hispanic (U.S. Census Bureau 2012c), it is reasonable to expect that a substantial portion of those who will gain insurance under ACA will be from URM groups. (It may be less than half, since many uninsured Hispanics may not be eligible for coverage under ACA due to their immigration status.) Where are these people going to find culturally competent doctors to take care of them?

Academic Programs to Increase Student Diversity

The issue of recruiting for URM students into the medical school "pipeline" has been under discussion for decades. In the period following the 1978 Bakke decision, attention shifted to programs to recruit more URM students into premedical studies at the university level and to support them in their preparation for medical school. As I have described elsewhere, these programs were of three principal types: (1) Programs to encourage high school students to consider a health professions career and offering academic support in pursuit of that goal, (2) programs to offer academic enrichment and other types of support to URM college students who are considering a health professions career, and (3) postbaccalaureate programs to offer students who have graduated from college additional instruction and support to encourage and enable their application to medical school (Barr, 2010, p. 142).

These programs, many initiated in the 1980s in the immediate aftermath of Bakke, contributed substantially to the gradual increase in the number of URM students attending medical school in the 1990s. However, as described earlier, that number largely leveled off after 2000. Despite the ongoing increase in the diversity of the U.S. population, the diversity of those trained in medicine has remained fairly flat.

A number of studies have identified the barriers to further increases in the number of URM students entering medicine. Principal among these seems to be the early college experience of students, especially those coming from lower socioeconomic backgrounds. I have studied this issue along with colleagues at Stanford University and the University of California, Berkeley (UC Berkeley). Both institutions have high standards, admitting only students who are fully prepared academically. However, we found that, somewhere between matriculation as a freshman and graduation as a senior, many of the URM students at these universities lose interest in a medical career. At Stanford, the average number of freshmen entering each year between 2001 and 2005 who expressed an initial interest in pursuing a medical career was 363, of whom 108 (30%) were URM students. However, between 2001 and 2008, the average number of students from Stanford who submitted an application to medical school was 294, of whom only 50 (17%) were URM students (Barr, 2010). The attrition of non-URM students was from 255 to 244, a loss of 4%, while that of URM students was from 108 to 50, for a loss of 54%.

In order to determine the cause or causes of this disproportionate attrition from premedical studies, we did a series of follow-up interviews with students in their junior year of college, all of whom had indicated interest as a freshman in becoming a physician. We found one clear answer to our question, "What were the factors that led to the decrease in your level of interest in being pre-med?" Overwhelmingly the students responded that the principal factor leading to the decrease in their level of interest was the courses they had taken. When asked if there were any specific courses that had been particularly discouraging, again students responded with one voice: chemistry. All of the 29 Stanford students whose interest had declined and who identified a specific course said chemistry. As described in the interview by one URM student, "Everyone says [chemistry courses are] more like a weeding-out process than anything [else], and I just ended up being one of those people" (Barr, Gonzalez, & Wanat, 2008, p. 507).

We did a parallel study at UC Berkeley and found similar results. Of the 29 URM students we interviewed at Berkeley, 26 identified chemistry courses as having discouraged their interest in remaining pre-med. Chemistry was mentioned three times more often than the next most frequent course, mathematics. As described by a URM student there, "I think having to drop Chem IA in the Spring of my first year made me question whether or not I could do it.... So it was just a matter of me not wanting to feel dumb around other people" (Barr, Matsui, Wanat, & Gonzalez, 2010, p. 50).

For more than 100 years, undergraduate science courses, especially chemistry courses, have been used to "weed out" undergraduates interested in becoming physicians but who have difficulty in undergraduate science courses. This "weed out" process has disproportionately affected two groups: URM students and students from socially or economically disadvantaged backgrounds. The theory behind using relatively lower grades or test scores in science in this manner is that performance in undergraduate sciences is an accurate predictor of success in medical school and as a physician. Unfortunately, the best available scientific evidence, gathered over a period of several decades, disproves this theory. While undergraduate science grades are correlated with grades in the initial preclinical science courses in medical school (e.g., biochemistry), they are almost wholly unrelated to subsequent clinical skills as a physician.

That is, unless you consider their inverse association with noncognitive characteristics such as empathy, sensitivity and communication ability. We have known for more than 40 years that students who do better in science classes as undergraduates tend to be "narrower in interests, less adaptable, less articulate, and less comfortable in interpersonal relationships" than their peers (Gough, 1978, p. 291). A study from the 1990s found that students who did the best in the premedical sciences also scored lower on standardized measures of empathy and tended to be "shy," "submissive," "withdrawn" or "awkward and ill at ease socially," characteristics the author suggested are "the antithesis of what most of us would want in a clinician" (Tutton, 1996, p. 185).

Expanding the Concept of Diversity in Selecting Students for Medical School

Diversity is not only about race or ethnicity. Diversity, especially as it pertains to the practice of medicine, also encompasses a wide range of cognitive and non-cognitive characteristics, gender diversity, as well as diversity of socioeconomic background and cultural experience. For patients to have access to high-quality care, the physicians and other providers they see for advice and treatment must represent this full range of diversity. Continuing to focus principally on performance in the undergraduate science classroom, as medical schools have done for nearly a century, will not provide the type of medical profession the United States will need in the 21st century.

Will de-emphasizing science performance weaken the clinical skills of the physicians we train? The answer comes from multiple research studies: "No." While students of differing science backgrounds may often have differing performance in the science-based components of the medical school curriculum, once they enter and then complete clinical training, their clinical skills become largely indistinguishable, leaving their interpersonal and noncognitive skills as distinguishing factors (Gough, Hall, & Harris, 1963; Price, Taylor, & Nelson, 1971; Wingard & Williamson, 1973).

Facing the challenge of providing financially accessible, culturally competent healthcare for the U.S. population in the 21st century will require a fundamental rethinking of how we choose students for training as physicians. For now, we will continue to use race and ethnicity as relevant factors in selecting students for admission, although their use will doubtlessly subside over time, either as the result of court rulings or as part of a growing recognition that diversity is a much broader concept. Socioeconomic diversity has become equally important as a factor to consider, for two important reasons.

The first is that the growing non-White population in the United States disproportionately includes those who have grown up experiencing socioeconomic disadvantage and who, as a consequence, can more easily establish a cultural understanding of the experiences of many of their patients. The second reason has to do with what type of medical practice different groups tend to prefer. As described earlier, for a number of years, medical students chosen based on traditional selection criteria have increasingly eschewed primary care practice in favor of subspecialty training and practice. The result has been a growing shortage of primary care physicians, especially those practicing in rural or urban underserved communities. Perhaps not surprisingly, when students who have grown up in these disadvantaged communities are given the opportunity to train as physicians, they disproportionately choose to return to practice in these communities, often in a primary care specialty (Lupton, Vercammen-Grandjean, Forkin, Wilson, & Grumbach, 2012).

Thus, finding culturally competent primary care physicians to provide care, first to the millions of Americans who will newly gain health insurance coverage under ACA and eventually to the increasingly diverse U.S. population, will require a fundamental rethinking of how we select qualified college students for training as physicians. The AAMC (2013) has adopted policies consistent with this approach, emphasizing on its website that, "The AAMC's commitment to diversity includes embracing a broader definition of 'diversity,' and supporting our members' diversity and inclusion efforts…." If medical schools are able to incorporate this perspective into their recruitment and selection processes, we can expect to see a narrowing of the inequality in access to high-quality healthcare, even if the level of economic inequality remains unchanged.

REFERENCES

Association of American Medical Colleges. (2006). *Diversity in the physician workforce: Facts & figures 2006.* Retrieved from https://members.aamc.org/eweb/upload/Diversity%20in%20the%20Physician%20Workforce%20Facts%20and%20Figures%202006.pdf

Association of American Medical Colleges. (2010). *Diversity in the physician workforce: Facts & figures 2010.* Retrieved from https://members.aamc.org/eweb/upload/Diversity%20in%20the%20Physician%20Workforce%20Facts%20and%20Figures%202010.pdf

Association of American Medical Colleges. (2012a). *AspiringDocs.Org. America needs a more diverse physician workforce.* Retrieved from https://www.aamc.org/download/87306/data/physiciandiversityfacts.pdf

Association of American Medical Colleges. (2012b). *Diversity in medical education: Facts and figures 2012.* Retrieved from https://members.aamc.org/eweb/upload/Diversity%20in%20Medical%20Education%20Facts%20and%20Figures%202012.pdf

Association of American Medical Colleges. (2013). *Diversity and inclusion.* Retrieved from https://www.aamc.org/initiatives/diversity/

Barr, D. A. (2010). *Questioning the premedical paradigm: Enhancing diversity in the medical profession a century after the Flexner report.* Baltimore, MD: Johns Hopkins University Press.

Barr, D. A., Gonzalez, M., Wanat, S. F. (2008). The leaky pipeline: Factors associated with early decline in interest in pre-medical studies among under-represented minority undergraduate students. *Academic Medicine, 83,* 503–511.

Barr, D. A., Matsui, J., Wanat, S. F., & Gonzalez, M (2010). Chemistry courses as the turning point for premedical students. *Advances in Health Sciences Education, 15*(1), 45–54.

Congressional Budget Office. (2013). How has CBO's estimate of the net budgetary impact of the Affordable Care Act's health insurance coverage provisions changed over time? Retrieved from http://cbo.gov/publication/44008

Cooper-Patrick, L., Gallo, J. J., Gonzales, J. J., Vu, H. T., Powe, N. R., Nelson, C., & Ford, D. E. (1999). Race, gender, and partnership in the patient-physician relationship. *The Journal of the American Medical Association, 282,* 583–589.

Cunningham, P. (2011). *State variation in primary care physician supply: Implications for health reform Medicaid expansions.* Center for Studying Health System Change. Retrieved from http://www.hschange.com/CONTENT/1192/

Cunningham, P., & May, J. (2006). *Medicaid patients increasingly concentrated among physicians.* Center for Studying Health System Change. Retrieved from http://www.hschange.com/CONTENT/866/866.pdf

Fisher v. University of Texas (2013). 09-50822.

Gaskin, D. J., Dinwiddie, G. Y., Chan, K. S., & McCleary, R. R. (2012). Residential segregation and the availability of primary care physicians. *Health Services Research, 47*(6), 2353–2376.

Gough, H. G. (1978). Some predictive implications of premedical scientific competence and preferences. *Journal of Medical Education, 53,* 291.

Gough H. G., Hall, W. B., & Harris, R. E. (1963). Admissions procedures as forecasters of performance in medical training. *Journal of Medical Education, 38,* 983–998.

Grutter v. Bollinger, 539 U.S. 306 (2003).

Kaiser Family Foundation. (2010). *Medicaid beneficiaries and access to care.* Retrieved from http://www.kff.org/medicaid/upload/8000-02.pdf

Kaiser Family Foundation. (2011). "Average Annual Premiums for Family Health Benefits Top $15,000 in 2011, Up 9 Percent, Substantially More than the Growth in Worker's Wages, Benchmark Employer Survey Finds." Retrieved from http://www.kff.org/insurance/092311nr.cfm

Kaiser Family Foundation. (2012). *A historical review of how states have responded to the availability of federal funds for health coverage.* Retrieved from http://www.kff.org/medicaid/upload/8349.pdf

LaVeist T. A., Nickerson, K. J., & Bowie, J. V. (2000). Attitudes about racism, medical mistrust, and satisfaction with care among African American and white cardiac patients. *Medical Care Research and Review, 57*(1), 146–161.

Lupton, K., Vercammen-Grandjean, C., Forkin, J., Wilson, E., & Grumbach, K. (2012). Specialty choice and practice location of physician alumni of University of California premedical postbaccalaureate programs. *Academic Medicine, 87*(1), 115–120.

Nivet, M. V. (2011). Diversity 3.0: A necessary systems upgrade. *Academic Medicine, 86*(12), 1487–1489.

Petterson, S. M., Liaw, W. R., Phillips, R. L. Jr., David, L., Rabin, D. L., Meyers, D. S. & Bazemore, A. W. (2012). Projecting US primary care physician workforce needs: 2010–2025. *Annals of Family Medicine, 10*(6), 503–509.

Price, P. B., Taylor, C. W., & Nelson, D. E. (1971). *Measurement and predictors of physician performance: Two decades of intermittently sustained research* (Report no. PB-224 543). Washington, DC: U.S. Department of Health, Education, and Welfare.

Regents of the University of California v. Bakke, 438 U.S. 265 (1978).

Saez, E. (2013). *Striking it richer: The evolution of top incomes in the United States.* Retrieved from http://elsa.berkeley.edu/~saez/saez-UStopincomes-2011.pdf

Schwartz, M. D. (2012). The US primary care workforce and graduate medical education policy. *The Journal of the American Medical Association, 308*(21), 2252–2253.

Shi, L., Lebrun-Harris, L. A., Daly, C. A., Sharma, R., Sripipatana, A., Hayashi, A. S., & Ngo-Metzger, Q. (2013). Reducing disparities in access to primary care and patient satisfaction with care: The role of health centers. *Journal of Health Care for the Poor and Underserved, 24,* 56–66.

Tutton, P. J. (1996). Psychometric test results associated with high achievement in basic science components of a medical curriculum. *Academic Medicine, 71,* 181–186.

U.S. Census Bureau. (2011). *Health insurance historical tables.* Retrieved from http://www.census.gov/hhes/www/hlthins/data/historical/HIB_tables.html

U.S. Census Bureau. (2012a). *Income, poverty, and health insurance coverage in the United States: 2011.* Retrieved from http://www.census.gov/hhes/www/poverty/data/incpovhlth/2011/index.html

U.S. Census Bureau. (2012b). *Population projections.* Retrieved from http://www.census.gov/population/projections/data/national/2012.html

U.S. Census Bureau. (2012c). *People without health insurance coverage by selected characteristics: 2010 and 2011.* Retrieved from http://www.census.gov/hhes/www/hlthins/data/incpovhlth/2011/Table7.pdf

U.S. Department of Health and Human Services Office of Minority Health. (2013a). National standards on culturally and linguistically appropriate services. Retrieved from http://minorityhealth.hhs.gov/templates/browse.aspx?lvl=2&dvlid=15

U.S. Department of Health and Human Services Office of Minority Health. (2013b). *What is cultural competency?* Retrieved from http://minorityhealth.hhs.gov/templates/browse.aspx?lvl=2&dvlid=11

Wingard, J. R., & Williamson, J. W. (1973). Grades as predictors of physicians' career performance: An evaluation literature review. *Journal of Medical Education, 48,* 311–322.

CHAPTER EIGHT

Gentrification AND Health

Patterns of Environmental Risk

RUSSELL LOPEZ

This chapter presents a three-part model of the sequence of gentrification and then looks at the potential environmental health effects that may occur during each of these. These effects are considered across the physical, built and social domains of the environment. During each phase of gentrification, the environment provides a potential mix of protected and harmful features. Together, and in concert with preexisting factors, the health of residents and displacement can be substantially affected. These varying effects suggest that interventions and policies be put into place to address the environmental impact of gentrification in advance of actually occurring in order to adequately protect the public's health.

DOMAINS OF THE ENVIRONMENT

While there are many ways to consider and parse out the various dimensions of the environment, one way is to think of the environment as having three broad and overlapping domains: the physical, built and social environments (Srinivasan, O'Fallon, & Dearry, 2003). The physical environment consists of the standard media: air, water, chemical, biological and other agents present or absent in an area that can affect health. Many of these factors are regulated by our various local, state and national environmental laws, and these are often the things that people think of when they first think of environmental factors and health. For example, the Clean Air Act, the

Clean Water Act and similar statutes were adopted to regulate factors in the physical environment (Copeland, 2001; Hubbell, Crume, Evarts, & Cohen, 2010).

The built environment includes all the various human-made features that shape much of daily living: our homes, streets, neighborhoods, businesses and so forth (Lopez, 2012). These features can provide vital health protection, housing and measures that protect against floods, for example, or they can foster poor health: neighborhoods without stores that sell healthy food, mold or dangerous housing and highways and high-traffic roads that each may have important health implications for the populations affected. The scope and scale of these factors can range from very small to continent wide or global (Frumkin, 2003). For example, contaminants or hazards in the home often reflect materials used in construction and maintenance as well as architectural features of a building: ventilation systems, fire protection mechanisms and so forth. Other aspects of the built environment are influenced by very broad aspects of it, such as the global transport of oil and other commodities, the interstate highway system or the movement of goods across international borders (Jackson, 2003).

Finally, there is the social environment, which includes all the various socially constructed, but very important, factors such as race, income, gender, legal structures, social capital and so forth (Gehlert et al., 2008). These factors tend to be socially constructed: They represent shared and individualized responses to society-wide norms. Social systems operate to define attributes such as race, income, poverty and wealth. These features have no biological, physical or chemical imperative; instead, they reflect how societies work to define individuals and groups. They arise from social processes, history, economics and power relations, and though they may be shaped by physical constraints, they represent socially defined outcomes (Giles-Corti & Donovan, 2003).

Even though these definitions and constructs can be fluid, they are nevertheless strong determinants of health. For example, race is a social construct with no underlying biological mechanisms that would allow us to categorically assign one person to a certain race that would be valid across all societies (Griffith, Johnson, Ellis, & Schulz, 2010). But labeling an individual to be of a certain race has important lifelong implications for health, income, housing opportunities and so forth (Caulfield et al., 2009).

These environmental domains are interrelated and they can influence exposures and health outcomes together. For example, a highway is a built environment feature often routed through a community because of the social environment: These highways have been disproportionately sited in low-income neighborhoods and communities of color (Bullard, 2004). The very political decisions that allocate more dollars for highways rather than transit are socially determined as well (Gutfreund, 2004). Whatever the factors behind highway construction, many of the negative health impacts of highways operate through exposures associated with the physical environment: particulates and other air pollutants, noise and so forth. Thus, none of these three domains exist in isolation (Hoffmann et al., 2009).

In addition, it is rare that any one of these three domains operates in purely healthy or unhealthy ways. In general, there are combinations of health-promoting and health-inhibiting factors operating simultaneously in each of these domains. Again, an example of this is the automobile. It provides access to employment, goods and services for millions, and being able to drive to healthcare or to a supermarket is associated with improved health outcomes. But automobile use also creates pollution, reduces walking behavior and increases the risk of accidents (Tatom, 2006).

Finally, it is important to remember that health is not merely the absence of disease. On the contrary, health represents how people live and should provide the context for lifelong possibilities or reaching one's potential. There are physical, mental and social aspects to health and there is a strong influence on health exerted by the environment (World Health Organization, 1946). There are important environmental impacts on obesity, for example, and exposure to nature may be essential for good mental and physical health (Fenton, 2005; Frumkin, 2001).

PHASES OF GENTRIFICATION

There are many ways to parse how gentrification affects a community, but for understanding its impacts on the environment, perhaps it is most useful to divide the process of gentrification into three broad phases: early or pre-gentrification, peak gentrification and post-gentrification or continuing gentrification. In general, the early or pre-gentrification period represents the time when gentrification is yet to be visibly present or has only minimally begun to affect a neighborhood. A few newcomers may be buying properties in an area, but their cumulative impact is yet to be apparent and their overall impacts on housing values, demographics or community resources are yet to be identified or measured. Peak gentrification is the period when widespread change is occurring in a neighborhood. Rents and property values may be rapidly rising, significant segments of the community are being displaced, government services and investments are being rapidly increased and the physical appearance of the neighborhood is being modified to reflect the affluence and political/social power of new residents. Long-term, new residents and outsiders are aware that the neighborhood is undergoing extensive change, to the extent that financial institutions reassess the economic health of the community, governments begin to improve services, the media may report on the new activity in the neighborhood and many residents may begin to fear or experience displacement. Finally, in the post-gentrification period, the process of change may continue as the few remaining affordable housing units are threatened. A new wave of gentrification, perhaps replacing middle- and upper middle-class households with the wealthy, may continue to change in the demographics of a neighborhood. Even though the physical appearance of a community may now appear static, there are still underlying changes that might be affecting the more vulnerable segments of the neighborhood.

Obviously, it is impossible to forecast how long any of these time periods may last or even accurately predict that an individual neighborhood will progress through all three. Some neighborhoods have started off as fairly well-to-do communities before undergoing gentrification, further complicating this reference typography. However, classic gentrification, where a community was mostly poor and working class, then experiences profound impacts and displacement caused by wealthier households moving into the area followed by a substantially altered social environment along with large-scale investments in the built environment and perhaps a renewed interest by government in providing services, is a phenomenon that has been described to have affected communities across the globe. It is this three-phased process that is discussed here.

Phase I: Pre-Gentrification

In one sense, this phase may last for decades as a capitalist society, along with neoliberal social policy and global economic forces, allow or encourage disinvestment in a neighborhood, inhibit access to capital and marginalize entire segments of the population (Saiz, 2010). As a consequence, housing quality deteriorates, and often this disinvestment is accompanied by a deliberate or unintentional decline in government services: Garbage may not be picked up, police protection declines, the quality of schools deteriorates, the streets and sidewalks may not be maintained and so forth (Rueben & Lei, 2010). This physical decline and socially mediated set of issues do not mean that the neighborhood will be totally lacking in features that promote health or does not have amenities that add to residents' quality of life. On the contrary, many residents will choose to live in the area because it may help them find housing they can afford; it offers close accessibility to family, friends and ethnic amenities; and the public transportation, clinics and other public services help them maintain their independence despite their scarce resources (Diez-Roux, 2004). Poor neighborhoods provide vital support to many of their residents, and along with social ties that can assist vulnerable people to live independently, these communities represent important places for the poor and the marginalized to live (Maantay, 2001). All of these may assist residents in accessing health.

Physical environment issues. Environmental justice research has identified many of the physical environment problems that have been disproportionately located in low-income neighborhoods and communities of color. These include increased siting of hazardous waste facilities, more brownfields, higher levels of air pollution and greater exposures to many other environmental toxins. The physical environment is more likely to be contaminated and have multiple pathways of exposures by different contaminants, thereby increasing health pressures on a population (Brulle & Pellow, 2006; Bullard, 1994).

These neighborhoods are more likely to have lead-contaminated housing, and many of the buildings have old plumbing, further increasing exposures to lead (Gaitens et al., 2009; Needleman, 2009). Other pollution sources may exacerbate asthma and other respiratory diseases, cardiovascular problems, immune disorders and other important causes of morbidity and mortality (Rose, Bodor, Hutchinson, & Swalm, 2010). Unfortunately, many residents may also have increased vulnerability to environmentally related illnesses, perhaps because they face a multitude of stressors and hazards or they may have a lower health status to start. These multiple health threats may be additive, a concept known as cumulative risk (Hynes & Lopez, 2007).

Built environment issues. The years of disinvestment in these communities can result in a range of problems and hazards associated with the built environment (Gee, Chen, See, Spencer, & Takeuchi, 2004). These include poor housing conditions, which potentially create risk through mold, pest infestations and various hazardous conditions. Disinvestment can increase the risk of arson and other fires at the same time the pull back of government services reduces fire protection and building inspection support (Drucker, 2010).

The lack of adequate government maintenance of infrastructure and enforcement of regulations can be a particular important health hazard. As sidewalks crack and deteriorate, it can reduce walking behavior, put seniors and people with disabilities at risk of falls and further reinforce deterioration. Street lights may not work, street trees may not be maintained or replanted and traffic conditions may erode, also reducing walkability of neighborhoods (Bedimo-Rung, Mowen, & Cohen, 2005). When schools, libraries and parks are allowed to decay, the ability of the built environment to promote health also declines. Research also suggests that these communities are less likely to have parks and other quality recreational opportunities and they are at increased risk of being targeted for highways and other problematic infrastructure (Cunningham & Michael, 2004).

However, there may be features of the built environment that are protective of health, and these may represent some of the reasons residents who have choices choose to live in these areas. Many of these neighborhoods have fine-grained street networks, access to stores and parks, and other infrastructure that supports physical activity, which makes it possible to walk to destinations despite the deterioration of street conditions. Many of these neighborhoods are in inner cities that are in first-ring suburbs that are close to public transit, facilitating access to employment, healthcare, places of worship, family and other necessary destinations. Many older communities provide shopping and other amenities needed by lower income residents (Atkinson, Sallis, Saelens, Cain, & Black, 2005).

Social environment issues. These pre-gentrification communities provide very important social amenities. These neighborhoods might have senior services, health clinics, government offices and other social service and community-based organizations that can assist low-income people, new immigrants and the elderly to live independently. There may be a wide web of social support and what is called social capital that allow people at risk to access the help of others in times of need. For many vulnerable households, the social networks institutions are vital. In addition, these neighborhoods may represent memories of the young and the elderly, of times with family and friends that promote good mental health (Sorensen et al., 2003).

Observed health problems. There are a range of health problems that can be found in pre-gentrifying communities that include various environmentally related diseases: asthma, lead poisoning, conditions associated with exposures to pollution and toxins (immune disorders, developmental conditions, cancers), cardiovascular and inflammatory issues and so forth. There are also a number of health concerns that are related to accidents and similar factors: death from fires, traffic accidents, pedestrian injuries and so forth. In addition, there can be physical and mental health issues associated with or aggravated by stress: immunological problems, depression, diabetes and so forth. In communities that lack access to fresh healthy food (food deserts) or are overly burdened by fast food restaurants and other sources of unhealthy food, obesity rates may be higher, and residents are at increased risk of the consequences of this excess weight: diabetes, cardiovascular disease, cognitive decline and so forth (Cohen, Farley, & Mason, 2003).

Phase II: Peak Gentrification

The peak phase is the time that most people consider to be gentrification. There are large-scale demographic shifts; as new people move in and residents are displaced, construction activity is widespread with a number of less obvious accompanying processes: Loans for property purchases and renovations become plentiful (for those with the economic means to take advantage of them), environmental problems in vacant and underutilized land are investigated, identified and remediated, and even buildings with no outward sign of renovation activity can change ownership or tenant mix. Pre-gentrification institutions may face dislocation by rising rents or departing clientele. Property taxes can rise, putting further pressure on low-income homeowners who may be forced to sell and move. The neighborhood is in flux and conditions rapidly change.

Physical environment. Though the renewed economic investment and government support that occurs in many gentrifying communities can ultimately result in a decreased burden of environmentally related issues in a community, the very act of remediation and cleanup can release contaminants into the environment and create hazardous exposures for neighbors. For example, improper lead remediation can create problems for neighbors in adjoining units and buildings, and soil removal at hazardous waste sites can release solvents and contaminants into water supplies and the air and put children and others at risk (Ding, 2008). Another potential problem is that rising populations and the movement of households with higher incomes into a neighborhood can increase air pollution, as total traffic in a community increases. This may result in increased exposure to air pollutants. Truck traffic associated with construction activity and deliveries of goods and materials may also exacerbate issues with air quality (Lloyd & Cackette, 2001).

Built environment. The impacts of gentrification on the built environment can be extensive, as housing and buildings are renovated and new retail moves into a neighborhood (Centers for Disease Control and Prevention, 2009). Unfortunately, the positive impacts of new investment may not include the housing of pre-gentrification residents. These households may experience net deterioration in their housing quality, as overcrowding increases in response to rising rents and property owners reduce maintenance in anticipation of selling their units to investors and others. Arson and other housing problems may increase, as unscrupulous owners use illegal methods to force out tenants or try to capitalize on rising property values. Construction activity can place neighbors at risk, as sidewalks can be closed to accommodate renovations and new buildings. This may force pedestrians to walk in the street, increasing risks of accidents. Retail that caters to low-income residents may be forced out, reducing affordable food, even as high-end grocers and specialty shops move in.

Social environment. The disruption of preexisting social networks can have large impacts on the social environment, particularly harming the social capital and personal connections of residents living in the community before gentrification begins. As residents are displaced, old ties are disrupted, potentially increasing neighborhood social disorder and reducing social capital. Though some of these connections may continue even as residents are dispersed, the loss of day-to-day interactions can severely stress relationships or end them altogether. Related to these interpersonal losses are those that result from the closing and/or relocation of long-standing institutions and services. Clinics may close, social services may relocate and other support mechanisms that help residents live independently may cease operation. This decline in institutional support may further reduce the social capital in a community. The distrust generated by relocation and displacement may also stress a neighborhood. These social capital effects may burden both residents who stay in the neighborhood and those who are forced to leave (Do et al., 2008; Keizer, Lindenberg, & Steg, 2008).

At particular risk to these disruptions of social capital are the elderly, who may suffer severe physical, mental and emotional effects from the loss of family, friends and community around them. They may lose their ability to function on their own, become physically withdrawn and shut in or become severely incapacitated or even face higher risks of mortality from the stresses of neighborhood change (Lopez & Goldoftas, 2009).

There are two types of social capital: bonding and bridging. Bonding reflects the ties between in-group or closely related people; bridging social capital includes the social ties that form between members of different groups and more distant people. While over time the degree of bridging social capital may increase, as existing neighbors develop connections with new residents, there is no reason to expect that these new ties will adequately replace the loss of long duration bonding-type connections (Coffé & Geys, 2007; Garcia & McDowell, 2010).

Community psychologist Mindy Fullilove suggests that the disruption and dislocation of a neighborhood can result in "root shock," long-lasting psychological harm and alienation. As residents are moved out, they end up suffering from conditions that are akin to grief and may mourn the loss of community and the loss of close ties with family and neighbors. As rent and expenses rise, the resulting housing and or food changes can put vulnerable people at risk, increasing stress, and there are potential mental health impacts including depression, anxiety and other related disorders (Fullilove, 2005).

Phase III: Post-Gentrification

The post-gentrification neighborhood may seem to be static with its former low income residents departed, but this is not necessarily the case. The rate of building activity may have declined from its earlier peaks, but renovation and new construction may continue across a long time frame, even including the re-renovation of previously converted buildings. Often, some of the pre-gentrification residents may remain in the community, but they may still be at risk of displacement, as property taxes continue to rise and expiring tax breaks and mortgage commitments for long-standing affordable units convince some property owners to convert to market-rate rentals or conversely buildings into condominiums. The physical environment may be vastly improved by renewed government investment in parks and street infrastructure, but the social environment may become more restrictive, and renewed police patrols may target youth of color. In addition, the physical and mental consequences of earlier phases of gentrification may persist.

Physical environment. In the post-gentrified neighborhood, there may be a marked improvement in the physical environment, as increased levels of political power and rising property values result in the closing of environmentally hazardous businesses and the cleanup of brownfields and other problematic properties is complete. There may be a substantial reduction in the burden of environmental contamination and a marked increase in environmental quality in the neighborhood. The look and feel of the neighborhood may seem to have improved (Ley, 1994).

These benefits can only accrue to those who can stay in the gentrified community, however. These net increases in environmental quality must be balanced against potential experiences with environmental degradation suffered by those who are forced to move out of a community by rising real estate prices or scarcities of affordable units (Smith, 1996). In many metropolitan areas, gentrification in select neighborhoods can be associated with increased rents and housing costs throughout the region, forcing lower-income households to pay more for shelter costs with lower quality to show for it (Wyly & Hammel, 1999). This can burden many of these households and may make them more likely to live near hazardous industries or places where the overall environmental quality is lower.

Built environment. There may be similar patterns of rising and declining built environment experiences as well. For those pre-gentrification residents who can stay in their communities post-gentrification, the renewed neighborhood may be improved by substantial investments in local infrastructure. Parks may be renovated and renewed, streets and sidewalks improved and new mixed-use developments may incorporate the many features that recent research suggests are health and social promoting. Again, these benefits need to be reconciled with the experiences of those who are displaced. If these relocated households end up in worse environmental conditions, then the process of gentrification has negative impacts on these former residents.

Social capital. There may be some long-term improvements in social capital, as the tensions and stresses of the peak gentrification period subside. Ongoing residents may accrue increased social interactions with their new neighbors, crime may decrease and overt signs of neighborhood disorder—graffiti, trash, abandoned buildings and broken windows—may subside. Ongoing residents may feel better living in their newly improved neighborhood. Again, these improvements must be considered alongside the negative experiences of former residents who end up living in worse conditions than they were previously surrounded by. New behavioral norms, enforced by official and unofficial means, may be a burden for certain residents. Gatherings of ethnic minorities accompanied by traditional music and cooking may now be disallowed in public spaces. Young people may be particularly vulnerable to these changes in social norms, and they may be targeted by the police or by other residents whenever they gather in groups or try to visit places they once frequented (Betancur, 2002; Lees, Slater, & Wyly, 2008).

With the departure of clinics and social service organizations, the elderly and other vulnerable residents may no longer be able to access the services they need to be independent. No longer knowing their neighbors, many may be less likely to leave their homes despite a drop in crime rates. This increased social isolation can lead them to be at risk for heat stress, cardiovascular disease and cognitive decline (Jackson & Scott, 2008).

The long-term effects of the disruption of social capital can be profound, and it may take decades for residents and communities to recover. Fortunately, most people are resilient and can adapt to new circumstances and environments. However, it is important to remember that those who are most likely to suffer the greatest effects by gentrification, such as those who have been forced to leave the neighborhood, are the least likely to be included in studies of the effects of gentrification or to have their problems recognized by their previous neighborhood institutions. Many of these people have simply dropped out of sight.

Mitigating the Environmental Impacts of Gentrification

Public health practice suggests a number of ways in which the environmental impacts of gentrification can be prevented and addressed. Most important is how public health considers prevention. Primary prevention means stopping a condition from happening before it begins. In the context of gentrification, this might include protections for residents, long-term funding of affordable housing, assistance with community organizing and the development of neighborhood institutions and other measures that might prevent gentrification from occurring in the first place. Secondary prevention involves the monitoring of population health to identify problems as they begin, in order to come up with solutions, interventions or treatments that might address the health consequences of these developing problems. In these changing communities, this might mean monitoring communities to identify when gentrification is occurring and when displacement is happening. Tertiary prevention involves getting the treatment to the people who are already experiencing symptoms or are in preclinical phases of an illness or condition. For gentrification, this would mean providing services to these neighborhoods so that people who are affected by the environmental problems such as exposures to hazardous materials or to dental health consequences of displacement receive the treatment they need (Brown et al., 2003).

Long-Term Research Needs

Despite this wide range of potential health impacts presented by the process and effects of gentrification, we know fairly little about the actual health consequences caused by gentrification in a particular neighborhood to a particular population. We know that voluntary moves can produce some improvements in physical and mental health, at least for women and girls. The few studies on the effects of forced relocation indicate that the mental health of many people is negatively impacted, particularly the elderly and those who are most vulnerable. Though there is some evidence that residents who can remain in gentrified neighborhoods report satisfaction with staying there, these studies suffer from the tremendous disadvantage of not having included those who have been forced to leave. Therefore, it is imperative that long-term, perhaps even generations-long studies, on the impacts of gentrification and neighborhood change be conducted.

CONCLUSIONS

The effects here represent the range of potential impacts of gentrification on health. None of them is inevitable; there is nothing mandated that certain individuals and groups must have their health put at risk in order to facilitate the transfer of real estate to others. What makes these risks potentially more widespread is the lack of action to identify and prevent them. Protecting residents will take a concerted effort by policymakers, advocates and residents to ensure that everyone has an equal ability to access health.

REFERENCES

Atkinson, J. L., Sallis, J. F., Saelens, B. E., Cain, K. L., & Black, J. B. (2005). The association of neighborhood design and recreational environments with physical activity. *American Journal of Health Promotion, 19*(4), 304–309.

Bedimo-Rung, A. L., Mowen, A. J., & Cohen, D. A. (2005). The significance of parks to physical activity and public health: A conceptual model. *American Journal of Preventive Medicine, 28*(2, Suppl 2), 159–168.

Betancur, J. (2002). The politics of gentrification: The case of West Town in Chicago. *Urban Affairs Review, 37*(6), 780–814.

Brown, P., Mayer, B., Zavestoski, S., Luebke, T., Mandelbaum, J., & McCormick, S. (2003). The health politics of asthma: Environmental justice and collective illness experience in the United States. *Social Science and Medicine, 57*(3), 453–464.

Brulle, R. J., & Pellow, D. N. (2006). Environmental justice: Human health and environmental inequalities. *Annual Review of Public Health, 27*, 103–124.

Bullard, R. (1994). Environmental justice for all: It's the right thing to do. *Journal of Environmental Law and Litigation, 9*(2), 281–308.

Bullard, R. (2004). Addressing urban transportation equity in the United States. *Fordham Urban Law Journal, 31,* 1183.

Caulfield, T., Fullerton, S. M., Ali-Khan, S. E., Arbour, L., Burchard, E. G., Cooper, R. S., & Daar, A. S. (2009). Race and ancestry in biomedical research: Exploring the challenges. *Genome Medicine, 1*(1), 8.

Centers for Disease Control and Prevention. (2009). *Health effects of gentrification.* Retrieved from http://www.cdc.gov/healthyplaces/healthtopics/gentrification.htm

Coffé, H., & Geys, B. (2007). Toward an empirical characterization of bridging and bonding social capital. *Nonprofit and Voluntary Sector Quarterly, 36*(7), 121–139.

Cohen, D. A., Farley, T. A., & Mason, K. (2003). Why is poverty unhealthy? Social and physical mediators. *Social Science and Medicine, 57*(9), 1631–1641.

Copeland, C. (2001). Clean Water Act. In V. A. Silyok (Ed.), *Environmental laws: Summaries of statutes administered by the environmental protection agency* (pp. 29–38). Huntington, NY: Nova Science Press.

Cunningham, G. O., & Michael, Y. L. (2004). Concepts guiding the study of the impact of the built environment on physical activity for older adults: A review of the literature. *American Journal of Health Promotion, 18*(6), 435–443.

Diez-Roux, A. (2004). Estimating neighborhood health effects: The challenges of causal inference in a complex world. *Social Science and Medicine, 58,* 1953–1960.

Ding, E. (2008). Brownfield remediation for urban health: A systematic review and case assessment of Baltimore, Maryland. *The Journal of Young Investigators, 14*(2), 596–600.

Do, D. P., Finch, B. K., Basurto-Davila, R., Bird, C., Escarce, J., & Lurie, N. (2008). Does place explain racial health disparities? Quantifying the contribution of residential context to the black/white health gap in the United States. *Social Science and Medicine, 67*(8), 1258–1268.

Drucker, J. (2010). *Risk factors of arson.* Newark, NJ: Rutgers Center for Public Security.

Fenton, M. (2005). Battling America's epidemic of physical inactivity: Building more walkable, livable communities. *Journal of Nutrition Education and Behavior, 37*(Suppl 2), S115–120.

Frumkin, H. (2001). Beyond toxicity: Human health and the natural environment. *American Journal of Preventive Medicine, 20*(3), 234–240.

Frumkin, H. (2003). Healthy places: Exploring the evidence. *American Journal of Public Health, 93*(9), 1451–1456.

Fullilove, M. (2005). *Root shock: How tearing up city neighborhoods hurts America, and what we can do about it.* New York, NY: Random House.

Fullilove, M., & Fullilove, R. (2000). Place matters. In R. Hofrichter (Ed.), *Reclaiming the environmental debate: The politics of health in a toxic culture* (pp. 77–92). Cambridge MA: The MIT Press.

Gaitens, J. M., Dixon, S. L., Jacobs, D. E., Nagaraja, J., Strauss, W., Wilson, J. W., & Ashley, P. J. (2009). Exposure of U.S. children to residential dust lead, 1999–2004: I. Housing and demographic factors. *Environmental Health Perspectives, 117*(3), 461–467.

Garcia, M., & McDowell, T. (2010). Mapping social capital: A critical contextual approach for working with low-status families. *Journal of Marital and Family Therapy, 36*(1), 96–107.

Gee, G., Chen, J., See, S., Spencer, M., & Takeuchi, D. (2004, November). *Sociohistorical determinants of neighborhood factors associated with the health of Filipino Americans.* Paper presented at the meeting of the American Public Health Association, Washington, DC.

Gehlert, S., Sohmer, D., Sacks, T., Mininger, C., McClintock, M., & Olopade, O. (2008). Targeting health disparities: A model linking upstream determinants to downstream interventions. *Health Affairs (Millwood), 27*(2), 339–349.

Giles-Corti, B., & Donovan, R. (2003). Relative influences of individual, social environmental, and physical environmental correlates of walking. *American Journal of Public Health, 93*(9), 1583–1589.

Griffith, D. M., Johnson, J., Ellis, K. R., & Schulz, A. J. (2010). Cultural context and a critical approach to eliminating health disparities. *Ethinicity Disease, 20*(1), 71–76.

Gutfreund, O. (2004). *20th-Century sprawl: Highways and the reshaping of the American landscape.* New York, NY: Oxford University Press.

Hoffmann, B., Moebus, S., Kroger, K., Stang, A., Mohlenkamp, S., Dragano, N., & Jockel, K. H. (2009). Residential exposure to urban air pollution, ankle-brachial index, and peripheral arterial disease. *Epidemiology, 20*(2), 280–288.

Hubbell, B. J., Crume, R. V., Evarts, D. M., & Cohen, J. M. (2010). Regulation and progress under the 1990 Clean Air Act amendments. *Review of Environmental Economics and Policy, 4*(1), 122–138.

Hynes, H., & Lopez, R. (2007). Cumulative risk and a call for action in environmental justice communities. *Journal of Health Disparities Research and Practice, 1*(2), 29–58.

Jackson, R. (2003). The impact of the built environment on health: An emerging field. *American Journal of Public Health, 93*(9), 1382–1384.

Jackson, V., & Scott, L. (2008). Bulldozed: Innovative strategies for addressing the mental health consequences of gentrification. *Coummunity Forum*, Paper 3. Retrieved from http://scholarworks.gsu.edu/ssw_forum/3

Keizer, K., Lindenberg, S., & Steg, L. (2008). The spreading of disorder. *Science, 322*(5908), 1681–1685.

Lees, L., Slater, T., & Wyly, E. (2008). *Gentrification.* New York, NY: Routledge.

Ley, D. (1994). Gentrification and the politics of the new middle class. *Environment and Planning D: Society and Space, 12*(1), 53–74.

Lloyd, A. C., & Cackette, T. A. (2001). Diesel engines: Environmental impact and control. *Journal of Air and Waste Management Association, 51*(6), 809–847.

Lopez, R. (2012). *The built environment and public health.* San Francisco, CA: John Wiley & Sons.

Lopez, R., & Goldoftas, B. (2009). The urban elderly in the United States: Health status and the environment. *Reviews on Environmental Health, 24*(1), 47–57.

Maantay, J. (2001). Zoning, equity, and public health. *American Journal of Public Health, 91*(7), 1033–1041.

Needleman, H. (2009). Low level lead exposure: History and discovery. *Annals of Epidemiology, 19*(4), 235–238.

Rose, D., Bodor, J. N., Hutchinson, P. L., & Swalm, C. M. (2010). The importance of a multi-dimensional approach for studying the links between food access and consumption. *Journal of Nutrition, 140*(6), 1170–1174.

Rueben, K., & Lei, S. (2010). *What the housing crisis means for state and local governments.* Cambridge, MA: Lincoln Land Insitute.

Saiz, A. (2010). The geographic determinants of housing supply. *Quarterly Journal of Economics*, Forthcoming, *125*(3), 1253–1296.

Smith, N. (1996). *The new urban frontier: Gentrification and the revanchist city.* New York, NY: Routledge.

Sorensen, G., Emmons, K., Hunt, M. K., Barbeau, E., Goldman, R., Peterson, K., & Berkman, L. (2003). Model for incorporating social context in health behavior interventions: Applications for cancer prevention for working-class, multiethnic populations. *Preventive Medicine, 37*(3), 188–197.

Srinivasan, S., O'Fallon, L., & Dearry, A. (2003). Creating healthy communities, healthy homes, healthy people: Initiating a research agenda on the built environment and public health. *American Journal of Public Health, 93*(9), 1446–1450.

Tatom, J. (2006). Urban highways and the reluctant public realm. In C. Waldheim (Ed.), *The landscape urbanism reader* (pp. 179–196). New York, NY: Princeton Architectural Press.

World Health Organization. (1946). *Preamble to the constitution.* Geneva, Switzerland: Author.

Wyly, E., & Hammel, D. (1999). Islands of decay in seas of renewal: Housing policy and the resurgence of gentrification. *Housing Policy Debate, 10*(4), 711–765.

CHAPTER NINE

Residential Segregation

Trends and Implications for Conducting Effective Community-Based Research to Address Ethnic Health Disparities

T. HENRY AKINTOBI, RONALD BRAITHWAITE AND ANIKA DODDS

BACKGROUND AND INTRODUCTION

Historically, race and ethnicity have been associated with a variety of health outcomes. Blacks have disproportionately higher morbidity and mortality rates associated with cardiovascular disease, cancer, diabetes and HIV/AIDS than other racial groups in the United States, despite national efforts and initiatives, most of these disparities have persisted over time (Braithwaite, Taylor, & Treadwell, 2010; National Center for Health Statistics, 2012; U.S. Department of Health and Human Services [HHS], 2013). These trends help explain the focus of public health research and practice agendas on racial health disparities. Efforts to eliminate such disparities can begin with avoidance of the term *race* and use of more appropriate labels for social and cultural population subgroups. In contrast to *race*, traditionally understood to represent perceived biological traits, *ethnicity* refers to social groups characterized by distinctive cultural traditions that are maintained across generations and a common history or origin (Last, 1995). For this reason, *ethnicity* is used instead of *race* in this essay.

The societal burden of ethnic health disparities in America manifests in a myriad of ways, including access to health care, inequalities in health and human services and neighborhood safety (HHS, 2013). Public health practitioners, clinicians and researchers generally agree that social, physical and environmental factors, beyond individual factors, are central to understanding and addressing increased mortality and morbidity risk across the life span. These factors, also called the social determinants of health, represent conditions in which people are born, grow, live, work, age and serve as barriers or facilitators to optimal health (World Health Organization, 2009). These factors are not static, but they change with population demographic shifts and the political will to invest in certain communities.

Healthy People, a nationally and internationally recognized compass in setting objectives to address the leading causes of morbidity and mortality in the United States, has prioritized social determinants of health as central to addressing ethnic health disparities (HHS, 2009). Residential segregation is among the established social determinants of health, contributing to patterns of neighborhood engagement, security and well-being (Acevedo-Garcia & Lochner, 2003; HHS, 2013). This chapter focuses on the significance of residential segregation, including demographic shifts and its empirical literature positioning in association with health outcomes and implications for research designed to address health disparities.

RESIDENTIAL SEGREGATION

Significance

Residential segregation, or the differentiation of two or more groups among dimensions of a given social space, has received increased attention, as neighborhood context has become a more formally investigated condition that may put individuals at risk of adverse health conditions (Acevedo-Garcia & Lochner, 2003; Diez-Roux, 2003; Kawachi & Berkman, 2003). [Early studies of residential segregation effects were examined among primarily Black minority groups. These studies demonstrated that higher residential segregation was associated with increased stress and poorer health (Collins & Williams, 1999; Frazier, 1957; Jiobu, 1972; Petersen & Krivo, 1999; LaVeist, 1993; Poldenak, 1993, 1996; Shihadeh & Flynn, 1996; Yankauer, 1950). Studies that have explored the effects of residential segregation on non-Black ethnic groups have yielded inconsistent findings (Collins & Williams, 1999; Fang, Madhaven, Bosworth, & Alderman, 1998).

Historical Context

Residential segregation reflects decades of individual and group-level discrimination rooted in racism and whose primary goal was to maintain social distance among defined groups (Pettigrew & Meertens, 1995). Historically, segregation was driven by majority perceptions of minority group inferiority with discriminatory policies led by the real estate industry, federal housing policy, banking institutions and organizations that sought to ensure the restriction of Blacks from housing choices relegating them to substandard residential areas (Williams & Collins, 2001). Politically, the concept of gerrymandering was born, broadly recognized as the geopolitical or social separation of communities so as to give one group or area an unequal advantage, thus disadvantaging minority groups even more.

Conceptualization and Operational Dimensions

The broad definition of residential segregation masks the different dimensions by which groups may be segregated. Massey and Denton (1988) conducted a seminal study that provided the foundation upon which the conceptualization and measurement of residential segregation have been operationalized. Residential segregation is conceptualized across five dimensions: centralization, concentration, clustering, evenness and exposure. Centralization is the degree to which members of an ethnic group live near an urban area. Concentration measures the physical space that is occupied by an ethnic group in a geographical area. Clustering measures the degree to which ethnic minority areas cluster together in space. Evenness represents the degree of spatial separation among ethnic groups. Exposure measures a group's experience of segregation by the degree to which members come into contact with one another or with another designated group. Population composition also has been widely recognized although broadly defined in operationalizing and measuring segregation (Fang et al., 1998; Jackson, Anderson, Johnson, & Sorlie, 2000). The majority of the aforementioned studies focused on large geographical areas as the unit of analysis or compared smaller areas across a broad geography.

Demographic Trends and Shifts

The United States has experienced a steady increase in residential segregation with declines that have only occurred within the last three decades. In the late 19th early 20th century, prior to the Great Migration of Blacks from the rural South to urban areas, segregation was modest (Glaeser & Vigdor, 2012). Between 1910 and 1960, Blacks moved to urban areas, where they experienced housing discrimination, including restrictive deed covenants, federally sponsored redlining in mortgage lending and discrimination by landlords, real estate agents or local public housing authorities, resulting in increased segregation. By midcentury, most African Americans in urban areas lived in communities where 80% of the Black population moved in order to achieve integration. By 2010, both dissimilarity and isolation had decreased to their lowest levels. Despite this reality, most urban African American dwellers live in a housing market where more than half the Black population would need to move in order to achieve complete integration (Glaeser & Vigdor, 2012).

While national trends indicate that residential segregation has decreased, investigation of trends by ethnicity, metropolitan statistical area (MSA) size and region are not homogeneous. An examination of residential segregation among Blacks, Hispanics, Asians and Pacific Islanders, American Indians and Alaska natives from 1990 to 2000 indicated that only Blacks experienced declines across all dimensions, as described by Massey and Denton (1988). Patterns of segregation among Blacks in the United States remain the highest of all racial/ethnic groups and higher than levels of economic segregation (Massey, Rothwell, and Domina, 2009). National trends suggest that patterns of Black/White segregation have declined between 1980 and 2000 (Iceland, Weinberg, & Steinmetz, 2002; Massey et al. 2009). However, these declines occurred in areas with small Black populations, such as Portland, Oregon; whereas levels of high segregation in Northwestern and Midwestern metropolitan statistical areas (MSAs) like New York City and Milwaukee have not abated over time.

Pathways Associated with Social Determinants of Health

Residential segregation's effects on health are purported to occur by creating vastly different economic, physical and social neighborhood environments (Kramer & Hogue, 2009). Economic effects are associated with more restricted employment, particularly for ethnic minorities (Collins & Williams, 1999; Mays, Cochran & Barnes, 2007; Schulz et al., 2002, 2005; Williams & Collins, 2001). Smaller real estate returns are another consequence of segregation, particularly for Blacks. Despite increases in housing equality in the United States, fewer Blacks have experienced this primary source of wealth for the American family. This is due, in part, to the increased number of minorities (particularly Black) in less desirable housing areas (Logan & Alba, 1993; Kramer & Hogue, 2009; Oliver & Shapiro, 1997; Schulz et al., 2002; Williams & Collins, 2001). Neglect and deterioration of physical environments and a disproportionate representation of undesirable land use also characterize highly segregated areas. Policymakers are more likely to cut important spending on social services in areas where residents are unlikely to mount political opposition (Shihadeh & Flynn, 1996). Withdrawal of fire and police services from such areas can trigger migration of arsonists, drug dealers and other individuals who may initiate high-risk, maladaptive behaviors (Greenberg & Schneider, 1994).

The role of discrimination and related stressors is important in discussions of ethnic health disparities and residential segregation. Discrimination has been a well-documented determinant of the marginalization of ethnic minority and low socioeconomic status (SES) groups (Gee, 2002; Krieger 2001). The stressors that result from individually experienced and institutionally sanctioned discriminatory practices are important to discussions of health disparities. Williams' (1996) framework conceptualized the influence of societal structures on ethnic discrimination. These social structures and institutions differentially influence Whites and ethnic minorities through social stratification, geographic isolation, decreased opportunities for social ties, lack of medical care and ethnic bias (Collins & Williams, 1999). The structural conditions in residentially segregated environments may induce cultural responses that weaken the commitment to norms and values that are important for socioeconomic mobility.

Associated Mortality Outcomes

While a broad range of health outcomes have been associated with residential segregation, the strongest associations are for Black infant and all-cause adult mortality (Cooper et al., 2001; Fang, Mandhavan, Bosworth, & Alderman, 1998; Collins & Williams, 1999; Guest, Almgren, & Hussey, 1998; Hart, Kunitz, Sell, & Mukamel, 1998; Jackson et al., 2000; LaVeist, 1993, 2003; LeClere, Rogers, & Peters, 1997; Massey, White, & Phua, 1996; Peterson & Krivo, 1999; Poldenak, 1993, 1996; Shihadeh & Flynn, 1996; Yankauer, 1950). Some studies, however, demonstrate segregation's protective effects (Blanchard et al., 2004; Fang et al., 1998, Hutchinson et al., 2009; Inagami et al., 2006). Both of these outcomes demonstrate local and national significance. The health status of mothers and infants are critical indicators of the nation's health (HHS, 2013). All-cause adult mortality, despite changing definitions in cause-specific deaths, serves as a broadly recognized measure by which the health of communities, states and nations are compared. Relatedly, nations are typically contrasted using the median household income.

Associated Morbidity Outcomes

Among health morbidity outcomes, infant outcomes are the most frequently investigated. *Healthy People* 2020 objectives to reduce low birth weight to 7.8 % and preterm deliveries to 11.4.% are critical because they represent perinatal health indicators that are the leading causes of neonatal death and compromised quality of life to infants and families (HHS, 2013). Morbidity associated with preterm birth, small for gestational age and low birth weight outcomes has been investigated (Bell, Zimmerman, Almgren, Mayer, & Huebner, 2006; Ellen, 2000; Grady, 2006; Grady & McLafferty, 2007; Grady & Ramirez, 2008; Acevedo-Garcia & Osypuk, 2008), but the direction of the association varies (White & Borell, 2011).

DISCUSSION

Limitations of Residential Segregation Research

The investigations detailed in this essay provide a new direction for research of the role of residential segregation in predicting risk factors for adverse health, but they suffer from a number of limitations. First, the effects of residential segregation often differ by ethnic group. Black-White socioeconomic disparities have clear spatial expressions, previously cited at the metropolitan area level along both central-city/suburban lines and across residential areas (Altshuler, Morrill, Wolman, Mitchell, & The Committee on Improving the Future of U.S. Cities through Improved Metropolitan Area Governances, 1999). Further, the significantly more adverse socioeconomic and residential quality measures among Blacks when compared to Whites have been attributed to differential effects of residential segregation (Massey, 2001; Massey & Denton, 1993; Williams & Collins, 2001). Second, while the influences of residential segregation are manifested at both individual (e.g., education, employment opportunities) and contextual levels (e.g., concentrated poverty), the majority of the research on the residential segregation and health relationship are based on single level, aggregate analyses (Acevedo-Garcia & Lochner, 2003; Acevedo-Garcia, Lochner, Osypuk, & Subramanian, 2003; Ellen, 2000). Third, the majority of studies have examined mortality rates as the outcome of interest. The myriad of social and environmental consequences of ethnic residential segregation infer the need for increased investigation of how specific characteristics of segregated areas may be associated with risk for cause-specific health outcomes. Finally, individual preferences and values associated with where they live are not accounted for in most studies. For example, a person may perceive benefits to living in an ethnically homogeneous neighborhood that are not captured by the data employed in many studies. It is anticipated that the contribution of this initial investigation is a foundation for future longitudinal examination of the meanings and perceptions of residential structures and contexts to more fully understand the risk or protective effects of residential segregation on health outcomes.

Associations between Residential Segregation and Gentrification

The demonstrated protective effects of residential segregation for some communities require further investigation, in light of the recognized stressors associated with gentrification. Gentrification can be thought of as "the rehabilitation of working-class and derelict housing and the consequent transformation of an area into a middle class neighborhood" (Smith & Williams, 1986, p. 573). The term was first coined by London-based sociologist, Ruth Glass in 1964. She used the term to explain the rise in middle-class residents who were displacing working- or lower-class residents (Glass, 1964). Gentrification also can be aptly called "equitable development," in which the creation and maintenance of economically and socially diverse communities are stable over the long term, through means that generate a minimum of transition costs that fall unfairly on lower-income residents (Kennedy & Leonard, 2001). With equitable development, city officials, developers, business owners and community advocates view gentrification as a means of economic revitalization (Kennedy & Leonard, 2001). Gentrification occurs in waves, often due to local and national trends within the real estate market and in conjunction with government "antipoverty" and economic policies (Kennedy & Leonard, 2001; Zukin, 1987). Positive examples of effects of gentrification include reduction in crime, increased viability of further development, stabilization of declining areas and increased social mix (Atkinson & Bridge 2005).

While positive reasoning and justification precede gentrification, the negative effects stretch far and wide. Negative consequences of previously segregated communities that experience gentrification include economic displacement through rent/price increases, community resentment and conflict, homelessness, increased cost and changes to services and housing demand pressures on surrounding poor areas (Lees, Slater & Wyly, 2008). Social displacement of neighborhood residents is the most common result. According to the Centers for Disease Control and Prevention (CDC), health effects due to gentrification can also include limited access or lack of availability of affordable health, housing, healthy food choices, transportation choices, social networks and stress levels (CDC, 2012).

With the Great Recession of the early 2000s and its negative effects amongst neighborhoods composed of primarily older and predominately African American and Hispanic descent, such areas not only experienced higher foreclosure rates but significant increases in new buyers who were not of the neighborhood ethnic minority descent or SES. Such shifts in demographics affect economic authority and political power of an ever-changing electorate (Toon, 2003). The research examined in this chapter demonstrates that the health effects of segregation are unique to ethnic groups and may be protective. Implications are that the effects of gentrification are equally unique, creating social contexts where groups become new minorities and socially and economically displaced. The stressors associated with such transitions may not be health promoting.

DIRECTIONS FOR FUTURE RESEARCH

The ethnic disparities that persist for the majority of health outcomes in the United States demonstrate the need for expanded investigation of a more comprehensive array of variables that influence health outcomes at multiple levels. While administrative and geopolitical boundaries most frequently govern the assessment of structural forces (e.g., isolation index, dissimilarity), the internalization and meanings associated with these social, cultural and political influences are critical to complete assessment of the determinants of and pathways to health outcomes. Psychosocial and biological assessments, context and duration of current and previous residential exposure and individuals' perceptions of their neighborhoods are among the measures that should be analyzed in each study of associations among space, place and health outcomes.

A number of theoretical issues that may provide fresh avenues of research include determining what constitutes an integrated neighborhood in "real life." The quantitative residential segregation indicators employed in the studies described in this essay are based on ideal values and thresholds of integration. A question that may be investigated is whether having an "equal" number of Blacks and Whites in neighborhoods will necessarily improve health. Investigation of voting records and city government initiatives that are focused on improving neighborhoods will further provide information on the political power of a community. The previous associations among measures of social capital, cohesion, integration and health (House, Landis, & Umberson, 1988, Kawachi et al., 1996; Wilkinson; 1996) call for assessment and incorporation of these contextual measures in order to strengthen studies in this research field.

The choice to live among those who are ethnically similar may more often mean that Blacks live among a greater percentage of poorer individuals than do Whites. However, the degree to which social support may buffer against the deleterious effects of residential segregation may help explain the current research findings. Blacks or Whites who choose to live in ethnically homogeneous neighborhoods may, in essence, be choosing residential solidarity over residential integration. Residential segregation may not be protective of health, but these key, health enhancing moderators may buffer against its anticipated, deleterious effects.

Community-based participatory research (CBPR) is a research approach that emphasizes neighborhood-researcher partnership and shared leadership in the planning, implementation, evaluation and dissemination of initiatives designed to address health disparities. Applied to the residential segregation research limitations and implication described in this report, among the advantages of CBPR are strengthened neighborhood-campus relationships, improved research question relevance, enhanced research recruitment and implementation, collective dissemination and mutual benefit for a diverse group of stakeholders (Blumenthal, DiClemente, Braithwaite, & Smith, 2013; Cargo & Mercer 2008; Israel et al., 1998, 2005; Jagosh et al., 2012; O'Fallon & Dearry 2002; Seifer & Cisco, 2006). Conceptually, it is anticipated that through utilizing CBPR, outcomes will include not only answering a research question and reaping associated benefits but also addressing community-identified social, economic or policy concerns (Blumenthal, 2006; Israel, Eng, Schulz, & Parker, 2005; Wallerstein & Duran, 2010), that are among constructs associated with residential segregation, in addition to other social determinants of health.

Development of sustained, mutually beneficial partnerships between communities and researchers are germane to ensuring sound CBPR practice. Also acknowledged are the differing priorities and cultures of these respective partners. This can be challenging when (a) academicians have not previously been guided by neighborhood experts in the evolution of a community's ecology, (b) community members have not led discussions regarding their health priorities, or (c) academic, agency and neighborhood experts have not historically worked together as a single body with established rules guiding roles and function (Henry Akintobi, Goodin, Trammel, Collins, & Blumenthal, 2011). The displacement of socially connected communities that may be considered residentially segregated, according to solely objective indices, may serve to not only threaten the health of communities but the partnerships developed to catalyze and sustain the collaborations built, over time, to address ethnic disparities within and across these neighborhoods.

CONCLUSION

The complex web of causation associated with variations in population health requires identification of the spiders, or fundamental causes, that instigate risk for risk factors (Krieger, 1994). In order to draw closer to identifying the determinants and pathways to adverse health outcomes, comprehensive data on biological, social and spatial markers are critical. Further, studies of the association between place and health should incorporate a thorough assessment of residential preferences and experiences. Until more is known about motivations for residential perceptions, our understanding of how place and space influence health is incomplete.

REFERENCES

Acevedo-Garcia, D., & Lochner, K. (2003) Residential segregation and health. In I. Kawachi & L. Berkman (Eds.) *Neighborhoods and health* (pp. 265–281). New York, NY: Oxford University Press.

Acevedo-Garcia, D., Lochner, K., Osypuk, T., & Subramanian, S. (2003). Future directions in residential segregation and health research: A multilevel approach. *American Journal of Public Health, 93*(2), 215–221.

Acevedo-Garcia, D., & Osypuk, T. (2008). Invited commentary: Residential segregation and health—the complexity of modeling separate social contexts *Am J Epidemiol*, 168, 1255–1258.

Altshuler, A., Morrill, W., Wolman, H., Mitchell, F., & The Committee on Improving the Future of U.S. Cities through Improved Metropolitan Governances. (1999). *Governance and opportunity in metropolitan America*. Washington, D.C., National Academy Press.

Atkinson, R., & Bridge, G. (2005). *Gentrification in a global context: The new urban colonialism*. Oxford, UK: Psychology Press.

Bell, J. F., Zimmerman, F. J., Almgren, G. R., Mayer, J. D., & Huebner, C. E. (2006). Birth outcomes among urban African-American women: A multilevel analysis of the role of racial residential segregation. *Soc Sci Med 63*(12), 3030–3045.

Blanchard T. C., Cossman J. S., & Levin M. L. (2004). Multiple meanings of minority concentration: Incorporating contextual explanations into the analysis of individual-level U.S. black mortality outcomes. *Population Research and Policy Review. 23*, 309–326.

Blumenthal, D. S. (2006). A community coalition board creates a set of values for community-based research. *Preventing Chronic Disease, 3*(1), A16.

Blumenthal, D. S., DiClemente, R. J., Braithwaite, R. L., & Smith, S. A. (2013). *Community-based participatory health research: Issues, methods and translation to practice*. New York, NY: Springer.

Braithwaite, R. L., Taylor, S. E., & Treadwell, H. (2010). *Health issues in the black community*. San Francisco, CA: Jossey-Bass.

Cargo, M., & Mercer, S. L. (2008). The value and challenges of participatory research: Strengthening its practice. *Annual Review Public Health, 29*, 325–350.

Centers for Disease Control and Prevention (CDC). (2012, September 25). *Health effects of gentrification*. Retrieved from http://www.cdc.gov/healthyplaces/healthtopics/gentrification.htm

Collins, C., & Williams, D. (1999). Segregation and mortality: The deadly effects of racism? *Sociological Forum, 14*(3), 495–523.

Cooper, H., Friedman, S., Tempalski, B., & Friedman. R. (2007). Residential segregation and injection drug use prevalence among black adults in US metropolitan areas. *American Journal of Public Health, 97*(2), 344–352.

Diez-Roux, A. V. (2003). A glossary of multilevel analysis. *Epidemiological Bulletin, 24*(3), 11–13.

Ellen, I. (2000). Is segregation bad for your health? The case of low-birth weight. In W. Gale & J. Pack (Eds.), *Papers in urban affairs* (pp. 203–238). Washington, DC: Brookings Institution.

Fang, J., Madhaven, S., & Alderman, M. (1997). Low birth weight: Race and maternal nativity-impact of community income. *Pediatrics*, 103, e5.

Fang, J., Mandhavan, S., Bosworth, W., & Alderman, M. H. (1998). Residential segregation and mortality in New York City. *Social Science and Medicine, 47*, 469–476.

Frazier, E. (1957). *The Black bourgeoisie*. New York, NY: Simon and Schuster.

Gee, G. (2002). A multilevel analysis of the relationship between institutional racial discrimination and health status. *American Journal of Public Health, 92*, 615–623.

Glaeser, E., & Vigdor, J. (2012). *The end of the segregated century: Racial separation in America's neighborhood, 1890–2010* (Civic Report No. 66, p. 3). Manhattan, NY: The Center for State and Social Leadership at the Manhattan Institute.

Glass, R. (1964). *Introduction. Aspects of change, in centre for urban studies*. London, UK: MacGibbon and Kee.

Grady, S. C. (2006). Racial disparities in low birthweight and the contribution of residential segregation: a multilevel analysis. *Social Science and Medicine 63*(12), 3013–3029.

Grady, S., & McLafferty, S. (2007). Segregation, nativity and health: Reproductive health inequalities for immigrant and native-born Black women in New York City. *Urban Geography, 28*(4), 377–397.

Grady, S. C., & Ramirez, I. J. (2008). Mediating medical risk factors in the residential segregation and low birthweight relationship by race in New York City. *Health Place. 14*(4): 661–677.

Greenberg, M., & Schneider, D. (1994). Violence in American cities: Young black males are the answer, but what is the question? *Social Science and Medicine, 39*, 179–187.

Guest, A., Almgren, G., & Hussey, J. (1998). The ecology of race and socioeconomic distress: Infant and working-age mortality in Chicago. *Demography, 35*, 23–34.

Hart, K., Kunitz, S., Sell, R., & Mukamel, D. (1998). Metropolitan governance, residential segregation, and mortality among African Americans. *American Journal of Public Health, 88*(3), 434–438.

Henry Akintobi, T., Goodin, L., Trammel, E., Collins, D., & Blumenthal, D. (2011). Challenges in improving community engaged research, clinical and translational science award community engagement key function committee task force on the principles of community engagement. In Clinical and Translational Science Award Community Engagement Key Function Committee Task Force on the Principles of Community Engagement (Eds.), *Principles of community engagement* (2nd ed.) pp. 136–138, Washington, DC: U.S. Department of Health and Human Services.

House, J., Landis, K., & Umberson, D. (1988) Social relationships and health. *Science, 214*, 540–545.

Hutchinson, R. N., Putt, M. A., Dean, L. T., Long, J. A., Montagnet, C. A., & Armstrong, K. (2009). Neighborhood racial composition, social capital and black all-cause mortality in Philadelphia. *Social Science & Medicine, 68*, 1859–1865.

Iceland, J., Weinberg, D., & Steinmetz, E. (2002). *Racial and ethnic residential segregation in the United States: 1980–2000.* Washington, DC: U.S. Census Bureau.

Inagami, S., Borrell, L., Wong, M., Fang, J., Shapiro, M., & Asch, S. (2006). Residential segregation and Latino, black and white mortality in New York City. *Journal of Urban Health, 83*(3), 406–420.

Israel, B., Schulz, A., Parker, E., & Becker, A. (1998). Review of community-based research: Assessing partnership approaches to improve public health. *Annual Review of Public Health*, 19, pp. 173–202

Israel B., Eng, E., Schultz, A. J., & Parker, E. A. (Eds). (2005). *Methods in community-based participatory research for health.* San Francisco: Jossey-Bass Publishers.

Jackson, S. A., Anderson, R. T., Johnson, N. J., & Sorlie, P. D. (2000). The relation of residential segregation to all-cause mortality: A study in black and white. *American Journal of Public Health, 90*, 615–617.

Jagosh J., Macaulay, A. C., Pluye, P., Salsberg, J., Bush, P. L., Henderson, J., …Greenhalgh T. (2012). Uncovering the benefits of participatory research: implications of a realist review for health research and, practice. *Milbank Quarterly, 90*(2), 311–46.

Jiobu, R. M. (1972). Urban determinants of racial differentiation in infant mortality. *Demography, 9*, 603–615.

Kramer, M. R., Hogue, C. R. (2009). Is segregation bad for your health? *Epidemiologic Reviews, 31*, 178–194.

Kawachi, I., & Berkman, L. F., eds. 2003. *Neighbourhoods and Health.* New York: Oxford Univ. Press.

Kawachi, I., Colditz, G., Ascherio, A., Rimm, E., Giovannucci, E., Stamfer, M., & Willet, W. C. (1996). A prospective study of social networks in relations to total mortality and cardiovascular disease in men in the U.S. *Journal of Epidemiology Community Health, 50*, 245–251.

Kennedy, M., & Leonard, P. (2001). *Dealing with neighborhood change: A primer on gentrification and policy changes*. Washington, DC: Brookings Institution.

Krieger, N. (1994). Epidemiology and the web of causation: Has anyone seen the spider? *Social Science and Medicine, 39*(7), 887–903.

Krieger, N. (2001). Theories for social epidemiology in the 21st century: An ecosocial perspective. *International Journal of Epidemiology, 30*(4), 608–677.

Last, J. M. (1995). *Dictionary of epidemiology* (3rd ed.). New York, NY: Oxford University Press.

LaVeist, T. (1989). Linking residential segregation and infant mortality disparity in U.S. cities. *Social Science Research*, 73, 90–94.

LaVeist, T. (1993). Segregation, poverty, and empowerment: Health consequences for African Americans. *The Milbank Quarterly, 71*(1), 41–66.

LaVeist, T. (2003). Racial segregation and longevity among African Americans: An individual-level analysis. *Health Services Research, 38(6), 1719*–1734.

LeClere, F., Rogers, R., & Peters, K. (1997). Ethnicity and mortality in the United States. Individual and community correlates. *Social Forces*, 76, 169–198.

Lees, L., Slater, T., & Wyly, E. K. (2008). *Gentrification*. New York, NY: Routledge.

Logan, J., & Alba, R. (1993). Locational returns to human capital: Minority access to suburban community resources. *Demography, 30*, 243–268.

Massey, D. (2001). Residential segregation and neighborhoods conditions in U.S. metropolitan areas. In N. Smelser, W. Wilson, & F. Mitchell (Eds.), *American becoming: Racial trends and their consequences, Volume I* (pp. 391–434). Washington, D.C.: National Academy Press.

Massey, D., & Denton, N. (1988). The dimensions of residential segregation. *Social Forces, 76*(2), 281–315.

Massey, D., & Denton, N. (1993). *American apartheid: Segregation and the making of the underclass*. Cambridge, MA: Harvard University Press.

Massey, D. S., Rothwell J., Domina T. (2009). The changing bases of segregation in the United States. *Annals of the American Academy of Political and Social Science*, 626, 74–90.

Massey, D., White, M., & Phua, V. (1996). The dimensions of segregation revisited. *Sociological Methods and Research*, 25, 172–206.

Mays, V. M., Cochran, S. D., & Barnes., N. W. 2007. Race, race-based discrimination, and health outcomes among African Americans. *Annual Review of Psychology, 58*, 201–225.

National Center for Health Statistics. Healthy People 2010 Final Review (2012). Hyattsville, MD.

O'Fallon, L. R., & Dearry, A. (2002). Community-based participatory research as a tool to advance environmental health sciences. *Environmental Health Perspective, 110*(Suppl 2), 155–159.

Oliver, M., & Shapiro, T. (1997). *Black wealth/White wealth: A new perspective on racial inequality*. New York, NY: Routledge.

Peterson, R., & Krivo, L. (1999). Racial segregation, the concentration of disadvantage, and Black and White homicide victimizations. *Social Forum, 14*(3), 465–493.

Pettigrew, T., & Meertens, R. (1995). Subtle and blatant prejudice in Western Europe. *European Journal of Social Psychology, 25,* 57–75.

Poldenak, A. (1993). Poverty, residential segregation and black/white mortality ratios in urban areas. *Journal of Health Care of the Poor and Underserved*, 4, 363–373.

Poldenak, A. (1996). Segregation, discrimination and mortality in U.S. blacks. *Ethnicity and Disease*, 6, 99–108.

Schulz, A. J., Kannan, S., Dvonch, J. T., Israel, B. A., Allen, A., James, S. A., Lepkowski, J. (2005). Social and physical environments and disparities in risk for cardiovascular disease: The healthy environments partnership conceptual model. *Enviromental Health Perspectives, 113*(12):1817–1825.

Schulz, A. J., Williams, D. R., Israel, B. A., & Lempert, L. B. (2002). Racial and spatial relations as fundamental determinants of health in Detroit. *Milbank Quarterly 80*(4), 677–707, iv.

Seifer, S., & Sisco, S. (2006). Mining the challenges of CBPR for improvements in urban health. *Journal of Urban Health, 83*(6), 981–984.

Shihadeh, E., & Flynn, N. (1996). Segregation and crime: The effect of Black isolation on the rates of Black urban violence. *Social Forces, 74,* 1325–1352.

Smith, N., & Williams, P. (1986). *Gentrification of the city.* London, UK: Routledge.

Toon, J. (2003, September 4). Tales of the city: Current gentrification in Atlanta contrast sharply to previous waves of urban restoration. *Georgia Institute of Technology Research News,* 33–34.

U.S. Department of Health and Human Services (2013). Office of Disease Prevention and Health Promotion. Healthy People 2020. Washington, DC. Retrieved from http://www.healthypeople.gov/2020/Consortium/HP2020Framework.pdf]. [URL accessed April 10, 2013.

Wallerstein, N., & Duran B. (2010). Community-based participatory research contributions to intervention research: The intersection of science and practice to improve health equity. *American Journal of Public Health, 100 Suppl 1*: S40-6. doi:10.2105/AJPH.2009.184036.

White, K., & Borrell, L. N. (2011). Racial/ethnic residential segregation: Framing the context of health risk and health disparities. *Health & Place, 17*(2), 438–448.

Wilkinson, R. (1996). *Unhealthy societies: The afflictions of inequality.* London, UK: Routledge.

Williams, D. (1996). Racism and health: A research agenda. *Ethnicity and Disease, 6,* 1–6.

Williams, D., & Collins, C. (2001). Racial residential segregation: A fundamental cause of racial disparities in health. *Public Health Reports, 116,* 404–416.

Wolff, E. N. (2012). The asset price meltdown and the wealth of the middle class. *National Bureau of Economic Research* (Working Paper No. 18559). New York, NY: New York University.

World Health Organization. (2009). *Social determinants of health.* Retrieved from http://www.who.int/social_determinants/sdh_definition/en/index.html

Yankauer, A. (1950). The relationship of fetal and infant mortality to residential segregation. *American Sociological Review, 15,* 644–648.

Zukin, S. (1987): Gentrification culture and capital in the urban core. *Annual Review of Sociology, 13,* 129–147.

PART FOUR

The Customer Is Always Right

Democracy under Threat in Public Schools

Gentrification and urban schooling are inextricably linked. No one seemed to care much about poor struggling urban schools until the middle class tired of long commutes and wanted back into the city. Suddenly, there was an epidemic of mass proportions—education (particularly urban schools) needed fixing because the new residents needed schools for their children. Central to the repair efforts is privatization (in the form of charter schools and online education) and accountability measures for students via standardized testing, teachers via teacher effectiveness evaluations, principals via principal effectiveness evaluations, schools via school report cards measuring Adequate Yearly Progress, and more recently, teacher education via the push to grade teacher education programs. However, these topics are not a major focus of public discourse these days because the attention has been directed elsewhere.

In a sleight of the hand magic trick, the magician (i.e., owners and power brokers of mass media outlets) directs the gaze of the public to the highly sensationalized and publicized violence in urban schools. While the need to address violence among youths is urgent, focus on what is happening *outside of* and *on the way to* school is so intense that many neglect what is happening *inside of* school. They fail to see what the magician's other hand is really doing, that is, how the trick happens. While the public focuses on what the magician wants us to see, that is the sensational, the authors in this section investigate the agitations that are the impetus to that which is on display.

When school closings force students to enter gang territories—risking safety for the promise of education—the historical memory of slaves risking life and limb to learn to read comes to mind. Ayers activates the memories of yesterday as he recounts the historical context of civil and educational court cases, noting the similarities to the present context of court case rulings that are regressive at best and obliterative at worst. These resemblances are harrowing.

Books' chapter delves into financial aspects of education, highlighting the effects of property-tax-based school funding and privatization on children residing in high-poverty communities. She cites the flawed funding structure as the primary antecedent to school failure and eventual death by closure, while privatization in the form of charter and online schooling is the beneficiary scavenging the remains.

Hankins and Henry look more closely at charter schooling, as they contrast parent activism in schooling postdesegregation gentrification to present-day gentrification. They conclude that the availability of charters allows White middle-class parents to transplant into neighborhoods as silos, making interaction with neighborhood traditional schools and institutions optional. This isolation *from* the community, while at the same time *within* the community, increases the rate at which the neighborhood changes as well as the racial inequality.

Finally, Giroux proposes what can be viewed as the rationale for the public's fixation on the display and the absence of inquiry into how and why. In sum, it is the near extinction of quality education that prepares a critical thinking, informed, engaged, socially conscious and active democratic citizenry. Let us watch closely as these authors demystify by taking us backstage, pulling back the curtain, and exposing the pulled strings and intricate preparations that makes it all possible.

CCR

CHAPTER TEN

Topsy-Turvy

Education at the End of Empire

WILLIAM AYERS

We must rapidly begin the shift from a "thing"-oriented society to a "person"-oriented society.... A civilization can flounder as readily in the face of moral and spiritual bankruptcy as it can through financial bankruptcy. (King, 1967, p. 186)

The paradox of education is precisely this—that as one begins to become conscious one begins to examine the society in which he is being educated. The purpose of education, finally, is to create in a person the ability to look at the world for himself, to make his own decisions, to say to himself this is black or this is white, to decide for himself whether there is a God in heaven or not. To ask questions of the universe, and then learn to live with those questions, is the way he achieves his own identity. (Baldwin, 1963, p. 43).

Emancipate yourselves from mental slavery/None but ourselves can free our minds. (Marley, 1980)

I felt the double blow to my gut before my head could reasonably catch up: The 1965 Voting Rights Act—a crowning achievement of the classical phase of the Black Freedom Movement—was emasculated on June 25th by the Supreme Court, 5-4, in Shelby County v. Holder (2013), and Trayvon Martin—the young Black man who by all accounts had done nothing wrong and was unarmed, walking home from a trip to the store for snacks on the night he was stalked, confronted and shot to death by an armed vigilante—was denied any semblance of justice on July 13th when his killer was acquitted in a Florida courthouse.

Every July 4th, it is worth rereading the abolitionist Frederick Douglass' angry and stirring 1852 speech, "What to the Slave Is the 4th of July?" and in 2013, one passage rings with particular urgency:

> Whether we turn to the declarations of the past, or to the professions of the present, the conduct of the nation seems equally hideous and revolting. America is false to the past, false to the present, and solemnly binds herself to be false to the future.

Let us look to the past: Trayvon Martin joins the long list of young Black men cut down by agents of the state or lynched by self-appointed enforcers: Emmett Till, Oscar Grant, Fred Hampton, Amadou Diallo—each a complex human being whose life was infinitely precious to himself, each singled out, sometimes randomly, as symbol and threat, each an open wound and an unresolved crime. None will be forgotten.

Go further: The 15th Amendment to the U.S. Constitution, ratified in 1870 at the height of Radical Black Reconstruction, stated the following: "The right of citizens of the United States to vote shall not be denied or abridged by the United States or by any State on account of race, color, or previous condition of servitude." This was a powerful victory brought about by decades of abolitionist action, the self-activity, resistance and general strike and flight of enslaved people, and finally a great Civil War, or the War of Liberation, a reminder—and we need to be reminded in good times as well as bad—of another provocative and incendiary statement from the incomparable Frederick Douglass: "Power concedes nothing and without a demand; it never has and it never will" (Douglass, 1857). With the withdrawal of federal troops in 1877 (the rotten compromise that won Rutherford Hayes the presidency, lost the Republican Party its soul and plunged the nation backward as it reversed the gains of that long and bloody Civil War), combined with a range of new laws and legal sanctions, as well as the opening of a mass campaign of terror against recently liberated Blacks—White supremacy reasserted itself with a vengeance.

Political power and the vast system of white-skin privilege it upheld rested firmly on the broad back of White supremacy, and cold reality made restricting the Black franchise a top priority and an urgent goal for the establishment. Symbols were everywhere—the logo of the Alabama Democratic Party was a white rooster with a banner above it reading, "White Supremacy"—but beyond codes and emblems, Whites built a broad and complex structure designed to maintain and strengthen White supremacy: the infamous "Black Laws," chain gangs, poll taxes, literacy tests combined with semi-official and sanctioned violence against the whole community—lynching, arson, banning and blacklisting—all played their roles in the campaign. And the law followed suit: In 1896 in *Plessy v. Ferguson*, the Supreme Court found segregation constitutional and enshrined the doctrine of "separate but equal" as settled law; in 1898 in *Mississippi v. Williams* and again in 1903 in *Giles v. Harris*, the court upheld laws that operated to disenfranchise African Americans. And the results were predictable: In 1896, there were 130,334 Blacks registered to vote in Louisiana; in 1904, there were only 1,342. Black turnout in the 1904 Presidential election in Virginia and South Carolina was exactly zero.

And America is false to the present: There are millions of African American men today (and growing numbers of African American women as well as men and women of other backgrounds) who have been barred from voting or seeking elective office or serving on juries based on their status as convicted felons. This is the system of mass incarceration which the brilliant lawyer and activist Michelle Alexander (2012) has named "the new Jim Crow." She points out that mass incarceration is a defining fact in the United States today, whether acknowledged or not, just as slavery was the fundamental reality in the 1800s. In fact, there are more Black men today held in prison or on probation or parole than there were held in slavery in 1850; there are significantly more people caught up in the system of incarceration and supervision in America today—over 6 million folks—than inhabited Stalin's Gulag at its height; the American Gulag is the second largest city in the United States, and while the United States constitutes less than 5% of the world's people, it holds over 25% of the world's combined prison population; on any given day, tens of thousands of men, overwhelmingly Black and Latino, are held in the torturous condition known as solitary confinement; and in the past 20 years, the amount states have spent on prisons has risen six times the rate spent on higher education. I could go on, but I will stop.

Mass incarceration and felony disenfranchisement constitute broader, more all-encompassing and more complex restrictions on voting rights than the Black Freedom Movement reckoned with when mass direct action for justice resulted in the passage of several far-reaching federal laws, including the Voting Rights Act of 1965. In 2006, Congress overwhelmingly reauthorized the act and named it the Fannie Lou Hamer, Rosa Parks, and Coretta Scott King Voting Rights Reauthorization Act and Amendments Act in recognition of the activism that created the law in the first place. Remember: "Power concedes nothing without a demand; it never has and it never will."

And now in *Shelby County v. Holder* (2013), the Supreme Court binds itself to be false to the future—it has gutted the Voting Rights Act and returned to form, aligning itself once more with *Mississippi v. Williams* (1898) and *Giles v. Harris* (1903). It cannot hold: We will quite soon become a majority non-White country, and another world is surely coming; whether that world is a better world depends more on our collective action for more democracy, more participation, more peace and justice now than it does on the good intentions, wisdom or sense of justice from on high. An authentically democratic society is as threatening and unacceptable to the establishment as emancipation was generations past. *Shelby* is one dying gasp for White power determined to block progress by any means necessary. It is time to get busy: "Power concedes nothing without a demand; it never has and it never will."

On October 26, 1992, the U.S. Congress designated Monroe Elementary School, one of the segregated Black schools in Topeka, Kansas, a National Historic Site because of its significance in the famous 1954 Supreme Court decision outlawing racial segregation in public schools, *Brown v. Board of Education*. The National Archives includes several documents from the case in its digital classroom.

Brown v. Board of Education overturned *Plessy v. Ferguson* and heralded the legal termination of racially segregated schools; it has become an icon in the popular story America tells itself about its inherent goodness and its inevitable upward trajectory: America the beautiful; *Brown* as icon. Americans are devoted to *Brown*, myth or symbol, just as we are dedicated to Superman's motto: Truth, Justice, and the American Way. But "Power concedes nothing without a demand; it never has and it never will."

Brown was decided in the wake of World War II, in the wash of that reenergized sense of freedom, and, critically, with the return of young Black veterans from Europe and Asia. Whenever you read or hear that *Brown* unleashed years of struggle for civil rights, flip the script and remind people that years of struggle for civil rights resulted in *Brown*. The decision followed incessant and increasingly intense demands and mobilizations by African Americans that the country live up to the promise of full citizenship encoded in the 14th Amendment:

> All persons born or naturalized in the United States, and subject to the jurisdiction thereof, are citizens of the United States and of the state wherein they reside. No state shall make or enforce any law which shall abridge the privileges or immunities of citizens of the United States; nor shall any state deprive any person of life, liberty, or property, without due process of law; nor deny to any person within its jurisdiction the equal protection of the laws.

And *Brown* coincided with clear White interests that had nothing to do with Black well-being: avoiding a revolution led and defined by subjugated African Americans, transforming the feudal South and integrating it into a repositioned capitalist juggernaut, removing a blatant hypocrisy and an embarrassing fact of American life that was effectively wielded against the United States in the United Nations and other international forums as an escalating Cold War raged on. White people needed *Brown*—but only a bit of *Brown*.

The language of *Brown* includes the language of justice. It repudiates racial segregation and says—correctly—that separate is inherently unequal. It affirms the full humanity of African Americans. It endorses core principles of democracy. It cries out for equality.

To take *Brown* to heart would require a hard look at the racial landscape we inhabit—a system with institutions operating at every level to construct Black "inferiority" and to deny full participation in social, political and economic life. That hard look could lead to an iron commitment, then, to smash the institution of White supremacy. No such luck—yet.

Brown also embodies a fundamental, even fatal flaw that runs deep in the American racial narrative. The argument in the case turns on the harm suffered by Black children and the feelings of inferiority that are a result of segregation, rather than the despicable, immoral and destructive system of White supremacy itself. Black people—not racism—became the exclusively acknowledged concern; Black pathology, not White privilege, was the focus of action.

And so *Brown*, the widely celebrated and lofty statement of principle, was followed immediately by its lesser known brother, the betrayer and assassin, *Brown II*. *Brown II* was the implementation, or remedy phase, of the decision, and here again—consistent with the long tradition of all things racial in America—the remedy fitted neither the crime nor the injury. In fact, *Brown II* gave local school districts, the parties defeated in *Brown*, the power and responsibility to construct the solution—to desegregate their schools *"with all deliberate speed."* The fox—far from being banished from the hen house—was given the only set of keys.

The Supreme Court had never in history issued an order to implement a constitutional right that was so vague, and so "all deliberate speed" turned out in practice to mean "never." The activity in the courts over the decades following *Brown* went decidedly south: Racially isolated communities of color were denied the right to draw students from adjoining White suburbs; children were denied the right to equal school funding; the concept of "neighborhood school" was reinforced and strengthened even if the result was resegregation, on and on and on. In 2007, the Supreme Court ruled against *voluntary* desegregation plans in Seattle, Washington, and Louisville, Kentucky, in which race was one of several factors used to maintain a diverse student body in public schools. Almost 60 years after *Brown*, school segregation is alive and well, more firmly entrenched than ever, and each year schools are more segregated. *Brown* is all but dead, and the structure of White supremacy rules.

As usual, White supremacy is hiding in plain sight. The most dissembling hypocrites argue that anyone who sees race is a racist, that race-conscious integration is the equivalent of Black-hating segregation—because both are based on skin color. This is an invented and wholly fictitious symmetry.

The problem in America is not and has never been race consciousness per se; the problem has always been White supremacy in fact. Anything that undermines White supremacy and fights for inclusion and equality sides with humanity; anything that excludes, segregates, subordinates is on the side of oppression and exploitation. And so, using the lofty language of *Brown*, ordinary White supremacists continue to herd Black children into unnatural and inferior schools, build walls and lock the gates.

Monroe Elementary—that iconic temple in Topeka elevated as a National Historic Site—may as well be turned into a mausoleum: Here is one more place where African American aspirations and the ongoing struggle for justice and liberation were laid to rest.

A dominant narrative in contemporary school reform is once again focused on exclusion and disadvantage, race and class, Black and White. "Across the US," the National Governor's Association declared in 2005, "a gap in academic achievement persists between minority and disadvantaged students and their white counterparts." This is the commonly referenced and popularly understood "racial achievement gap," and it drives education policy at every level. Once again, whether heartfelt or self-satisfied, the narrative never mentions the monster in the room: White supremacy.

It is true, of course, that standardized test scores reveal a difference between Black and White test-takers: 26 points in one area of comparison—fourth-grade reading—20 points in another, 23 in a third. But the significance of those differences is wildly disputed. Some argue—as Charles Murray and Richard Herrnstein (1994) did in their popular and incendiary book *The Bell Curve*—that genetic differences account for the gap, and there is little that can be done to lift up the poor inferior Black folks. An alternate theory—popular since the 1960s with progressives and liberals—holds that Blacks are not inherently inferior to Whites, but merely "culturally deprived," and that fixing the "massive pathologies" in the family and community will require social engineering on a grand scale.

Each of these explanations has its large and devoted following—the first, while difficult for many Whites to endorse publicly, carries the reflected power of eugenics and the certainty that what they had always secretly suspected (that Whites are indeed superior beings in so many intimately experienced ways—"I did nothing wrong, and I feel *great* about being *me!*") is true. The second has the advantage of giving a bit more than a pig's eye for the well-being of Black people while disturbing none of the pillars of White privilege. Either theory can live comfortably beneath the obsessive focus on the so-called achievement gap.

Clearly the second theory is in ascendency: The guys with all the money, the media, the armed forces and the super-sized megaphones are the autocrats and the authoritarians, the plutocrats, the patricians and all their various professional political allies—the troglodytes in Congress and the Broad Foundation, of course, but also the smart liberals and data-driven "scientific" progressives (Arne Duncan, Bill Gates, Michael Bloomberg, Rahm Emanuel) who are absolutely certain of their righteousness and their own beneficent intentions, who swell with pride when considering the gleaming architecture of their own specific talents that they are so generously willing to share with their inferiors in the service of general betterment and who are always a bit taken aback and then deeply resentful when the objects of their attention and affection do not have the good sense to comply with their plans for social uplift (school closings, a longer school day, privatization, cutting sports and arts).

Gloria Ladson-Billings (2006) upended all of this nonsense with an elegant reversal: There is no achievement *gap*, she argued, merely a glancing reflection of something deeper and more fundamental—America has a profound education *debt*. The educational inequities that began with the attempted annihilation of Native peoples and the enslavement of Africans, the conquest of a continent and the importation of both "free" labor (in chains) and serfs, have transformed into apartheid education something anemic, inferior, inadequate and oppressive. Over decades and then centuries, the debt has accumulated and has passed from generation to generation, and it continues to grow and pile up. Jonathan Kozol (2005) has documented that the debt—far from being ameliorated—grows year by year: Chicago serves 86% Black and Latina/o students and spends around $8,000 per pupil per year, while a few miles away in the tony suburb of Highland Park, 90% White, the school district spends $17,000 per student; New York City, 72% Black and Latina/o, spends around $12,000 per pupil annually, while suburban Manhasset, 91% White, spends over $22,000. In most states, the highest poverty districts receive far fewer resources, and, according to Ladson-Billings, in "30 states, high minority districts receive less money for each child than low minority districts."

Ladson-Billings (2006) imagined what could be done if the political powers took the "achievement gap" seriously: immediate reassignment of the best teachers in the country to schools for poor children of color, guaranteed places for those students in state and regional colleges and universities, smaller classes, a Marshall Plan-type effort to rebuild school infrastructure.

Ladson-Billings (2006) argued that the United States also owes a moral debt to African Americans, a debt that "reflects the disparity between what we know is right and what we actually do."

Will America educate African American youngsters? In 1933, Carter Woodson published *The Mis-Education of the Negro*, and he answered the question this way:

> When you control a man's thinking you do not have to worry about his actions. You do not have to tell him to stand here or go yonder. He will find his "proper place" and will stay in it. You do not need to send him to the back door. He will go without being told. In fact, if there is no back door, he will cut one for his special benefit. His education makes it necessary.

Woodson had in mind the way education serves the social order, the way American schools satisfy a society with identifiable structures of privilege and oppression based on race and reflect and promote that racial stratification perfectly. When there is, for example, a pervasive sense that there is nothing about the presence of African American youngsters, especially Black boys, that is deemed valuable or desirable or important—their presence always a problem, a deficit, an impediment—that gets manifested on the street and in the classroom.

Education, of course, is never neutral. It always has a value, a position, a politics. Education—teaching and schooling—either reinforces or challenges the existing social order. The largest, most generous purpose of education is always human enlightenment and human liberation, and the driving and undergirding principle is the unity of all humanity: Every human being is of incalculable value, entitled to decent standards concerning freedom and justice and education, and any violations, deliberate or inadvertent, must be fought against, testified to and resisted.

But because schools serve societies—in fact, in many ways all schools are microcosms of the societies in which they are embedded—every school is both mirror and window onto a specific social reality. If one understands the schools, one can see the whole of society; if one fully grasps the intricacies of society, one will know something about how its schools are organized. In a totalitarian society, for example, schools are built for obedience and conformity; in a kingdom, the schools teach fealty; in a racialized society, educational privileges and oppressions are distributed along the color line. In an authentic democracy, we would expect to find schools defined by a spirit of cooperation, inclusion and full participation, places that honor diversity while building unity.

And that takes us to the necessary and challenging task of naming our moment—*necessary* because if we fail to analyze our conditions concretely we are rudderless; *challenging* because this moment, like every other moment, is dense with possibilities and packed with energy—it refuses to stand still and it will not be nailed down. Every analysis is partial, contingent and unfinished. But without at least an attempt to understand in a systematic way the world around us—the apparent as well as the hidden forces at work—we are at sea.

The word "moment" is itself elastic, wobbling at the edges between a sense of the airy and the ephemeral, on one hand, and a claim to the momentous on the other. The moment is passing and profound, transient and memorable: It can be an event or a happening and it can, as well, define an epoch, a period or an era. So we must try to name and illuminate it with uncertainty and humility. We reach for the magnificent, knowing that we can never capture it and pin it to the board and, if we did, we would kill it.

We are living in the midst of an historic sea change—a dramatic and irreversible cultural, economic and political shift—in terms of global power. The financial crisis and the cyclical economic adjustments of the day grab the headlines and draw most of the attention just below the surface, roiling and churning; more profound upheaval is well under way: the decline of the U.S. empire and the eclipse of the "American Century," which, in all likelihood (but not necessarily—it depends *in part* on us), will be as messy as the end of the British, French, Japanese or Spanish empires; the turn from an economy with industry in the imperial centers to one where major production is in the colonial and postcolonial regions, which may well be (again, not inevitably) as murderous as the great leap from agriculture to industry; an unprecedented ecological dislocation that is already redrawing all existing maps and propelling millions of environmental refugees out of their homes and into a shrinking world. The center cannot hold, and we are—each and all of us, whether we recognize it or not—in the mix and on the move, witnessing and participating in the end of empire and the creation of a new social order in one way or another.

Today, however, the imperial dream of an unchallenged and grotesquely lopsided world is coming to a painful end. This is not the heralded "end of history," that weird ideology manufactured by the intellectual servants of power to explain and justify the imbalance and the injustice; rather it is the end of the arrogant hope for a thousand year *Pax Americana*. The evidence of terminal rot at the center is everywhere and the accompanying collapse is all around us: an economic and financial emergency based on deep structural problems; an environmental crisis that cannot and will not be ignored; the demographic changes caused by globalization and immigration leading to the fateful narrowing of a European American majority in the United States and challenges to White supremacy in Europe and elsewhere; the "Arab Spring" and "Occupy" openings combined with the stalemate and impending defeat of Western military forces in Iraq, Afghanistan and the entire region; and the various challenges to U.S. hegemony from a number of directions including Europe, North Asia, North Africa, the Middle East, Latin America and oddly jerry-rigged entities such as the BRIC alliance (Brazil, Russia, India, China).

Another world is possible, as the hopeful slogan from the World Social Forum has it, and in fact another world is inevitable. But will it be a better world? Not necessarily—a world of permanent war and sprawling work camps, massive prisons and constant control, environmental disaster on an even more gargantuan scale is also a possibility. Nothing is guaranteed, and nothing is settled once and for all. Everything is dynamic, in motion, on the move and on the make, incomplete, unfinished. And as we must continue to remind ourselves: "Power concedes nothing without a demand; it never has and it never will."

This is precisely where a focus on education—on reason and evidence and argument—becomes essential. This historic moment, this epoch, could surely be increasingly violent and horrifying or it could be a time of new hope, beauty and unforeseen possibilities. This is in part up to us: it depends on how we think and how we act. In education, this moment challenges us to reconsider every assumption and to reexamine first and fundamental principles.

An essential step is to reimagine the project of schooling, teaching and learning, curriculum and instruction, in radically new ways. Education at the end of empire is inevitably where our identity and our destiny will be developed and worked out. Education is one of the key pillars of the superstructure of capitalist society, the arena of politics and ideology, where humans become conscious of class conflict and fight it out. Education is a site of class sorting, the development of ideological hegemony and the debate over what it means to be human, where we are on the clock of the universe and what kind of future we mean to create.

Education at the end of empire is education in crisis and contestation. The outlines of the agenda of the powerful are increasingly apparent: shrinking of public discourse and liberties, privatization and drastically lowered expectations for students and families, the demonization of teachers, zero tolerance as a cat's paw for surveillance and control, sort-and-punish curricula, a culture of obedience and conformity, a narrowing definition of learning as job training and education as a product to be bought and sold in the market, the school-to-prison pipeline. On the other side there is a growing fight back based on the principle that all human beings are of incalculable value and that life in a just and free society must be geared toward and powered by a profoundly radical idea: The fullest development of all human beings regardless of race or ethnicity, origin or background, gender identity, ability or disability is the necessary condition for the full development of each person; and, conversely, the fullest development of each is the condition for the full development of all. On this side are those who recognize that access to education, the development of skills and critical capacity, make citizens and residents not just "college ready" or work prepared but also ready to become leaders of struggles for a humane future.

This points to the importance of opposing the hidden curriculum of obedience and conformity in favor of foregrounding and teaching initiative, questioning, doubt, skepticism, courage, imagination and creativity. These are central and not peripheral to an education based on principles of equality, justice and basic human rights. These are the qualities educators must struggle to model and nourish, encourage and defend in our communities and our classrooms.

In a free society, students are able to think for themselves and develop minds of their own, to make judgments based on evidence and argument and to build capacities for exploration and invention. They are encouraged then to ask the most fundamental and essential questions that are, like the young themselves, always in motion, dynamic and never twice the same: Who in the world am I? How did I get here and where am I going? What in the world are my choices and my chances? What did I learn that the teacher didn't know? What's my story, and how is it like or unlike the stories of others? What is my responsibility to those others?

Teachers and students who long for schooling as something transcendent and powerful find ourselves locked in institutions that glorify sorting students into winners and losers, reduce learning to a mindless and irrelevant routine of drill and skill and teaching to a kind of glorified clerking, passing along a curriculum of received wisdom and predigested (and often false) bits of information. This is unlovely in practice and it is unworthy of our deepest dreams.

The dominant neoliberal metaphor of the rich and powerful posits schools as businesses, teachers as workers, students as products and commodities, and it leads rather simply to thinking that school closings and privatizing the public space are natural events, relentless standardized test-and-punish regimes sensible, zero tolerance a reasonable proxy for justice. This is what the true believers call "reform."

The hijacking of school reform by the neoliberal corporate planners, the U.S. Chamber of Commerce and the Business Roundtable, U.S. government strategists and the education elites intensifies, an attack on teachers, unions, teacher education, schools and the kids themselves. The aim is to recreate the privileges of the powerful while forging a generation of technicians and passive followers and disciplining the lower classes to accept their place in the matrix. The gravitational pull of this narrative is so great that even radical reformers find themselves revoicing the deceptive goals and the phony frames. If we are to take a thorough and honest look at the educational landscape before us, we cannot accept the standards and benchmarks established by the elite, from the acceptance of capitalist development, meaningless and wasteful work, and ecological depredations as the only way forward, to the normalizing of White, middle-class discourse as the gold standard of excellence, anointed with titles like "Standard English" or "Academic English."

Schools for obedience and conformity are characterized by passivity and fatalism and infused with anti-intellectualism and irrelevance. They turn on the little technologies for control and normalization, the elaborate schemes for managing the crowd, the knotted system of rules and discipline, the exhaustive machinery of schedules and clocks, the laborious programs of sorting the crowd into winners and losers through testing and punishing, grading, assessing and judging, all of it adding up to a familiar trap, an intricately constructed hierarchy, everyone in a designated place and a place for everyone. In schools as they are, knowing and accepting one's pigeonhole on the towering and barren cliff becomes the only lesson one really needs.

Educators who are today oriented toward justice and liberation and enlightenment as living forces and powerful aspirations focus their efforts not on the production of things but on the production of fully developed human beings who are capable of controlling and transforming their own lives, citizens and residents who can participate actively in public life, people who can open their eyes and awaken themselves and others as they think and act ethically in a complex and ever-changing world. This kind of teaching encourages students to develop initiative and imagination, the capacity to name and constantly interrogate the world, the wisdom to identify the obstacles to their full humanity and to the humanity of others and the courage to act upon whatever the known demands. Education, then, is changed from rote boredom and endlessly alienating routines into something that is transformative, always opening doors and opening minds as students forge their own pathways into a wider world.

Teaching in this political moment is both a challenge and a gift, for this moment embodies what educators, beginning with early childhood teachers, have always called "a teachable moment." Teachable moments are times of disequilibrium and dislocation, times when lesson plans are thrown into doubt and newness can enter, times when the predictable and the commonplace are recognized as inadequate and fresh and startling winds can blow, for teachers no less than for students. The teachable moment aligns neatly with a certain kind of pedagogy, one that does not know the answers and is compelled to improvise with the unfinished, the contingent and the surprising/unforeseen.

In the schools we need, education is constructed as a fundamental human right geared toward the fullest development of the human personality and the reconstruction of society around basic principles of equality and justice and recognition.

"Power concedes nothing without a demand; it never has and it never will."

REFERENCES

Alexander, M. (2012). *The new Jim Crow*. New York, NY: The New Press.
Baldwin, J. (1963, December 21). A talk to teachers. *Saturday Review*, pp. 42–44.
Brown v. Board of Education, 347 U.S. 483 (1954).
Douglass, F. (1852). *What to the slave is the fourth of July?* Retrieved from http://teachingamericanhistory.org/library/document/what-to-the-slave-is-the-fourth-of-july/
Douglass, F. (1857). If there is no struggle, there is no progress. Retrieved from http://www.blackpast.org/1857-frederick-douglass-if-there-no-struggle-there-no-progress
Fannie Lou Hamer, Rosa Parks, and Coretta Scott King Voting Rights Reauthorization Act and Amendments Act (2006).
Giles v. Harris, 189 U.S. 475 (1903).
King, M. L. Jr. (1967). *Where do we go from here?* New York, NY: Harper Row.
Kozol, J. (2005). *The shame of the nation: The restoration of apartheid*. New York, NY: Crown.
Ladson-Billings, G. (2006). From the achievement gap to the education debt: Understanding achievement in U.S. schools. *The American Education Research Association*. Retrieved from http://brokersofexpertise.net/cognoti/content/file/resources/documents/e1/e1f36433/e1f364339be984a86674b4d7a4f055033df43956/Ladson_Billings.pdf
Marley, B. (1980). Redemption song. *Uprising*. Kingston: Island Records.
Mississippi v. Williams, 170 U.S. 213 (1898).
Murray, C., & Herrnstein, R. (1994). *The bell curve: Intelligence and class structure in American life*. New York, NY: Free Press.
National Governor's Association. (2005). Annual Meeting. Des Moines, IA. Retrieved from http://www.nga.org/files/live/sites/NGA/files/pdf/2005NGAAnnualMeeting.pdf
Plessy v. Ferguson, 163 US 537 (1896).
Shelby County v. Holder 570 U.S. (2013).
U.S. Const. amend. XIV.
U.S. Const. amend. XV.
Woodson, C. G. (1933). *The mis-education of the Negro*. Washington, DC: The Associated Publishers.

CHAPTER ELEVEN

Cultural Studies IN Dark Times

Public Pedagogy and the Challenge of Neoliberalism

HENRY A. GIROUX

INTRODUCTION

As the right wages a frontal assault against all remnants of the democratic state and its welfare provisions, left progressives are in disarray. Theoretical and political impoverishment feed off each other as hope of a revolutionary project capable of challenging the existing forces of domination appears remote. Militarism increasingly engulfs the entire social order as matters of "war and national security" become "consuming anxieties" that provide the "memories, models, and metaphors that shape broad areas of national life" as well as drive American foreign policy (Sherry, 1995, p. xi). As U.S. military action expands its reach into Iraq, Afghanistan and possibly Iran and Syria under the guise of an unlimited war against terrorism, public spaces on the domestic front are increasingly being organized around values supporting a bellicose, patriarchical and jingoistic culture that is undermining "centuries of democratic gains" (Buck-Morss, 2003, p. 33). As politics is separated from economic power, the state surrenders its obligation to contain the power of corporations and financial capital, reducing its role to matters of surveillance, disciplinary control and order. Market fundamentalism and the militarization of public life mutually reinforce each other to displace the promise, if not the very idea, of the Great Society—with its emphasis on the common good, basic social provisions for all, social justice and economic mobility. Fuelled by dreams of empire as well as the desire to mask the shape political power is taking in a period of economic and social decline, militarism and neoliberalism cloak themselves in the discourse of democracy in order to hide the barbarism being reproduced in the torture prisons in Iraq and Afghanistan, in the spread of wage slavery in the interest of capital accumulation and in the carceral surveillance and disciplinary measures being imposed on the nation's public schools. Democratic political projects appear remote and give rise to either cynicism, solipsism or reductionistic ideologies on the part of many progressives within and outside of the academy. The crucial task of theorizing a politics suitable for the 21st century has fallen on hard times. Economistic theories return to dominate much of the left, reducing politics to a reflection of economic forces, interests and measures. Within the university, critically engaged intellectuals appear in short supply as most academics, especially in the humanities and social sciences, bid a hasty retreat to arcane discourses, retrograde notions of professionalism or irrelevant academic specialties (Agger, 1989; Said, 2004). Rather than reinventing and rethinking the challenge of an oppositional politics within a global public sphere, the academic left appears to be withdrawing from the demands of civic engagement by retreating into what Susan Buck-Morss (2003) called "theoryworld," a space where the "academic freedom of critical theorists coincides with our lack of influence in public and political debate" (p. 68). Hope, once embodied in the politics of persuasion, the drive for instituting critical education in a diverse number of public spheres, collective efforts to organize struggles within major institutions and the attempt to build international social movements, seems, at best, a nostalgic remnant of the 1960s. The naturalness and commonsense appeal of the neoliberal economic order produces a crisis of political and historical imagination, on the one hand, and an educational crisis on the other. It is in opposition to the current turn away from matters of history, culture and politics that I begin with a quote from Susan George (1999), a powerful critic of neoliberalism and a leading voice in the antiglobalization movement:

> In 1945 or 1950, if you had seriously proposed any of the ideas and policies in today's standard neo-liberal toolkit, you would have been laughed off the stage or sent to the insane asylum. At least in the Western countries, at that time, everyone was a Keynesian, a social democrat, or a social-Christian democrat or some shade of Marxist. The idea that the market should be allowed to make major social and political decisions; the idea that the state should voluntarily reduce its role in the economy, or that corporations should be given total freedom, that trade unions should be curbed and citizens given much less rather than more social protection—such ideas were utterly foreign to the spirit of the time. Even if someone actually agreed with these ideas, he or she would have hesitated to take such a position in public and would have had a hard time finding an audience. (para 2)

Times have changed, and altered historical conditions posit new problems, define different projects and often demand fresh discourses. The complex theoretical discourses fashioned in the academy in the 1980s and 1990s seem hopelessly disconnected, if not irrelevant, in the current moment. And the space of democratic political and social thought now appears exhausted by a panoply of military, religious and market fundamentalisms that refuse to question their own assumptions and appeal to the naturalness and inevitability of their ascendency historical struggles that produced it. George's comments are instructive because, in resurrecting historical memory, they not only point to a current period in American history in which the seemingly impossible has become possible (Giroux, 2004), but they also gesture toward those forces that must be named in order to become the object of resistance and refusal. The impossible in this case is the specter of authoritarianism replacing a weakened and damaged liberal democracy. With the election of George W. Bush to the presidency in 2000, the United States found itself in the midst of a revolution in which the most basic underlying principles of democracy had begun to unravel. The nature of this right-wing revolution resides in the lived relations of the contemporary social order and the ways in which such relations exacerbate the material conditions of inequality, undercut a sense of individual and social agency, hijack democratic values—such as egalitarianism and dissent—and promote a deep sense of hopelessness and cynicism. Resuscitating a deeply antimodernist past as a way to command the future, the Bush administration has evoked the cult of traditionalism, religious fundamentalism and the absolute reign of the market as central features of an emerging authoritarianism designed to "roll back the twentieth century quite literally" (Greider, 2003, p. 11). The alliance of militant neoconservatives, extremist evangelical Christians and free-market fundamentalists imagine a social order modeled on the presidency of William McKinley and the values of the robber barons. The McKinley presidency, which spanned from 1897 to 1901, "had a consummate passion to serve corporate and imperial power" (Moyers, 2004). This was an age when Blacks, women, immigrants and minorities of class "knew their place," big government served the exclusive interests of the corporate monopolists, commanding institutions were under the sway of narrow political interests, welfare was a private enterprise and labor unions were kept in check by the repressive forces of the state—all while an imperialist war raged in the Philippines. With the geographic shift to Iraq, all of these conditions are being reproduced under the leadership of an extremist element of the Republican Party that holds sway over all branches of government.

One of the central elements of the new authoritarianism is a structural relationship between the state and the economy that produces rigid hierarchies, concentrates power in relatively few hands, unleashes the most brutal elements of a rabid individualism, destroys the welfare state, incarcerates large numbers of its now disposable populations, economically disenfranchises large segments of the lower and middle classes and reduces entire countries to pauperization (Giroux, 2003; Harvey, 2003). Neoliberalism not only dissolves the bonds of sociality and reciprocity, it also undermines the nature of social obligations by defining civil society exclusively through an appeal to market-driven values.

At the same time, neoliberalism feeds a growing authoritarianism steeped in militarism, Christian fundamentalism and jingoistic patriotism, encouraging intolerance and hate as it punishes critical engagement and questioning, especially if it is at odds with the reactionary religious and political agenda being pushed by the Bush administration.

Increasingly, education appears useful only to those who hold political and economic power, and issues regarding how the academy might contribute to the quality of democratic public life on a national and global level are either ignored or dismissed. On the right, neoliberal cheerleaders are pushing hard to turn the university into another outpost of corporate learning and training. On the left, education as a site of dialectical struggle, persuasion and critical engagement is all too often reduced to ritual debunking and demystification, revealing the political logic of a debased capitalists system. But revelation guarantees nothing, and in this case, it substitutes a limited form of reportage for the hard pedagogical work connecting empowering forms of knowledge to the realities and social forms that bear down on students' everyday lives (Freire, 1998). The collective struggle to widen the reach and quality of education as a basis for creating critical citizens—so alive in the 1960s—is rendered defunct within the corporate drive for efficiency, downsizing, profits and an utterly instrumentalist notion of excellence. Cornel West (2004) has argued persuasively that just as we need to analyze those dark forces shutting down democracy, "we also need to be very clear about the vision that lures us toward hope and the sources of that vision" (p. 18). I want to act on West's utopian call by recapturing the vital role that an expanded notion of critical education might play for educators, students, cultural studies advocates and other progressives by providing a language of critique and possibility that addresses the growing threat of free market fundamentalism to an inclusive democracy and the promise of a cultural politics in which pedagogy occupies a formative role in shaping both critical agency and the radical imagination.

But before I make that case, I want to address in more detail neoliberalism as one of the most powerful antidemocratic ideologies now threatening both the idea and formation of a critically informed citizenry, a viable notion of social agency and the idea of the university as a democratic public sphere. In doing so, I hope to establish a context for analyzing the importance of cultural studies as a theoretical, pedagogical and political intervention that makes clear both the responsibility of academics to understand and engage neoliberalism within the rising tide of authoritarianism in the United States and elsewhere and what it might mean to offer students and others the hope and tools necessary to revitalize the culture of politics as an ethical response to the demise of democratic public life. At the very least, such a challenge demands that educators and other cultural workers struggle to preserve and revitalize those institutional spaces, forums and public spheres that support and defend critical education, help students come to terms with their own power as individual and social agents and reclaim those nonmarket values such as caring, community, trust, conviction and courage that are vital to a substantive democracy.

THE POLITICS OF NEOLIBERALISM

Neoliberalism not only exerts unparalleled influence on the global economy but also redefines the very nature of politics and society. Free market fundamentalism rather than democratic idealism is now the driving force of economics and politics in most of the world. It is a market ideology driven not just by profits but by an ability to reproduce itself with such success, that to paraphrase Fred Jameson (1994, p. xii), it is easier to imagine the end of the world than it is to imagine the end of capitalism, even as it creates vast inequalities and promotes human suffering throughout the globe. Wedded to the belief that the market should be the organizing principle for all political, social and economic decisions, neoliberalism increasingly drives the meaning of citizenship and social life while waging an incessant attack on democracy, public goods, the welfare state and noncommodified values.

Neoliberal economics has dominated American society since the 1970s and has been embraced by both New Democrats and conservatives. Both political parties in the United States embrace the defining principles of neoliberalism, especially the notion that the market is self-regulating and should be free of interference by the government, that choice is defined as an economic prerogative and that "economic transactions can subordinate and [in] many cases replace political democracy" (Newfield, 2002, p. 314). While there is some political opposition among the established parties to the brutalizing policies of neoliberalism, both political parties generally buy into a corporate-driven legislative agenda, which includes

> deregulation of business at all levels of enterprises and trade; tax reduction for wealthy individuals and corporations; the revival of the near-dormant nuclear energy industry; limitations and abrogation of labor's right to organize and bargain collectively; a land policy favoring commercial and industrial development at the expense of conservation and other pro-environment policies; elimination of income support to chronically unemployed; reduced federal aid to education and health; privatization of the main federal pension program, social security; limitations on the right of aggrieved individuals to sue employees and corporations who provide services. (Aronowitz, 2003, p. 102)

Under neoliberalism, everything either is for sale or is plundered for profit. One might also add to Aronowitz's (2003) list the attack on institutions dedicated to critically informing the public; politicians willingly hand the public's airwaves over to a handful of powerful broadcasters and large corporate interests without a dime going into the public trust; entire populations, especially those of color who are poor, are now considered disposable, schools more closely resemble either jails or high-end shopping malls, depending on their clientele and teachers are forced to get revenue for their school by hawking everything from hamburgers to pizza parties. And attending the university in an era of drastic cutbacks and spiraling tuition becomes once again the near exclusive preserve of the upper-middle classes (Giroux & Giroux, 2004).

Corporations more and more design not only the economic sphere but also shape legislation and policy affecting all levels of government, and with limited opposition. As corporate power lays siege to the political process, the benefits flow upward to the rich and the powerful. In Bush's ownership society, government policy now works to benefit the biggest corporations. For example, Bush's 2006 budget contained drastic cuts for many of the major regulatory agencies, not only compromising everything from emission standards to drug safety programs, but also presenting the "possibilities—indeed, probability—that these public agencies will become captives of private corporations they are supposed to regulate" (Drutman & Cray, 2005, p. 17). It gets worse. Included in such benefits are reform policies that shift the burden of taxes from the rich to the middle class, the working poor and state governments, as can be seen in the shift from taxes on wealth (capital gains, dividends and estate taxes) to a tax on work, principally in the form of a regressive payroll tax (Collins, Hartman, Kraut, & Mota, 2004). During the 2002–2004 fiscal years, tax cuts delivered $197.3 billion in tax breaks to the wealthiest 1% of Americans (i.e., households making more than $337,000 a year), while state governments increased taxes to fill a $200 billion budget deficit (Gonsalves, 2004). Equally alarming, a 2004 Congressional study revealed that 63% of all corporations in 2000 paid no taxes, while "[s]ix in ten corporations reported no tax liability for the five years from 1996 through 2000, even though corporate profits were growing at record-breaking levels during that period" (Woodard, 2004, para. 11).

As neoliberal policies dominate politics and social life, the breathless rhetoric of the global victory of free market rationality is invoked to cut public expenditures and undermine those noncommodified public spheres that serve as the repository for critical education, public dialogue and collective intervention. Public services such as healthcare, child care, public assistance, education and transportation are now subject to the rules of the market. Social relations between parents and children, doctors and patients, teachers and students are reduced to that of supplier and customer, just as the laws of market replace those noncommodified values capable of defending vital public goods and spheres. Forsaking the public good for the private good and hawking the needs of the corporate and private sector as the only source of sound investment, neoliberal ideology produces, legitimates and exacerbates the existence of persistent poverty, inadequate healthcare, racial apartheid in the inner cities and growing inequalities between the rich and the poor

(Krugman, 2003; Phillips, 2003; Street, 2004). Under neoliberalism, the state now makes a grim alignment with corporate capital and transnational corporations, legitimating the dangerous presupposition that corporations should be planning our future and that progress should be defined almost exclusively in economic and technological terms rather than in social and ethical terms. Corporations, in turn, are not designed to be responsible citizens. On the contrary, their sole purpose is to make money and by default, accumulate power. Unfortunately, when left unregulated, "they begin to overwhelm the political institutions that can keep them in check, eroding key limitations on their destructive capacities. Internationally, of the 100 largest economies in the world, 51 are corporations and 49 are nations" (Drutman & Cray, 2005, p. 17).

In its capacity to de-historicize and naturalize such sweeping social change, as well as in its aggressive attempts to destroy all of the public spheres necessary for the defense of a genuine democracy, neoliberalism reproduces the conditions for unleashing the most brutalizing forces of capitalism (Derber, 2002). Social Darwinism, with its ruthless indifference to human suffering, has risen like a phoenix from the ashes of 19th-century pseudoscience and can now be seen in full display on most reality TV programs and in the unfettered self-interest that now drives popular culture and fits so well with the spirit of authoritarianism. There is no public politics in this discourse, only the private domain of market identities, values and practices (Giroux, 2004). As social bonds are replaced by unadulterated materialism and narcissism, public concerns are now understood and experienced as utterly private miseries, except when offered up on *Jerry Springer* as fodder for entertainment. Where public space—or its mass mediated simulacrum—does exist, it is the backdrop for a highly orchestrated and sensational confessional for private woes, a cutthroat game of winner-take-all replacing more traditional forms of courtship, as in *Who Wants to Marry a Millionaire?* or as advertisement for crass consumerism, like MTV's *Cribs*.

Conscripts in a relentless campaign for personal responsibility, Americans are now convinced that they have little to hope for—and gain from—the government, nonprofit public spheres, democratic associations, public and higher education or other nongovernmental social forces. With few exceptions, the project of democratizing public goods has fallen into disrepute in the popular imagination, as the logic of the market undermines the most basic social solidarities. The consequences include not only a weakened social state but a growing sense of insecurity, cynicism and political retreat on the part of the general public. The incessant calls for self-reliance that now dominate public discourse betray an eviscerated and refigured state that neither provides adequate safety nets for its populace, especially those who are young, poor or racially marginalized, nor gives any indication that it will serve the interests of its citizens in spite of constitutional guarantees.

In fact, the reconfigured state is increasingly becoming a carceral enterprise more concerned with punishing and policing than with nurturing and investing in the public good. Situated within an expanding culture of fear, market freedoms seem securely grounded in a defense of national security, capital and property rights. When coupled with a media-driven culture of panic and hyped-up levels of insecurity, surviving public spaces are increasingly monitored and militarized. For example, when the media alerted the nation's citizenry to new terrorist threats in New York, New Jersey, and Washington, DC, CNN ran a lead story on their impact on tourism—specifically on the enthusiastic clamor by tourist families to get their pictures taken among U.S. paramilitary units now lining city streets, fully flanked with their imposing tanks and massive machine guns. The accouterments of a police state now vie with high-end shopping and museum visits for the public's attention, with only the occasional murmur of protest. But the investment in surveillance and carceral containment is hardly new. Since the early 1990s, state governments have invested more in prison construction than in education, and prison guards and security personnel in public schools are two of the fastest growing professions.

NEOLIBERALISM AS PUBLIC PEDAGOGY

Within neoliberalism's market-driven discourse, corporate power marks the space of a new kind of public pedagogy, one in which the production, dissemination and circulation of ideas emerge from the educational force of the larger culture. Public pedagogy in this sense refers to a powerful ensemble of ideological and institutional forces whose aim is to produce competitive, self-interested individuals vying for their own material and ideological gain. Under neoliberalism, pedagogy has become thoroughly reactionary as it operates from a variety of education sites producing forms of pedagogical address in which matters of personal agency, social freedom and the obligations of citizenship conceive of political and social democracy as a burden, an unfortunate constraint on market relations, profit making and a consumer democracy (Newfield, 2002). Corporate-driven public pedagogy and culture largely cancel out or devalue gender, class-specific and racial injustices of the existing social order by absorbing the democratic impulses and practices of civil society within narrow economic relations. Knowledge has become capital to invest in the economy but has little to do with the power of self-definition or the capacities needed to expand the scope and operations of freedom and justice. Similarly, corporate public pedagogy has become an all-encompassing cultural horizon for producing not only megacorporate conglomerates but also market identities, values and atomizing social practices. As politics increasingly becomes privatized, some neoliberal advocates argue that the answer to solving the healthcare and education crises faced by many states is to sell off public assets to private interests, just as they insist the problem of social security can be solved through private investment accounts. The Pentagon even considered, if only for a short time, turning the war on terror and security concerns over to futures markets, subject to online trading. Neoliberalism utterly privatizes politics and offers absurd solutions to collective problems, such as in suggesting that water pollution can be solved by buying bottled water. Thus, noncommodified public spheres are replaced by commercial spheres, as the substance of critical democracy is emptied out and replaced by a democracy of goods available to those with purchasing power, amidst the increasing expansion of the cultural and political might of corporations throughout the world.

Under neoliberalism, dominant public pedagogy, with its narrow and imposed schemes of classification and limited modes of identification, uses the educational force of the culture to negate the basic conditions for critical agency. What becomes clear in the new information age, or what Zygmunt Bauman (2000) called liquid modernity, is that the power of the dominant order is not just economic but ideological—rooted in the ability to mobilize consent, define a particular notion of agency, impose narrow visions of the future and decouple politics from both social radical notions of agency and democratic visions of freedom and social justice.

Within neoliberal public pedagogy, individuality has nothing to do with self-development or self-empowerment. On the contrary, self-development is now refashioned as the endless pursuit of personal interests. A belief in the power of a brutalizing self-interest replaces any notion of shared responsibility or social justice. Misfortune in this discourse does not arouse the obligations of citizenship but is relegated to the status of an individual weakness. Public goods are now transformed into sites for individual financial gain and social problems dissolve into the discourse of pathology. Poverty is now viewed as a crime. Racism is viewed as a personal prejudice (more often than not victimizing Whites), and unemployment is a mark of weak character. Power, inequality and social justice disappear from the language of the social just, as the individual increasingly lives in a world in which private interests take precedence over social concerns.

As collective agents recede under neoliberalism, market forces incessantly attempt to privatize or commercialize public space. One consequence is that those noncommodified spaces capable of providing individuals with the discourses, values and subject positions crucial to identifying and struggling over institutions vital to the life of democracy begin to disappear from the political vernacular. Under such circumstances, matters of agency become even more crucial to viable democratic politics, as those spaces capable of producing critical modes of agency increasingly slip into the black hole of commercialized space. As public spaces disappear, it becomes more difficult to develop a democratic discourse for educating collective social agents capable of raising critical questions about the limits of a market-driven society, as well as what it might mean to theorize about the future of public institutions central to the development of truly substantive democratic society.

In the absence of public spaces that promote shared democratic values, a new authoritarian politics and culture emerge in which the state makes a grim alignment with corporate capital, neoconservative visions of empire and Christian fundamentalism. Political power is now accumulated behind an alliance of economic, political and religious fundamentalists who recognize that "military-like discipline abroad requires military-like discipline at home" (Harvey, 2003, p. 193). Repressive legislation is used to sacrifice civil liberties in the cause of national security, the government promotes a culture of fear to implement neoliberal policies at home and neoconservative visions of empire abroad, dissent is labeled as unpatriotic and the media and political parties increasingly become adjuncts of official power (Barber, 2003; Giroux 2003; Robin, 2004).

As neoliberal economics is accorded more respect than democratic politics, the citizen has been abandoned and the consumer becomes the only viable model of agency. As public spending decreases, education is divorced from democratic politics and the political state increasingly becomes the corporate state (Hertz, 2003). All the more reason to take seriously Hannah Arendt's (1965) claim that, "Without a politically guaranteed public realm, freedom lacks the worldly space to make its appearance" (p. 149). And it is precisely within such a realm that subjects are socialized into forms of individual and social agency in which they learn how to govern rather than be governed, to assume the responsibilities of engaged citizens rather than be reduced to consumers or investors. Arendt understood quite clearly that democracy can only emerge, if not flourish, within political organizations in which education is viewed both as a site of politics and the foundation that provides the pedagogical conditions in which individuals can learn the knowledge, skills and values necessary for those forms of citizenship, leadership and social engagement that deepen and extend the realities of an inclusive democracy. Politics often begins when it becomes possible to make power visible, to challenge the ideological circuitry of hegemonic knowledge and to recognize that "political subversion presupposes cognitive subversion, a conversion of the vision of the world" (Bourdieu, 2001, p. 128). But another element of politics focuses on where politics happens, how proliferating sites of pedagogy bring into being new forms of resistance, raise new questions and necessitate alternative visions regarding autonomy and the possibility of democracy itself. Neoliberal ideology and pedagogy have been reproduced and reinforced within the advanced countries of the West through the development of new sites of pedagogy and new technologies that penetrate spaces that historically have been beyond the reach of the logic of commercialism and commodification. Hence, it is all the more necessary for educators and other cultural workers to take seriously both the proliferating sites of these new forms of ideological address and the work they do within the social order to create agents and subject positions that become complicitous with the brutalizing logic of the market.

At this point in American history, neoliberal capitalism is not simply too overpowering; on the contrary, "democracy is too weak" (Barber, 2002, p. A23). Profound transformations have taken place in the public space, producing new sites of pedagogy marked by a distinctive confluence of new digital and media technologies, growing concentrations of corporate power and unparalleled meaning-producing capacities. Unlike traditional forms of pedagogy, knowledge and desire are inextricably connected to modes of pedagogical address mediated through unprecedented electronic technologies that include high-speed computers, and new types of digitized film. Such sites operate within a wide variety of social institutions and formats, including sports and entertainment media, cable television networks, churches and channels of elite and popular culture such as advertising. The result is a public pedagogy that plays a decisive role in producing a diverse cultural sphere that gives new meaning to education as a political force.

While John Dewey, Paulo Freire and various other leading educational theorists in the last century understood the important connection between education and democracy, they had no way in their time of recognizing that the larger culture would extend beyond, if not supercede, institutionalized education, particularly schools, as the most important educational force over developed societies. In fact, education and pedagogy have long been synonymous with schooling in the public mind. Challenging such a recognition does not invalidate the importance of formal education to democracy, but it does require a critical understanding of how the work of education takes place in such institutions as well as in a range of other knowledge- and meaning-producing spheres such as advertising, television, film, the Internet, video game culture and the popular press. Rather than invalidate the importance of schooling, it extends the sites of pedagogy and in doing so broadens and deepens the meaning and importance of public pedagogy. What is being suggested here is that educators, cultural studies theorists and others take seriously the role that culture plays, as Raymond Williams (1967, p. 15) put it, as a form of "permanent education."

The concept of public pedagogy as a form of permanent education underscores the central importance of formal spheres of learning that unlike their popular counterparts—driven largely by commercial interests that more often mis-educate the public—must provide citizens with those critical capacities, modes of literacies, knowledge and skills that enable them to both read the world critically and participate in shaping and governing it. Put differently, formal spheres of learning provide one of the few sites where students can be educated to understand, critically engage and transform those institutions that are largely shaping their beliefs and sense of agency. I am not claiming that public or higher education is free from corporate influence and dominant ideologies but that such sites of education, at best, have historically provided the spaces and conditions for prioritizing civic values over commercial interests and for conveying the important role critical education plays in fostering democratic life. In spite of its present embattled status and contradictory roles, higher education, in particular, remains uniquely placed—though also under attack by the forces of corporatization—to prepare students to both understand and influence the larger educational forces that shape their lives. Needless to say, those of us who work in such institutions by virtue of our privileged positions within a rather obvious division of labor have an obligation to draw upon those traditions and resources capable of providing a critical education to all students in order to prepare them for a world in which information and power have taken on new and significant dimensions. In fact, the critique of information cannot be separated from the critique of power itself, providing a substantial new challenge for how we are to theorize politics for the 21st century. One way to take up this challenge is to address the theoretical contributions that a number of radical educators and cultural studies theorists have made in engaging not only the primacy of culture as a political force but also *how* the relationship between culture and power constitutes a new site of politics, pedagogy and resistance.

CULTURAL STUDIES AND THE QUESTION OF PEDAGOGY

Of course, my position on the civic obligations of the academy is not without its critics. It is not a position that supports traditional views of humanistic education, its canons or its implicit demand for reverence rather than engagement. Consider, by way of the counterexample, Jeffrey Hart (1996), Dartmouth professor and a senior editor with the *National Review* (the right-wing magazine founded William F. Buckley, a founder of American conservatism and a former employee of the Central Intelligence Agency). Echoing the central concerns of the culture wars that conservatives have been waging in full force since the 1980s, his claim is twofold: Higher education has been taken over by radicals who are a product of the 1960s, and conservative students are being mistreated because they are overwhelmingly subjected to political indoctrination or harassment. Sounding the alarm on the disciplinary and theoretical advances of the last several decades—like cultural studies and women's studies—Hart responded to the question of how to get a decent college education as follows:

> Select the ordinary courses. I use ordinary here in a paradoxical and challenging way. An ordinary course is one that has always been taken and obviously should be taken—even if the student is not yet equipped with a sophisticated rationale for so doing. The student should be discouraged from putting his money on the cutting edge of interdisciplinary cross-textuality.... If the student should seek out those ordinary courses, then it follows

> that he should avoid the flashy come-ons. Avoid things like Nicaraguan Lesbian poets. Yes, and anything listed under "Studies," any course whose description uses the words "interdisciplinary," "hegemonic," "phallocratic," or "empowerment," anything that mentions "keeping a diary," any course with a title like "Adventures in Film." Also, any male professor who comes to class without a jacket and tie should be regarded with extreme prejudice unless he has won a Nobel Prize. (p. 34)

Unlike Hart, who believes that cultural studies is the enemy of not only higher education but also what he would term the "disinterested" mind, I believe that cultural studies, for all of its diversity and contradictions, is one of the few theoretical traditions within the academy that links learning to social change and education to the imperatives of a critical and global democracy.

My own interest in cultural studies emerges out of its early concern with adult education, exemplified in the work of Richard Hoggart (1957), Raymond Williams (1958), Stuart Hall (1992) and Paul Willis (1981), and more recently in the work of Lawrence Grossberg (1997), bell hooks (1994), Stanley Aronowitz (2000) and Nick Couldry (2001), who focus on education more broadly. This tradition, often ignored today, views cultural studies as an empowering practice that "acts directly upon the conditions of culture to change them" (Couldry, 2001, p. 66), engages the politics of cultural studies as part of a broader project related to democracy and views matters of pedagogy as central to the project of cultural studies itself. Within this perspective, intellectual work and practice within the university are articulated as a matter of democracy. Defining the task of cultural studies, Raymond Williams (1989) argued as follows:

> It has been about taking the best we can in intellectual work and going with it in this very open way to confront people for whom it is not a way of life, for whom it is not in any probability a job, but for whom it is a matter of their own intellectual interest, their own understanding of the pressures on them, pressures of every kind, from the most personal to the most broadly political—if we are prepared to take that kind of work and revise the syllabus and discipline as best we can...then Cultural Studies has a very remarkable future indeed. (pp. 161–162)

Such a project calls for intellectual work that is theoretically rigorous, radically contextual, interdisciplinary and self-critical about its motivating questions and assumptions. This project engages culture through a wide variety of social forms and material relations of power, views theory as a resource and historical memory as a series of ruptures rather than a totalizing narrative. Cultural studies in this perspective is not only deconstructive but also willing, to quote Hall (1992), "to address the central, urgent, and disturbing questions of a society and a culture in the most rigorous intellectual way we have available" (p. 11). Such a discourse points to the hard work of providing a language of critique and possibility, of imagining different futures and addressing the pedagogical conditions that make possible the agents, politics and forms of resistance necessary to reclaim the promise of a truly global, democratic future.

My commitment to cultural studies emerges out of an ongoing project to theorize the diverse ways in which culture functions as a contested sphere over the production, distribution and regulation of power and how and where it operates both symbolically and institutionally as an educational, political and economic force. In this perspective, cultural studies recognizes the primacy of the pedagogical as a critical practice through which politics is pluralized, understood as contingent and open to many formations. But cultural studies is also crucial for resisting those mutually informing material and symbolic registers in which matters of representation and meaning work to secure particular market identities, legitimate dominant relations of power and privatize spaces of dialogue and dissent, especially as neoliberalism attempts to undermine the very meaning and practice of a substantive democracy.

Against the neoliberal attack on all things social, cultural studies can play an important role in producing narratives, metaphors images, and desiring maps that exercise a powerful pedagogical force over how people think about themselves, engage with the claims of others, address questions of justice and take up the obligations of an engaged citizenship. Within a cultural studies discourse, culture is the primary sphere/space/location in which individuals, groups and institutions learn to translate the diverse and multiple relations that mediate between private life and public concerns (Bauman, 1999). Far from being exclusively about matters of representation and texts, culture becomes a site, event and performance in which identities and modes of agency are configured through the mutually determined forces of thought and action, body and mind and time and space. Culture offers a site where common concerns, new solidarities and public dialogue refigure the fundamental elements of democracy. Culture is also the pedagogical and political ground in which a global public sphere can be imagined to confront the now planetary inequities of symbolic and material power, just as it promotes the possibilities of shared dialogue and democratic transformation. Culture as an emancipatory force affirms the social as a fundamentally political space, just as neoliberalism attempts within the current historical moment to deny culture's relevance as a public sphere and its centrality as a political necessity.

Central to any viable notion of cultural studies, then, is the primacy of culture and power, organized through an understanding of how private issues are connected to larger social conditions and collective forces, that is, how the very processes of learning constitute the political mechanisms through which identities are shaped, desires mobilized and experiences take on form and meaning within those collective conditions and larger forces that constitute the realm of the social. This suggests the necessity on the part of cultural theorists to be particularly attentive to the connections between pedagogy and political agency. Yet, unfortunately, the much-needed emphasis on making the political more pedagogical has not occupied a central place in the work of most cultural studies theorists, as it did in the field's earliest formations. Pedagogy in most cultural studies work is either limited to the realm of schooling, dismissed as a discipline with very little academic cultural capital or is rendered reactionary through the claim that it simply accommodates the paralyzing grip of governmental institutions that normalize all pedagogical practices.

FROM A PEDAGOGY OF UNDERSTANDING TO A PEDAGOGY OF INTERVENTION

In opposition to these positions, I want to reclaim a tradition in radical educational theory and cultural studies in which pedagogy as a critical practice is central to any viable notion of agency, inclusive democracy and a broader global public sphere. Pedagogy as both a language of critique and possibility looms large in these critical traditions not as a technique or a priori set of methods but as a political and moral practice. As a political practice, pedagogy is viewed as the outgrowth of struggles and illuminates the relationship among power, knowledge and ideology, while self-consciously, if not self-critically, recognizing the role it plays as a deliberate attempt to influence how and what knowledge and identities are produced within particular sets of social relations. As a moral practice, pedagogy recognizes that what cultural workers, artists, activists, media workers and others teach cannot be abstracted from what it means to invest in public life, presuppose some notion of the future or locate oneself in a public discourse. The moral implications of pedagogy also suggest that our responsibility as intellectuals for the public cannot be separated from the consequences of the knowledge we produce, the social relations we legitimate and the ideologies and identities we offer up to students as well as colleagues.

Refusing to decouple politics from pedagogy means, in part, creating those public spaces for engaging students in robust dialogue, challenging them to think critically about received knowledge and energizing them to recognize their own power as individual and social agents. Pedagogy has a relationship to social change in that it should not only help students frame their sense of understanding, imagination and knowledge within a wider sense of history, politics and democracy but should also enable them to recognize that they can do something to alleviate human suffering, as the late Susan Sontag (2003) has suggested. Part of this task necessitates that cultural studies theorists and educators anchor their own work, however diverse, in a radical project that seriously engages the promise of an unrealized democracy against its really existing and radically incomplete forms. Of crucial importance to such a project is rejecting the assumption that theorists can understand social problems without contesting their appearance in public life. More specifically, any viable cultural politics needs a socially committed notion of injustice if we are to take seriously what it means to fight for the idea of the good society. Bauman (2002) is right in arguing that, "If there is no room for the idea of *wrong* society, there is hardly much chance for the idea of good society to be born, let alone make waves" (p. 170).

Cultural studies theorists need to be more forceful, if not more committed, to linking their overall politics to modes of critique and collective action that address the presupposition that democratic societies are never too just, which means that a democratic society must constantly nurture the possibilities for self-critique, collective agency and forms of citizenship in which people play a fundamental role in shaping the material relations of power and ideological forces that affect their everyday lives. Within the ongoing process of democratization lies the promise of a society that is open to exchange, questioning and self-criticism, a democracy that is never finished and one that opposes neoliberal and neoconservative attempts to supplant the concept of an open society with a fundamentalist market-driven or authoritarian one.

Cultural studies theorists who work in higher education need to make clear that the issue is not whether higher education has become contaminated by politics, as much as recognizing that education is already a space of politics, power and authority. At the same time, they can make visible their opposition to those approaches to pedagogy that reduce it to a set of skills to enhance one's visibility in the corporate sector or an ideological litmus test that measures one's patriotism or ratings on the rapture index. There is a disquieting refusal in the contemporary academy to raise broader questions about the social, economic and political forces shaping the very terrain of higher education—particularly unbridled market forces, fundamentalist groups and racist and sexist forces that unequally value diverse groups within relations of academic power.

There is also a general misunderstanding of how teacher authority can be used to create the pedagogical conditions for critical forms of education without necessarily falling into the trap of simply indoctrinating students. For instance, many conservative and liberal educators believe that any notion of critical pedagogy that is self-conscious about its politics and engages students in ways that offer them the possibility for becoming critical—what Lani Guinier (2003, p. 6) called the need to educate students "to participate in civic life, and to encourage graduates to give back to the community, which through taxes, made their education possible"—leaves students out of the conversation or presupposes too much or simply represents a form of pedagogical tyranny. While such educators believe in practices that open up the possibility of questioning among students, they often refuse to connect the pedagogical conditions that challenge how and what they think at the moment to the next task of prompting them to imagine changing the world around them so as to expand and deepen its democratic possibilities. Teaching students how to argue, draw on their own experiences or engage in rigorous dialogue says nothing about why they should engage in these actions in the first place. How the culture of argumentation and questioning relates to giving students the tools they need to fight oppressive forms of power, make the world a more meaningful and just place and develop a sense of social responsibility is missing in contemporary progressive frameworks of education.

While no pedagogical intervention should fall to the level of propaganda, a pedagogy that attempts to empower critical citizens cannot and should not avoid politics. Pedagogy must address the relationship between politics and agency, knowledge and power, subject positions and values, and learning and social change while always being open to debate, resistance and a culture of questioning. Liberal educators committed to simply raising questions have no language for linking learning to forms of public-minded scholarship that would enable students to consider the important relationship between democratic public life and education or what it would mean to encourage students pedagogically to enter the sphere of the political, enabling them to think about how they might participate in a democracy by taking what they learn into new locations and battlegrounds—a fourth-grade classroom, a church, the media, a politician's office, the courts, a campus—or for that matter taking on collaborative projects that address the myriad of problems citizens face on a local, national and global level in a diminishing democracy.

In spite of the professional pretense to neutrality, academics in the field of cultural studies need to do more pedagogically than simply teach students how to argue and question. Students need much more from their educational experience. Democratic societies need educated citizens who are steeped in more than the skills of argumentation. And it is precisely this democratic project that affirms the critical function of education and refuses to narrow its goals and aspirations to methodological considerations. As Amy Gutmann (1999) argued, education is always political because it is connected to the acquisition of agency, the ability to struggle with ongoing relations of power, and it is a precondition for creating informed and critical citizens who act on the world. This is not a notion of education tied to the alleged neutrality of the academy or the new conservative call for "intellectual diversity" but to a vision of pedagogy that is directive and interventionist on the side of producing a substantive democratic society. This is what makes critical pedagogy different from training. And it is precisely the failure to connect learning to its democratic functions and goals that provides rationales for pedagogical approaches that strip what it means to be educated from its critical and democratic possibilities.

Cultural studies theorists and educators would do well to take account of the profound transformations taking place in the public sphere and reclaim pedagogy as a central element of cultural politics. In part, this means once again recognizing, as Pierre Bourdieu and Günter Grass (2003) have insisted, that the "power of the dominant order is not just economic, but intellectual—lying in the realm of beliefs" (p. 66), and it is precisely within the domain of ideas that a sense of utopian possibility can be restored to the public realm. Such a task suggests that academics and other cultural workers actively resist the ways in which neoliberalism discourages teachers and students from becoming critical intellectuals by turning them into human data banks. Educators and other cultural workers need to build alliances across differences, academic disciplines and national boundaries as part of a broader effort to develop social movements in defense of the public good and social justice. No small part of this task requires that such groups make visible the connection between the war at home and abroad. If the growing authoritarianism in the United States is to be challenged, it is necessary to not only oppose an imperial foreign policy but also the shameful tax cuts for the rich, the dismantling of the welfare state, the attack on unions and those policies that sacrifice civil liberties in the cause of national security.

Opposing the authoritarian politics of neoliberalism, militarism and neoconservatism means developing enclaves of resistance in order to stop the incarceration of a generation of young Black and Brown men and women, the privatization of the commons, the attack on public schools, the increasing corporatization of higher education, the growing militarization of public life and the use of power based on the assumption that empire abroad entails tyranny and repression at home. But resistance needs to be more than local or rooted in the specificity of particular struggles. Progressives need to develop national and international movements designed to fight the new authoritarianism emerging in the United States and elsewhere. In part, this means revitalizing social movements such as civil rights, labor, environmental and antiglobalization on the basis of shared values and a moral vision rather than simply issue-based coalitions. This suggests organizing works, intellectuals, students, youth and others through a language of critique and possibility in which diverse forms of oppression are addressed through a larger discourse of radical democracy, a discourse that addresses not only what it means to think in terms of a general notion of freedom capable of challenging corporate rule, religious fundamentalism and the new ideologies of empire but also what it might mean to link freedom to a shared sense of hope, happiness, community, equality and social justice. Democracy implies a level of shared beliefs, practices and a commitment to build a more humane future. Politics in this sense points to a struggle over those social, economic, cultural and institutional forces that make democracy purposeful for all people. But this fundamentally requires something prior—a reclaiming of the social and cultural basis of a critical education that makes the very struggle over democratic politics meaningful and understandable as part of a broader affective, intellectual and theoretical investment in public life (Couldry, 2004).

As the U.S. administration spreads its legacy of war, destruction, commodification, privatization, torture, poverty and violence across the globe, we need a new language for politics, justice and freedom in the global public sphere. We need a new vocabulary for talking about what educational institutions should accomplish in a democracy and why they fail; we need a new understanding of public pedagogy for analyzing what kind of notions of agency and structural conditions can bring a meaningful democracy into being. Most important, we need to make pedagogy and hope central to any viable form of politics engaged in the process of creating alternative public spheres and forms of collective resistance. The question of agency cannot be separated from a concern about where democratic struggles can take place and what it might mean to create the affective conditions for students and others to want to engage in such struggles in the first place. Hope, as a precondition for agency and resistance, is a crucial element of democratic politics, because it not only rests on a promise of a better world, but it views the future as something more than a repeat of the present. Hope is central to political change and must find a way out of the manufactured cynicism that accompanies current forms of neoliberalism and religious fundamentalism. We need to recognize, as Zygmunt Bauman points out, that the real pessimism is quietism—falsely believing in not doing anything because nothing can be changed (Bunting, 2003). Most significantly, we need a new understanding of how culture works as a form of public pedagogy, how pedagogy works as a moral and political practice, how agency is organized through pedagogical relations, how individuals can be educated to make authority responsive, how politics can make the workings of power visible and accountable and how to reclaim hope in dark times through new forms of pedagogical praxis, global protests and collective resistance.

REFERENCES

Agger, B. (1989). *Fast capitalism: A critical theory of significance.* Urbana: University of Illinois Press.
Arendt, H. (1965). *Between past and future.* New York, NY: Meridian Books.
Aronowitz, S. (2000). *The knowledge factory.* Boston, MA: Beacon Press.
Aronowitz, S. (2003). *How class works.* New Haven, CT: Yale University Press.
Barber, B. R. (2002, July 29). A failure of democracy, not capitalism. *The New York Times,* p. A23.
Barber, B. R. (2003). *Fear's empire: War, terrorism, and democracy.* New York, NY: Norton.
Bauman, Z. (1999). *In search of politics.* Stanford, CA: Stanford University Press.
Bauman, Z. (2000). *Liquid modernity.* London, UK: Polity Press.
Bauman, Z. (2002). *Society under siege.* Malden, MA: Blackwell.
Bourdieu, P. (2001). *Language and symbolic power.* Cambridge, MA: Harvard University Press.
Bourdieu, P., & Grass, G. (2003). The "progressive" restoration: A Franco-German dialogue. *New Left Review, 14,* 63–77.
Buck-Morss, S. (2003). *Thinking past terror: Islamism and critical theory on the left.* London, UK: Verso.

Bunting, M. (2003, April 4). Zygmunt Bauman: Passion and pessimism. *The Guardian*. Retrieved from http://books.guardian.co.uk/print/0,3858,4640858-110738,00.html

Collins, C., Hartman, C., Kraut, K., & Mota, G. (2004, April 20). *Shifty tax cuts: How they move the tax burden off the rich and onto everyone else*. United for a Fair Economy. Retrieved from http://www.faireconomy.org

Couldry, N. (2001). Dialogue in an age of enclosure: Exploring the values of cultural studies. *The Review of Education/Pedagogy/Cultural Studies, 23*, 166.

Couldry, N. (2004). In the place of a common culture, what? *The Review of Education/Pedagogy/Cultural Studies, 26*, 1–19.

Derber, C. (2002). *People before profit*. New York, NY: Picador.

Drutman, L., & Cray, C. (2005). The people's business: Controlling corporations and restoring democracy. *In These Times, 14*, 16–28.

Freire, P. (1998). *Pedagogy of freedom*. Lanham, MD: Rowman & Littlefield.

George, S. (1999). A short history of neo-liberalism: Twenty years of elite economics and emerging opportunities for structural change. Retrieved from http://www.globalexchange.org/campaigns/econ101/neoliberalism.html

Giroux, H., & Giroux, S. (2004). *Take back higher education: Race, youth, and the crisis of democracy in the post-civil rights era*. New York, NY: Palgrave Macmillan.

Giroux, H. A. (2003). *Public spaces/private lives: Democracy beyond 9/11*. Lanham, MD: Rowman & Littlefield.

Giroux, H. A. (2004). *The terror of neoliberalism*. Boulder, CO: Paradigm.

Gonsalves, S. (2004). How to skin a rabbit. *The Cape Cod Times*. Retrieved from http://www.commondreams.org/views04/0420-05.htm

Greider, W. (2003). The right's ambition: Rolling back the 20th century. *The Nation, 12*, 11–19.

Grossberg, L. (1997). *Bringing it all back home: Essays on cultural studies*. Durham, NC: Duke University Press.

Guinier, L. (2003). Democracy tested. *The Nation, 5*, 6–7.

Gutmann, A. (1999). *Democratic education*. Princeton, NJ: Princeton University Press.

Hall, S. (1992). Race, culture, and communications: Looking backward and forward at cultural studies, *Rethinking Marxism, 5*, 10–18.

Hart, J. (1996). How to get a college education. *The New Republic*, September, 34–40.

Harvey, D. (2003). *The new imperialism*. New York, NY: Oxford University Press.

Hertz, N. (2003). *The silent takeover: Global capitalism and the death of democracy*. New York, NY: Harper Business.

Hoggart, R. (1957). *The uses of literacy*. London, UK: Chatto and Windus.

hooks, b. (1994). *Teaching to transgress*. New York, NY: Routledge.

Jameson, F. (1994). *The seeds of time*. New York, NY: Columbia University Press.

Krugman, P. (2003). *The great unraveling: Losing our way in the new century*. New York, NY: W.W. Norton.

Moyers, B. (2004). This is your story—The progressive story of America. Pass it on. Retrieved from http://www.commondreams.org/views03/0610-11.htm

Newfield, C. (2002). Democratic passions: Reconstructing individual agency. In R. Castronovo & D. Nelson (Eds.), *Materializing democracy* (pp. 314–338). Durham, NC: Duke University Press.

Phillips, K. (2003). *Wealth and democracy: A political history of the American rich*. New York, NY: Broadway.

Robin, C. (2004). *Fear: The history of a political idea*. New York, NY: Oxford University Press.

Said, E. (2004). *Humanism and democratic criticism.* New York, NY: Columbia University Press.
Sherry, M. S. (1995). *In the shadow of war: The United States since the 1930s.* New Haven, CT: Yale University Press.
Sontag, S. (2003). Courage and resistance. *The Nation, 5,* 11–14.
Street, Paul. 2004. *Empire and inequality: America and the world since 9/11.* Boulder, CO: Paradigm.
West, C. (2004). Finding hope in dark times. *Tikkun, 19*(4), 18–20.
Williams, R. (1958). *Culture and society.* London, UK: Chatto and Windus.
Williams, R. (1967). *Communications.* New York, NY: Barnes & Noble.
Williams, R. (1989). *The politics of modernism.* London, UK: Verso.
Willis, P. (1981). *Learning to labor.* New York, NY: Columbia University Press.
Woodard, C. (2004, April 15). *Who really pays taxes in America? Taxes and politics in 2004.* AskQuestions.org. Retrieved from http://www.askquestions.org/articles/taxes/

CHAPTER TWELVE

Disparity, Austerity AND Public Schooling IN THE United States

Why Quentin Can't Read

SUE BOOKS

Let me start not with Quentin (for whom the chapter is titled), but with a man I'll call Bernie—the proverbial drunk who looks for his lost car keys under the nearest lamppost. Bernie is searching under the post not because that's where he lost his keys, but rather because that's where the light is best. Like Bernie, policymakers have continued to search for "what works" in education in the wrong places, but unlike Bernie, they are declaring that, lo and behold, they've "found" their keys just where they were looking! The widely proclaimed problem is a broken system of public schooling, and the purported "keys" to a solution are competition and privatization.

Like Bernie's scrambling, the search for a solution to the alleged problems of public education seems misguided, if not disingenuous, as decades of research have linked a host of school-related challenges with poverty (e.g., Berliner, 2014; Coleman, 1966; Rothstein, 2004). Poverty looms behind low academic achievement (Lacour & Tissington, 2011), a rising income-achievement gap (Reardon, 2011), student disengagement rooted in (often astute) perceptions of educational inequality and unfairness (e.g., Kozol, 2005), disproportionate learning disabilities (Cortiella, 2011), severe mental illness (Berliner, 2014) and severe behavioral and emotional problems (Turner & Kaye, 2006). Low-income children and youth also drop out of (Hernandez, 2011) or are "pushed out" of high school in disproportionately high numbers (Advancement Project, 2010), enter college in disproportionately low numbers (Bailey & Dynarski, 2011) and too often become entrapped in the shameful "school to prison" pipeline (Kim, Losen, & Hewett, 2010).

Concentrated poverty, generally defined as census tracts with poverty rates of 30% or more, is linked with school difficulties even more closely than is individual family poverty (Rothstein, 2004). Families in high-poverty neighborhoods are far more likely than others to struggle to meet children's basic needs for food, housing and healthcare and to protect them from chronic distress, often prompted by fear and pervasive violence. More than 1 in every 10 children in the United States (11%) now lives in an area of concentrated poverty (Kids Count, 2012). Between 2007 and 2009, the number of high-poverty school districts increased sharply as well (Baker, Sciarra, & Farrie, 2012). These high-poverty schools are almost always starkly segregated. About 40% of all Hispanic and black students attend schools where 90% to 100% are minority and where 70% to 100% are living in poverty (Orfield, 2009). Indeed, "A map of schools attended by the average black or Hispanic student would almost perfectly match a map of high-poverty schools" (Gary Orfield, quoted in Schemo, 2001).

This chapter documents some of the causes and consequences of educational unfairness in the United States and considers how what Alex Molnar has called the "financialization of education" (Saul, 2011) is affecting children and youth living in poverty and usually attending impoverished schools as well. As a primary cause of persistent disparities in resources, our property-tax-based approach to school funding is a good place to start.

SCHOOL FUNDING IN THE UNITED STATES

"We wouldn't play Little League this way," a parent told Jonathan Kozol (2005), reflecting on school funding in Ohio. "We'd be embarrassed. We would feel ashamed" (p. 55). Although funding schemes vary from state to state, public schools generally rely on three primary sources: state revenues, local property taxes and federal funds. State revenues constitute, on average, 48.3% of the total pie, and property taxes, on average, 43.5%. The federal contribution now amounts to only 8.2% (Baker et al., 2012). Significant disparities exist at every level—among the states, among the 15,000-plus school districts spread across the nation and among individual schools within single districts.[1]

The constitutionality of this system, which virtually ensures disparities, was challenged in 1973 in a U.S. Supreme Court case, *San Antonio v. Rodriguez*. This class-action lawsuit was prompted by significant inequities in funding in neighboring school districts in San Antonio, Texas, and questioned whether such large gaps in funding violate the U.S. Constitution. By a 5-4 vote, the justices reversed a Texas court and said no. As long as some semblance of public schooling is provided—that is, as long as there is no "absolute [educational] deprivation"—even egregious disparities are constitutional. The

Rodriguez court concluded that education is not a fundamental right protected by the U.S. Constitution and students attending poorly funded schools do not constitute a "protected class." Consequently, the Court required a lower level of justification for the acknowledged disparities in funding and found that justification in the idea of local control: the notion that he or she who pays the piper calls the tune.

How do disparities in school funding arise? The Equity and Excellence Commission (2013) offered a particularly clear explanation:

> Imagine two towns: Town A has $100,000 in taxable property per pupil; Town B has $300,000. If Town A votes to tax its property at 4%, it raises $4,000 per pupil. But Town B can tax itself at 2% and raise $6,000 per pupil. Town B's tax rate is half as high as Town A's, but its public schools enjoy 50 percent more resources per student. (p. 17)

While these numbers are simplified, the scenario is real. Many poorer districts tax themselves at higher *rates* than wealthier districts (U.S. General Accounting Office, 1997), which suggests that high-poverty communities end up with struggling public schools not because they "don't care about education," as is popularly believed. Rather, they end up with less, at least in part, because low property values do not generate adequate revenue, even with a significant tax effort.

In a national study designed to bring more conceptual clarity to the question of fairness in school funding, Baker et al. (2012) asked, "Is school funding fair?" and concluded, essentially, "no." Acknowledging the well-documented correlation between student and school poverty and the costs of providing equal educational opportunity, the researchers used as a key measure of fairness whether states are providing "a sufficient level of funding distributed to districts within the state to account for additional needs generated by student poverty" (p. 5). States differ markedly on this measure in part because they differ with respect to the total funding provided for public schools as well as the weight given to poverty in the distribution of revenues. At the extremes, a student in Tennessee receives only about 37% of the funding that a student in Wyoming receives. Only 17 states provide more funding to high-poverty districts (with poverty rates of 30% or more). Fifteen states have "flat" systems, and 16 favor low- or no-poverty districts. In terms of sensitivity to the educational significance of poverty, the spread ranges from Utah, where high-poverty school districts receive 159% of the funding of districts with no child poverty, to Nevada, where high-poverty districts receive only 76% of the funding of no-poverty districts. Looking at school funding from a different angle, Heuer and Stullich (2011) found that in 2008–2009, almost half (46%) of all high-poverty (Title I) elementary schools spent less per pupil (in state and local revenues) than low-poverty (non-Title I) schools in the same district.

Private donations to schools create a second layer of disparity. Gifts from parents, alumni and corporations fund a wide range of needs and wants for recipient schools, including equipment, supplies and artists-in-residence. Other schools have no significant donors. Concerned about this fundamental unfairness, school administrators in Greenwich, Connecticut, capped the amount a single school could receive but often then waived caps when large donations were offered. An analysis by the school board a few years later showed continuing disparities: $17,000 for one elementary school compared to more than $50,000 each for eight others. To try to share the beneficence, 18 school districts in New York State joined together in the late 1990s to create a nonprofit foundation. However, the foundation faltered when "some of the well-heeled participants expressed interest in creating foundations that catered exclusively to their schools" (Cowan, 2007).

Tying school funding to district wealth—a practice, among industrialized nations, that only the United States follows—ensures that families who can afford million-dollar homes also can provide their children with a top-dollar education. The corollary holds true as well: Families that cannot afford anything close to a million-dollar home, or maybe any home at all, must watch their children cope as best they can with whatever they are given. Cogs in a wheel of social immobility, far too many of these young people grow up hearing in one way or another, "More money (for you) is not the answer."

WHY SCHOOL FUNDING MATTERS

The consequences of inequitable school funding are well known. "Point to a group of toddlers in an upper-middle-class neighborhood in America, and it's a good bet that they will go to college, buy nice houses and enjoy white-collar careers," Kristoff (2013) wrote in *The New York Times* (p. A27). "Point to a group of toddlers in a low-income neighborhood, and—especially if they're boys—they're much more likely to end up dropping out of school, struggling in dead-end jobs and having trouble with the law" (p. A27). Good teachers, small classes, useful libraries, up-to-date technology, extracurricular opportunities, school nurses and guidance counselors and other in- and out-of-school supports all carry a price tag. If some districts can afford much more of these resources than others, some students clearly will enjoy much greater educational opportunity than others. A survey of 33 schools in New York with high percentages of low-income, struggling or disabled students found that all but two of the schools could not afford enough teachers to meet standards in core subjects or to offer extra help for struggling students and that deep budget cuts had forced some school principals to take on the duties of secretaries, social workers or custodians (Rebell, Wolff, & Rogers, 2012). Beyond a bare-bones curriculum, many districts cannot offer the courses that college-bound students need.

> In a recent, glaring case [in New York State], the valedictorian of a rural school district outside Rochester was rejected by a nearby State University of New York campus—not because her grades were too low, but because her high school didn't offer the courses needed to compete for college admission. (Easton, 2012, p. A21)

In a market economy, teachers, like almost everyone else, respond to financial incentives. Experienced educators tend to seek positions in more affluent schools where they can earn more, which leaves students in poor neighborhoods with "low-paid rookie teachers who move on as they gain experience and rise up the salary scale" (Dillon, 2011, p. A29). Across the nation, high-poverty schools can neither compete for the best teachers and principals nor offer the rigorous academic and enrichment programs other schools provide (Equity and Excellence Commission, 2013). Worse, many students and teachers in high-poverty schools spend their days in ill-equipped classrooms in dilapidated and sometimes dangerous buildings with leaky roofs, faulty heating systems, broken toilets, moldy classrooms, cracked windows and missing equipment (Kozol, 2005).

In San Francisco, California, in a case focused on particularly horrendous conditions, *Williams v. State* (2000), the American Civil Liberties Union and other civil rights groups filed a class-action lawsuit on behalf of students attending awful schools. The complaint cited a lack of qualified teachers, libraries and instructional materials; overcrowding resulting in staggered classes and a shortened school year; and school conditions so "appalling" they "shock the conscience," including "filthy toilets and leaky roofs, sweltering classrooms (sometimes said to exceed 100 degrees), falling ceiling tiles, and schoolhouse vermin" (Schrag, 2003, p. 98). Less sensational, but more tragic, is the predictable record of academic failure. Since the early 1990s, California students have ranked near the bottom among all states in math and reading achievement, including in cross-state comparisons of the same ethnic groups (Schrag, 2003). The *Williams* case was settled out of court. However, eight years later, the state had yet to allocate even half the $800 million promised for school repairs, which left more than 700 schools waiting to fix broken toilets, battered walls and clogged sewer lines (National Education Access Network, 2013).

Money matters in educational reform as well. "Closing achievement gaps, increasing college and career readiness, and improving teacher quality are all laudable goals, but are goals that cannot be achieved and sustained without the fundamental base of a fair school funding" (Baker et al., 2012, p. 28). The achievement gap between students in rich and poor families has widened significantly over the last 25 years and is now more than twice as large as the Black-White achievement gap. In a review of 19 national studies, Reardon (2011) found that the spread in achievement between students at the 90th and 10th percentiles of family income is 40% to 50% *larger* for the cohort born in 2001 than for the cohort born 25 years earlier. In 2010, among 491 schools in Illinois, where 90% of the students are poor and 90% are minorities, only one, a magnet school with 200 students, could show that 90% of the students had met basic state standards (WBEZ, a Chicago radio station, cited in Berliner, 2014).

International comparisons reaffirm the link between income and achievement. An analysis of the 2009 scores on the Programme for International Student Assessment (PISA) shows that if U.S. students in schools where 10% or fewer of the students are eligible for free or reduced-price lunch were a nation unto themselves, their reading scores would have been *the highest in the world*, and their math and science scores well above average. However, if students in schools where 75% or more of the students qualify for free or reduced-price lunch were a nation, their scores would have been next to the bottom, above only Mexico, among the Organization for Economic Cooperation and Development (OECD) countries (Berliner, 2014). U.S. students from high-income families scored on average 100 points higher on the most recent PISA reading test than students from low-income families—"about the same disparity that exists between the average scores in the US and Tunisia" (Porter, 2013, p. B1).

As Kristoff (2013) argued, "Something is profoundly wrong when we can point to 2-year-olds in this country and make a plausible bet about their long-term outcomes—not based on their brains and capabilities, but on their ZIP codes" (p. A27). Sadly, we can make such predictions. Students in the top income quartile comprised two-thirds of the entering freshman class of 2010 at the 193 most selective colleges and universities in the United States while students in the bottom half of the family-income distribution made up only 15% (Leonhardt, 2011). Among high-achieving high school seniors in the bottom income quartile, only one-third in the graduating class of 2008 went on to enroll in one of the country's most selective colleges and universities, whereas among their similarly high-achieving peers in the top income quartile, more than three-quarters enrolled in these schools (Hoxby & Avery, 2012).

In a course I teach on comparative and international education, I often show students a documentary film focused on a school in a township outside Cape Town, South Africa, entitled *Testing Hope: Grade 12 in the New South Africa* (O'Brien Gardner & Blank, 2007). The film captures what my students always regard as a tragedy: The school's top student is rejected by the University of Cape Town not because his grades are too low, but rather because his school did not offer the upper-level courses the university admissions committee wanted to see. From this perspective, in the 60 years (as of 2014) since the U.S. Supreme Court declared in *Brown v. Board of Education* that education "must be made available to all on equal terms," the United States has progressed no further than post-apartheid South Africa.

BALANCING BUDGETS ON SCHOOLCHILDREN'S BACKS

Although the U.S. Constitution does not affirm education as a right, almost every state constitution includes language that assigns the state responsibility for public schooling. For decades, advocates have turned to the courts in an effort to hold states accountable for the funding inequities and inadequacies that have crippled so many schools, with devastating consequences for students. This advocacy has been an uphill battle, in part because the idea of local control has held such legal sway, but also because states have been so quick to look to the public schools to balance budgets or find resources for other priorities. In at least 30 states, state education funding was less in 2011 than in 2008 (Baker et al., 2012), and in at least 23 states, significantly less (Rebell & Wolff, 2011), with high-poverty schools taking the biggest hit. In Pennsylvania, for example, the poorest 150 school districts sustained cuts of $581 per student, while the wealthiest 150 districts had cuts of only $214 per student (Poor Schools Hit Hardest, 2011). In New York State, budget cuts for schools in poor and middle-class communities have been double or triple those for schools in wealthy communities. For example,

> Poughkeepsie, with a student poverty rate of 80%, has cut its full-day kindergarten to a half day, while wealthy Jericho offers high school classes in fashion design and civil engineering. Scarsdale offers 22 Advanced Placement courses, while poor and rural Massena, in New York's North Country, offers only two, even though many colleges now give AP courses greater weight than SAT scores in admissions. (Easton, 2012, p. A21)

Budget cuts have cost 30,000 New York State educators their jobs and have required class-size increases at two-thirds of the state's schools.

One "solution" to the shortfalls resulting from budget cuts has been to charge fees—for workbooks and other learning materials or to use lab equipment or participate in extracurricular activities. In Medina, Ohio, it now costs $660 to play a high school sport, $200 to join the concert choir and $50 to act in the school's spring play. The Palmyra schools in Pennsylvania are charging $20 lab fees for many science, art and music courses (Simon, 2011). Other stopgap measures have included lopping days off the school year (as some California districts have done), cancelling afterschool programs (which has affected 4,500 students in Miami) and cutting prekindergarten programs (as Texas has done for 100,000 toddlers; Rebell & Wolff, 2011).

Manchester, New Hampshire's largest school district with some of the state's lowest per-pupil spending, fired 95 full-time teachers in 2012 and increased class sizes significantly. "We had students sitting on the floor with a clipboard," said Jim O'Connell, president of a parent-teacher organization in the district. "It's one degree separated from a 1700s classrooms with chalk and a slate" (quoted in Bidgood, 2012, p. A12). The district's next step likely will be to invest about

$80,000 in "blending learning labs," where students will take courses online during the school day, with only a "facilitator" (not a teacher) present. "It's a high-tech babysitter," an exasperated parent told *The New York Times*. "The intent is to maximize the learning and to minimize the financial impact on Manchester School District," countered the superintendent. "I don't want people to think we're trying to divert resources so that we don't have to hire teachers" (quoted in Bidgood, 2012, p. A12).

Closing schools—as has happened on a large scale in Chicago, Detroit, Kansas City, Newark, Philadelphia, and Washington, DC—also makes it unnecessary to pay teachers. After a $419 million cut in state education funding in 2012, the Philadelphia School District (which has had among the lowest per-pupil funding of any major city) faced a budget deficit of $1.35 billion over the next five years. In response, a state commission overseeing the schools voted in March 2013 to close 23 schools, about 10% of the city's total (Hurdle, 2013). In Chicago, where 100 schools had already been closed since 2001, district officials announced in May 2013 that 49 more (about 8% of the city's total) would be shut down. As a result, hundreds of thousands of students, mostly Black and Hispanic in low-income families, will be educationally uprooted, for good or ill (Yaccino, 2013, p. A20).

Faced with budget cuts, often exacerbated by declining enrollments, some school districts, especially in rural areas, have been recruiting tuition-paying students from abroad. According to one estimate, more than 1,000 public schools have taken steps to recruit international students (Simon, 2012). The Newcomb (NY) Central School District in the Adirondack Mountains almost doubled its enrollment between 2006 and 2013 by bringing in international students. Newcomb Superintendent Clark Hults is now working with more than a dozen school districts in New York and Vermont that want to replicate his program while lobbying against visa restrictions, which at the moment limit public school visas to one year (Casey, 2013).

Perhaps the most alarming revenue-generating scheme, given its long-term consequences, is the use of capital appreciation bonds to pay for classroom renovations, athletic facilities and other building improvements. Fearful that voters will balk at any tax increase, school districts in recent years have borrowed millions through these bonds, which defer payment sometimes for decades but then balloon.

> In 2009, the Santa Ana Unified School District borrowed $35 million.... Not a cent is owed until 2026. But taxpayers will eventually have to pay $340 million to retire that $35 million debt.... In San Diego, property owners owe $630 million on a $164 million bond. For the Folsom Cordova Unified School District, a $514,000 bond will cost $9.1 million. And in the most expensive case yet, the Poway Unified School District borrowed $105 million to finish modernizing older school buildings, which local property owners will be paying off until four decades from now at an eventual cost of nearly $1 billion. (Lovett, 2013, p. A25)

While today's taxpayers may benefit from this "creative" financing—what the California state treasurer calls "the school district's version of printing money"— the real winners are the financial advisors who have netted millions of dollars in these deals (Lovett, 2013, p. A25).

DISPARITY AND AUSTERITY IN CONTEXT

Persistent disparities in school funding and radical responses to budget shortfalls occur in a broader ideological and policy context. The discussion below highlights three dimensions of this context: anti-tax fervor, urban reform initiatives and the growth of online charter schools.

Tax Caps and Credits

The flip side of the coin of strapped school districts resorting to drastic austerity measures is tax policy: a proliferation of caps and credits that siphon funds from public schools directly or indirectly.

Caps on state property taxes, tax breaks for corporate "job creators" or for individual "scholarship" donors and other initiatives marketed as school reforms all drain resources from public schools and exacerbate inequalities. New York and California provide instructive cases of how property tax caps affect school quality. Citing budgetary exigencies, New York cut education resources each year from 2009 to 2013, despite a promise two years earlier to devote $5.5 billion more to classrooms, with 72% going to the highest-need schools (Easton, 2012). In 2011, over the objections of educators, school district officials and unions, the state passed a 2% property tax cap. Because the allowed increase raises much less revenue for schools in low-wealth communities than in high-wealth communities, it hits poor communities the hardest. Arguably, the vote for a tax cap was not well informed. The cap is "popular because somebody comes up to you and says, 'Hey, we're going to put a cap on your property taxes,'" John Whiteley of the New York State Property Tax Reform Coalition told *The New York Times*. "Who's going to say no?" (quoted in Hu, 2011, p. A19).

California's experience should have informed conversations in New York. After the California Supreme Court found in *Serrano v. Priest I* (1971) and *II* (1976) that California's school funding violated both the state constitution and, in the first decision, the federal constitution, the state legislature discouraged property-tax increases in high-wealth districts and sought gradually to raise state aid to poor districts. However, before this could happen, voters passed Proposition 13, which rolled back local tax assessments, allowing them to increase only when property was sold and capped the tax rate at 1% of that reduced value. The fallout was predictable:

> Overnight...the state became the major source of school funding....That made it fairly easy to equalize California's school spending...but instead of leveling it up, it was effectively leveled down, and dramatically so, sinking from the top ten among the states in the 1960s to the bottom ten 30 years later....The results were the largest class sizes in the country; deep cuts in counseling staffs, nurses, and music and arts programs; and in many districts, including the suburbs, rotting buildings and overcrowded classrooms graced by little except emergency maintenance. (Schrag, 2003, p. 78)

At least a dotted line can be drawn from the tax-cap maneuver in California to the conditions that the *Williams* lawsuit documented 22 years later and to the state's current dismal student achievement levels. *Education Week*'s 2013 State Report Card gives California a D on K–12 achievement.

A host of state and federal tax breaks go to corporate "job creators," many of whom subsequently create few good jobs or sometimes none at all. These practices—arguably, forms of corporate extortion—further strap states and municipalities and therefore public schools. A 2011 study of state economic development subsidy programs (Mattera, Cafcas, McIlvanine, Seifter, & Tarczynsky, 2011) found that these deals cost states and cities $11 billion a year in lost revenues, in many cases with no promise of long-term, well-paying jobs. "That money could go a long way to improving education, transportation and other public services that would have a far better shot at promoting real economic growth," *The New York Times* editorialized (Race to the Bottom, 2012, p. A34). The researchers wondered "whether states [were] getting something for their money" (Mattera et al., 2011, p. ii). Pennsylvania seemingly was not. The state found that many businesses in its Keystone Opportunity Zone program were neither creating jobs nor generating capital investment. New Jersey found that $2.17 billion invested in its Urban Enterprise Zone program over six years had little economic impact. New York changed its Empire Zone program, intended to create jobs in high-poverty areas, after auditors found that some businesses that received tax breaks subsequently cut jobs or hired only in low-poverty areas (Cooper, 2011).

A little-known federal tax break, the New Markets Tax Credit, kicks in for wealthy investors and major banks that finance new charter school construction. Through the program, a bank or private equity firm that lends money to a nonprofit to build a charter school can receive a 39% federal tax credit over seven years, can piggyback this credit with other tax breaks (for example, for historic preservation or job creation), and by combining all the credits with the interest from the loan itself, almost double the investment in seven years (Gonzales, 2010).

Finally, consider the various voucher and scholarship programs through which states increasingly are redirecting public school funds to families and, through families, to private schools. Sixteen states now offer 32 programs that allow parents to use taxpayer money to pay private school tuition (American Federation for Children, 2013). Voucher programs provide parents with scholarships to enroll their children in eligible private schools. Other programs provide tax credits for donations for private school scholarships. Eight states now offer these tax credits, which in 2012 funneled to private schools almost $350 million that would have gone to public schools.

Georgians who donate to nonprofit scholarship programs receive dollar-for-dollar tax credits, up to $2,500 per couple. Although the $50-million-a-year program has been marketed as a way to help low-income children "escape" struggling public schools, there are no income caps for scholarship recipients. Furthermore, most of the donations seemingly are subsidizing students who were already attending private schools. From 2007 (the year before the Georgia tax credits started) through 2009, private-school enrollment increased by only 0.3% in the metropolitan counties that included most of the private schools in the program (Saul, 2012). Once the credits became available, "parents of children in private schools began flooding public school offices to officially 'enroll' their children... even though they had no intention of ever sending their children to public schools" (Saul, 2012, p. A1). Although school voucher and tax-credit programs often are touted as public school alternatives that give parents more "choice," there is no solid evidence that students who attend private schools subsidized by these programs are receiving a better education (Santos & Rich, 2013). Many private schools neither administer the annual tests public school students take nor face the increasingly stringent accountability measures to which public schools are subjected.

"School Reform" Initiatives

Public schools in high-poverty urban and rural neighborhoods have borne the brunt not only of chronic disparities in funding, state budget cuts and regressive tax policy, but also of the "financialization of education" (Molnar, quoted in Saul, 2011, p. A1). I use this term to describe a collection of neoliberal policies and initiatives that drain much-needed resources from public schools, especially in high-poverty neighborhoods. Such developments include a classic blame-the-victim practice of first creating "failure" among public schools, then using a stamp of "failed" or "broken" to justify closing the schools and opening alternatives that benefit middle-class students (Lipman, 2011). Steady disinvestment in public schools often precedes a declaration of academic failure, facilitated by the predictably unreachable No Child Left Behind (NCLB) targets. This label provides a public rationale for closing schools, especially in gentrified and gentrifying urban neighborhoods, then opening "better" schools, generally with admission hurdles.

Consider Chicago's Midsouth area, where 20 of the area's 22 schools were slated for closure as part of the Renaissance 2010 (Ren2010) initiative. At school board meetings in the area, Lipman (2011) heard persistent pleas for help: schools without libraries, staff cuts even as displaced students were enrolled from other closed schools and major infrastructure problems. For the most part, these pleas fell on deaf ears and the schools were closed as planned. New schools were opened, but not neighborhood public schools open to all. In 2008 and 2009, during the most contentious period of school closings, the "winners" and "losers" were blatantly clear:

> One school in a gentrifying area with a student population that was mainly Latino/a and low-income was replaced by a highly selective, prestigious public magnet school branded to appeal to affluent Whites. Another school with a similar population was phased out to make way for a new K-12 Renaissance 2010 "international"-themed school lobbied for by parents in the nearby Gold Coast, one of the wealthiest areas of Chicago. The new public school required a $2,500 parent "donation" for each incoming kindergarten child, while CPS eliminated the school's old boundaries, excluding the children from the neighborhood. (Lipman, 2011, pp. 68–69)

Along with "failure," low enrollment is a popular rationale for closing regular public schools, as was the case in Philadelphia, where the percentage of students attending charter schools almost doubled in the years before the school closings, rising from 12% to 23% between 2004–2005 and 2011–2012 (Hurdle, 2013). When public school students are enticed into alternative schools, charter schools or schools outside their district, the often underfunded schools "left behind" lose enrollment. In almost half of the nation's largest school districts, enrollment has declined steadily over the last five years, "triggering school closings that have destabilized neighborhoods, caused layoffs of essential staff and concerns in many cities that the students who remain are some of the neediest and most difficult to educate" (Rich, 2012, p. A1). All too often, those left behind are children with disabilities, children living in poverty, English-language learners and children of color.

Because school funding is generally allocated per pupil, enrollment declines set up a predictable domino effect of hardship: Teachers are laid off, and classes (typically foreign language, music and art) are cancelled. These distressed schools then struggle to compete with suburban and charter schools with lower teacher-student ratios and broader educational programs. Some of the largest school systems are now poised to become "the schools that nobody wants," said education historian Jeffrey Mirel, and the creation of schools no one wants portends even greater stratification and segregation, warned AFT President Randi Weingarten (quoted in Rich, 2012, p. A1). Consider, for example, Walter Payton College Preparatory High School, located within walking distance of the Cabrini-Green housing project in Chicago. The school was built in 2000 and enrollment was made selective. As of 2008, White students (37% of the student body) were the largest demographic group, and only 28.4% of the students were from low-income families. By contrast, at an elementary school nearby, 98.3% of the students are Black and 97.2% low income (Paperson, 2010).

Chicago's urban renewal project has neither desegregated highly segregated schools, improved students' educational opportunities, nor ameliorated funding disparities. Most of the lowest-income students who were displaced by school closings were reassigned to high-minority, low-achieving schools very similar to those they left. Eighty-four percent of these students ended up in schools with below-average district test scores and 44% in schools on probation for low test scores (Catalyst Chicago, 2007). Of the 5,445 elementary students displaced between 2001 and 2006, only 6% moved to academically strong schools (in the top quartile based on test scores; De la Torre & Gwynne, 2009).

Some students swept up in Chicago's urban renewal project were transferred to four schools in three years, as one school after another was closed. Looking at this socially constructed chaos, one wonders: What—besides the creation and protection of privilege—could justify such social, emotional and educationally unsound disruption? Schools, at the very least, provide breakfast, lunch, heat in winter, adult supervision and some stability for children. Until other systems are established to meet these critical needs, there is no justification for yanking away this support (Noguera, 2013).

For-Profit Online Charter Schools

Arguably even more educationally unsound than the push-some-out, pull-others-in strategy of school "reform" is the proliferation of highly profitable online charter schools that offer students no more than can be provided by underpaid, overworked, nonunionized teachers operating from their homes without ever interacting face-to-face with their students. These schools are now "educating" an estimated 200,000 students and are growing in number (Molnar, 2013). A *New York Times* expose of K12 Inc. and other purveyors of online charter schools found that student test scores in these taxpayer-subsidized schools are low, online class sizes are high and teachers are poorly paid. For *The New York Times*, at least,

> A look at the company's operations…raises serious questions about whether K12 schools—and full-time online schools in general—benefit children or taxpayers, particularly as state education budgets are being slashed. Instead, a portrait emerges of a company that tries to squeeze profits from public school dollars by raising enrollment, increasing teacher workload and lowering standards. (Saul, 2011, pp. 1)

In terms of student achievement, online charters have failed miserably. Only 27% of privately managed online schools met NCLB achievement targets in 2010–2011 (Miron & Urschel, 2012), and on-time graduation rates for full-time virtual schools (less than 38%) are less than half the national average. On both measures, "full-time virtual schools lag significantly behind traditional brick-and-mortar schools" (Molnar, 2013, p. ii).

Nevertheless, K12 Inc., the largest operator of for-profit virtual schools, took in $708 million in 2012 while employing 39 lobbyists working around the country to create a "policy friendly" environment for virtual learning (Davis, 2013). Thanks in part to the lobbying, there has been a "significant upsurge in legislative activity" to expand online schooling, regulate virtual education, and modify existing regulations (Molnar, 2013, p. 3). Because high-poverty districts in some states receive larger state subsidies, they are particularly attractive to companies like K12, which foresees a $15 million market for its schools. It's hard to argue with Molnar's observation: "These folks are fundamentally trying to do to public education what the banks did with home mortgages" (quoted in Saul, 2011, p. 1). But trying, they are. The 10 largest for-profit online school firms spent $94.4 million on advertising between 2007 and 2012, with K12 Inc. investing $21.5 million in the first eight months of 2012 alone (Toppo, 2012).

CONCLUSION

Let me conclude with what I fear is a cautionary tale of our times. A civil rights complaint brought in 2012 by the American Civil Liberties Union (ACLU) of Michigan on behalf of students in a high-poverty school district attempts to hold the state accountable for ensuring that all students receive the instruction they need to learn to read at least well enough to pass state proficiency exams. The case focuses on the Highland Park School District, which borders Detroit. Most of the students in the district (88%) qualify for free or reduced-price lunch, and state tests show two-thirds are reading below grade level and in many cases several grades below.

One student in the district, 14-year-old Quentin, just finished seventh grade, but he reads at a first-grade level, according to an expert hired by the ACLU:

> When asked to compose a letter to [the governor] to describe his school, Quentin misspelled his own name, writing, "My name is Quemtin ... and you can make the school gooder by getting people that will do the jod that is pay for get a football tame for the kinds mybe a baksball tamoe get a other jamtacher for the school get a lot of tacher." (*S.S. v. State of Michigan*, 2012).

During seventh grade, Quentin was enrolled in a regular language arts class and in Read 180, an online program designed to help struggling readers. Quentin could decide each day whether to attend his regular class or to participate in Read 180. In the Read 180 classroom, "the teacher did not provide any instruction while the students read books on their own, or in groups, or completed self-directed work on the computer.... The longest writing assignment Quentin had to complete...was a three paragraph summary of a book," according to the complaint (*S.S. v. State of Michigan*, 2012). "Kids are getting plopped in front of computers with no teacher in the classroom or the teacher is just sitting there, not engaged," said Kary Moss, executive director of the ACLU of Michigan. "There's no individualized assessment of what [the students] need, how they're doing, or monitoring of what's going on" (quoted in Layton, 2012). Court papers note that the state has taken over the Highland Park schools, which it plans to turn over to a charter operator—who can apply for a charter without providing any assurance that its schools will be able to address the severe literacy deficits in Highland Park.

In this sad, sad case we see much of what is wrong on a broader scale: poverty, inadequate funding, state irresponsibility and private capitalizing on the entirely predictable consequences of all this: seventh graders like Quentin who can barely read and write. The societal response finally must be *to own up to and address the heart of the problem.* The United States is unique among developed nations in sustaining such a high level of child poverty and educational inequality, as "no other developed nation has inequities nearly as deep or systemic" nor odds "so thoroughly stacked...against so many of its children" (Equity and Excellence Commission, 2013, p. 15). We must drastically reduce, if not eliminate, child and family poverty; ensure that high-poverty schools have the funding they need to educate students well (or, better yet, eliminate the economic segregation that creates high-poverty schools); and outlaw "reforms" that enrich others at the disproportionate expense of low-income students and students of color. The achievement gap must be seen for what it is: an educational manifestation of social inequality (Noguera, 2013). We have been looking for lost keys in the wrong places not because the light is better there, but worse, because profits are higher and privileges more protected there, just not for the students pushed to the bottom tier of a multi-tiered educational system.

NOTE

1. This section and the next revise and update similar discussions in Books (2013).

REFERENCES

Advancement Project. (2010). *Test, punish, and push out: How "zero tolerance" and high-stakes testing funnel youth into the school-to-prison pipeline.* Retrieved from http://b.3cdn.net/advancement/d05cb2181a4545db07_r2im6caqe.pdf

American Federation for Children. (2013). *Facts.* Retrieved from http://www.federationforchildren.org/facts

Bailey, M. J., & Dynarski, S. M. (2011). Inequality in postsecondary education. In G. J. Duncan & R. J. Murnane (Eds.), *Whither opportunity? Rising inequality, schools, and children's life chances* (pp. 117–132). New York, NY: Russell Sage.

Baker, B., Sciarra, D., & Farrie, D. (2012). *Is school funding fair? A National Report Card.* Retrieved from http://www.schoolfundingfairness.org/National_Report_Card_2012.pdf

Berliner, D. (2014). Effects of inequality and poverty v. teachers and schooling on America's youth. *Teachers College Record, 116*(1). Retrieved from http://www.tcrecord.org/Content.asp?ContentID=16889

Bidgood, J. (2012, December 8). Strapped district plans to add online classes. *The New York Times,* p. A12.

Books, S. (2013). Funding the right to equal educational opportunity: An overview and call to follow the money. In J. Hall (Ed.), *Children's human rights and public schooling in the United States* (pp. 193–211). Boston, MA: Sense.

Brown v. Board of Education. (1954). 347 U.S. 483.

Casey, D. C. (2013). Rural leader buoys school with foreign students. *Education Week, 32*(20), 12–14.

Catalyst Chicago. (2007). *School autonomy all over the map.* Retrieved from http://www.catalyst-chicago.org/assets/20070201/extra/0207reportcard.pdf

Coleman, J. S. (1966). *Equality of educational opportunity.* Washington, DC: U.S. Department of Health, Education, and Welfare.

Cooper, M. (2011, December 14). With states desperate to keep jobs, companies have upper hand, report shows. *The New York Times,* p. A23.

Cortiella, C. (2011). The state of learning disabilities. *New York: National Center for Learning Disabilities.* Retrieved from http://www.ncld.org/images/stories/OnCapitolHill/PolicyRelatedPublications/stateofld/2011_state_of_ld_final.pdf

Cowan, A. (2007, June 3). Schools' deep-pocketed partners. *The New York Times.* Retrieved from http://www.nytimes.com/2007/06/03/nyregion/nyregionspecial2/03Rschool.html?pagewanted=print

Davis, M. R. (2013, April 22). Ed. companies exert public-policy influence. *Education Week.* Retrieved from http://www.edweek.org/ew/articles/2013/04/24/29ii-politicalpower.h32.html

De la Torre, M., & Gwynne, J. (2009). When schools close: Effects on displaced students in Chicago Public Schools. *Consortium on Chicago School Research at the University of Chicago.* Retrieved from http://ccsr.uchicago.edu/sites/default/files/publications/CCSRSchoolClosings-Final.pdf

Dillon, S. (2011, November 30). Districts pay less in poor schools, report says. *The New York Times,* p. A29.

Easton, B. (2012, May 26). Albany's unkindest cut of all. *The New York Times,* p. A21.

Education Week. (2013, April 1). *Quality counts 2013: Code of conduct.* Retrieved from http://www.edweek.org/ew/qc/2013/state_report_cards.html?intc=EW-QC13-LFTNAV

Equity and Excellence Commission. (2013). *For each and every child: A strategy for education equity and excellence.* Alexandria, VA: U.S. Department of Education.

Gonzalez, J. (2010, May 6). Albany charter cash cow: Big banks making a bundle on new construction as schools bear the cost. *Daily News.* Retrieved from http://www.nydailynews.com/new-york/education/albany-charter-cash-big-banks-making-bundle-new-construction-schools-bear-cost-article-1.448008

Hernandez, D. J. (2011). Double jeopardy: How third-grade reading skills and poverty influence graduation. *Annie E. Casey Foundation*. Retrieved from http://www.aecf.org/~/media/Pubs/Topics/Education/Other/DoubleJeopardyHowThirdGradeReadingSkillsandPovery/DoubleJeopardyReport040511FINAL.pdf

Heuer, R., & Stullich, S. (2011). Comparability of state and local expenditures among schools within districts: A report from the study of school-level expenditures. *U.S. Department of Education Office of Planning, Evaluation and Policy Development*. Retrieved from http://www2.ed.gov/rschstat/eval/title-i/school-level-expenditures/school-level-expenditures.pdf

Hoxby, C. M., & Avery, C. (2012). The missing "one-offs": The hidden supply of high-achieving, low income students (Working Paper 18586). *Cambridge: National Bureau of Economic Research*. Retrieved from http://www.brookings.edu/~/media/projects/bpea/spring%202013/2013a_hoxby.pdf

Hu, W. (2011, January 24). Unlikely allies fight Cuomo's plan for property tax cap. *The New York Times*, p. A19.

Hurdle, J. (2013, March 8). Philadelphia officials vote to close 23 schools. *The New York Times*, p. A16.

Kids Count. (2012). Kids count data book: State trends in child well being. *Annie E. Casey Foundation*. Retrieved from http://datacenter.kidscount.org/DataBook/2012/OnlineBooks/KIDSCOUNT2012DataBookFullReport.pdf

Kim, C. Y., Losen, D. J., & Hewitt, D. T. (2010). *The school-to-prison pipeline: Structuring legal reform*. New York, NY: New York University Press.

Kozol, J. (2005). *The shame of the nation: The restoration of apartheid schooling in America*. New York, NY: Crown.

Kristoff, N. (2013, January 24). For Obama's new term, start here. *The New York Times*, p. A27.

Lacour, M., & Tissington, L. D. (2011). The effects of poverty on academic achievement. *Educational Research and Reviews*, 6(7), 522–527.

Layton, L. (2012, July 12). ACLU alleges Michigan school district violated students, right to learn to read. *The Washington Post*. Retrieved from http://www.washingtonpost.com/local/education/aclu-alleges-michigan-school-district-violated-students-right-to-learn-to-read/2012/07/11/gJQArf1jeW_print.html

Leonhardt, D. (2011, May 24). Top colleges, largely for the elite. *The New York Times*, p. B1.

Lipman, P. (2011). *The new political economy of urban education: Neoliberalism, race, and the right to the city*. New York, NY: Routledge.

Lovett, I. (2013, February 10). California schools finance upgrades by making the next generation pay. *The New York Times*, p. A25.

Mattera, P., Cafcas, T., McIlvanine, L., Seifter, A., & Tarczynsky, K. (2011). *Money for something: Job creation and job quality standards in state economic development subsidy programs*. Washington, DC: Good Jobs First. Retrieved from http://www.goodjobsfirst.org/sites/default/files/docs/pdf/moneyforsomething_noappdx.pdf

Miron, G., & Urschel, J. L. (2012). *Understanding and improving full-time virtual schools: A study of student characteristics, school finance, and school performance in schools operated by K12 Inc*. Boulder, CO: National Education Policy Center. Retrieved from http://nepc.colorado.edu/publication/understanding-improving-virtual

Molnar, A. (Ed.). (2013). *Virtual schools in the US. 2013: Politics, performance, policy, and research evidence*. Boulder, CO: National Education Policy Center. Retrieved from http://nepc.colorado.edu/publication/virtual-schools-annual-2013/

National Education Access Network. (2013). School funding cases in California. Retrieved from http://schoolfunding.info/2011/10/school-funding-cases-in-california/

Noguera, P. (2013, April 30). Education, racial inequality, and the future of American democracy. distinguished lecture. San Francisco, CA: American Educational Research Association.

O'Brien Gardner, E. (Producer), & Blank, M. (Director). (2007). *Testing hope: Grade 12 in the new South Africa* [DVD]. United States: York Zimmerman.

Orfield, G. (2009). Reviving the goal of an integrated society: A 21st century challenge. *Civil Rights Project*. Retrieved from http://civilrightsproject.ucla.edu/research/k-12-education/integration-and-diversity/reviving-the-goal-of-an-integrated-society-a-21st-century-challenge/orfield-reviving-the-goal-mlk-2009.pdf

Paperson, L. (2010). The post-colonial ghetto: Seeing her shape and his hand. *Berkeley Review of Education, 1*(1), 5–34.

Poor Schools Hit Hardest by Budget Cuts in Pennsylvania. (2011, August 7). *Patriot-News/Associated Press*. Retrieved from http://www.pennlive.com/midstate/index.ssf/2011/08/poor_schools_hit_hardest_by_bu.html

Porter, E. (2013, April 3). Investments in education may be misdirected. *The New York Times*, p. B1.

Race to the Bottom. (2012, December 6). *The New York Times*, p. A34.

Reardon, S. (2011). The widening academic achievement gap between the rich and the poor: New evidence and possible explanations. In G. J. Duncan & R. J. Murnane (Eds.), *Whither opportunity? Rising inequality, schools, and children's life chances* (pp. 91–116). New York, NY: Russell Sage.

Rebell, M. A., & Wolff, J. R. (2011, August 26). When schools depend on handouts. *The New York Times*, p. A27.

Rebell, M. A., Wolff, J. R., & Rogers, J. (2012). Deficient resources: An analysis of the availability of basic educational resources in high needs schools in eight New York state school districts. *Campaign for Educational Equity, Teachers College, Columbia University*. Retrieved from http://www.tc.columbia.edu/i/a/document/25804_DeficientResources2-21-13.pdf

Rich, M. (2012, July 24). Enrollment off in big districts, forcing layoffs. *The New York Times*, p. A1.

Rothstein, R. (2004). *Class and schools: Using social, economic, and educational reform to close the black-white achievement gap*. Washington, DC: Economic Policy Institute.

San Antonio Independent School Dist. v. Rodriguez. (1973). 411 U.S.1.

Santos, F., & Rich, M. (2013, March 28). States shifting aid for schools to the families. *The New York Times*, p. A1.

Saul, S. (2011, December 12). Profits and questions at online charter schools. *The New York Times*, p. A1.

Saul, S. (2012, May 22). Public money finds back door to private schools. *The New York Times*, p. A1.

Schemo, J. (2001, July 20). US schools turn more segregated, a study finds. *The New York Times*. Retrieved from http://www.nytimes.com/2001/07/20/us/us-schools-turn-more-segregated-a-study-finds.html

Schrag, P. (2003). *Final test: The battle for adequacy in America's schools*. New York, NY: New Press.

Serrano v. Priest. (1971). 5 Cal.3d 584 (*Serrano I*).

Serrano v. Priest. (1976). 18 Cal.3d 728 (Serrano II).

Simon, S. (2011, May 25). Public schools charge kids for basics, frills. *The Wall Street Journal*. Retrieved from http://online.wsj.com/article/SB10001424052748703864204576313572363698678.html

Simon, S. (2012, March 8). Insight: Public schools sell empty classroom seats abroad. *Reuters*. Retrieved from http://www.reuters.com/article/2012/03/08/us-usa-schools-foreign-idUS BRE8270A120120308?feedType=RSS&feedName=domesticNews

S.S. v. State of Michigan. (2012). Complaint and Petition for Writ of Mandamus. Circuit Court of the County of Wayne.

Toppo, G. (2012, November 28). Online schools spend millions to attract students. *USA Today*. Retrieved from http://www.usatoday.com/story/news/nation/2012/11/28/online-schools-ads-public-/1732193/

Turner, M. A., & Kaye, D. R. (2006). How does family well-being vary across different types of neighborhoods? (Low-Income Working Families Paper 6). Washington, DC: The Urban Institute. Retrieved from http://www.urban.org/UploadedPDF/311322_family_wellbeing.pdf

U.S. General Accounting Office. (1997). *State efforts to reduce funding gaps between poor and wealthy districts* (GAO/HEHS-97-31). Washington, DC: U.S. General Accounting Office. Retrieved from http://www.gao.gov/assets/230/223749.pdf

Williams v. State of California. (2000). First Amended Complaint for Injunction and Declaratory Relief, No. 312236. Cal. Super. Ct., S.F. County, filed May 17.

Yaccino, S. (2013, May 23). Protests fail to deter Chicago from shutting 49 schools. *The New York Times*, p. A20.

CHAPTER THIRTEEN

School Activism AND THE Production OF Urban Space IN Atlanta, Georgia

KATHERINE HANKINS AND ELIZABETH EGAN HENRY

> We were so thrust into starting the school that my whole experience with the neighborhood as it was changing was through that experience because we moved in and a year later my son was born. We had a porch party when he was six weeks old...we ended up with about thirty people on our porch that were either pregnant or had newborns or real little ones. That's how we built our group in the community...our gentrification process was connected to the school.
> (E. HENRY, INTERVIEW WITH MICHELLE BLACKMON, NOVEMBER 8, 2007)

For Atlanta mother Michelle Blackmon, the movement by community activists to open the city's first grassroots, parent-run charter school was tied to the gentrification of her intown Atlanta neighborhood, Grant Park. Young, middle-class couples, like the Blackmons, who were first attracted to the intown neighborhoods by the historic architecture and design of the city's early streetcar suburbs, chose to move to these areas to enjoy the cultural, dining and shopping amenities of an urban lifestyle and to avoid the ever-increasing commute times of sprawling metropolitan Atlanta. As they renovated and restored their bungalow, they quickly got to know other couples who had moved into Grant Park. When their first child was born, they made the decision not to move to the northern arc of White, middle-class suburbs or to send their children to one of the city's private schools. Committed to their neighborhood, they began asking themselves what they needed to do to create a neighborhood school for Grant Park. Reaching out through the neighborhood association and the Grant Park Parents Network electronic mailing list, they invited friends and neighbors to their home. Gathered on the Blackmon's front porch, as they so often did on weekends for parties and get-togethers, the couples discussed their options. What emerged from that meeting galvanized the gentrifiers who were moving into Atlanta's historic, close-in southeast neighborhoods, and, in the process, catalyzed the transformation of the neighborhood through the creation of a charter school, a publicly funded, privately managed neighborhood school.

What the Grant Park parents likely did not realize was how their neighborhood-based school activism mirrored that of the work of parents two decades prior in neighborhoods to the northeast of downtown, although the institutional framework and level of the activists' political engagement, as we explore later, were quite different. The gentrification of pockets of central city Atlanta neighborhoods began in the 1970s and initially concentrated in the neighborhoods of Morningside-Lenox Park, Virginia-Highland, Ansley Park and the BOND (Bass Organization for Neighborhood Development) community, which included Inman Park and Candler Park (and eventually Lake Claire; see Figure 1). During the 1970s and 1980s, activists from these neighborhoods engaged in a myriad of political fights through organizing and working with established political channels to halt the construction of two planned highways, elect pro-neighborhood mayoral, city council and school board candidates and influence zoning and education policy.

In this chapter we historicize the gentrification waves in Atlanta, demonstrating the key role that parent activism played in the gentrification process in two regions of the city that had been, in the tumultuous decades of the 1960s and 1970s, increasingly African American and working class. By the 1990s, the neighborhoods on the northeastern side of the city had become more affluent and mostly White, and by the 2000s, neighborhoods in the southeastern part of the city had followed suit. The activism around schooling, we argue, produced middle-class and increasingly White spaces of the city, a city that like so many others across the United States, cannot be understood without attention to the racialized policies and practices that have shaped it. What distinguishes these two moments in neighborhood activism is the institutional framework that shaped the *ways* in which activists engaged in neighborhood transformation. The Grant Park parents became involved in school activism shortly after charter school legislation passed in Georgia, thereby giving them instant control over a key neighborhood institution. As we demonstrate here, middle-class parents in the northeastern neighborhoods spent years building political coalitions, serving on the board of the Atlanta public schools and becoming active in a variety of neighborhood concerns. The Grant Park parents, on the other hand, were able to create their own school, given the neoliberalizing policies in the U.S. education system that increasingly offer "local control" to parents and private corporations. This institutional framework, we argue, provides the potential for exacerbating the well-documented effects of gentrification, including the displacement of working-class residents and people of color from central urban neighborhoods. Our empirical portrait is based on almost 10 years of research in these Atlanta neighborhoods and includes interviews with key activists and archival analysis of activists' documents and newspapers.

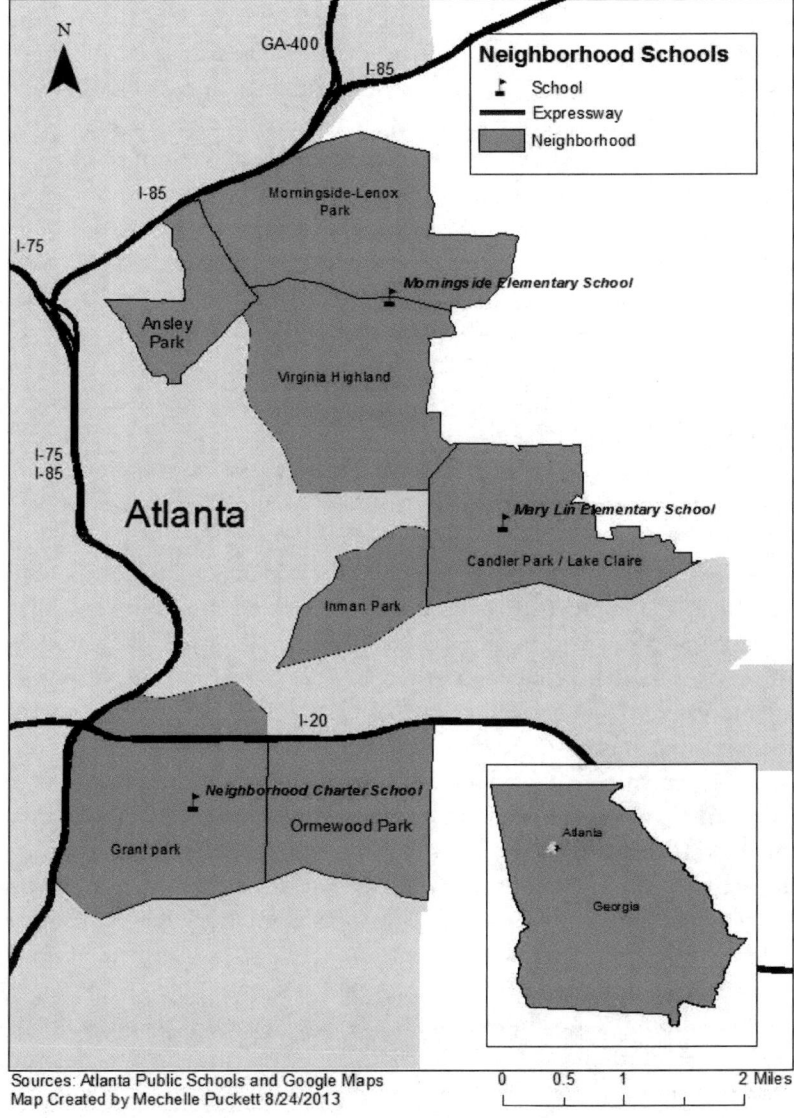

Fig 1. Schools in Atlanta's east-side neighborhoods.

GENTRIFICATION AND SCHOOLING

Gentrification, as is widely understood in urban geography and sociology, refers to the "class remake of the central city" (Smith, 1996, p. 39), wherein older, devalued properties in neighborhoods close to the core of a city become desirable spaces for more affluent residents. While debates about the "emancipatory" potential of gentrification (Ley, 1996) or the ways in which it captures the deleterious march of capitalist exploitation (Smith, 1996) are widely rehearsed in the literature (e.g., Lees, Slater, & Wyly, 2008), scholars are increasingly turning to understand the nuances of the "on-the-ground" experiences, subjects and spaces of gentrification processes (Brown-Saracino, 2009, 2010). For example, the early gentrification waves were seen as ushered in by the likes of artists and bohemians, who were willing to overlook decaying infrastructure for edgy and affordable housing (Brown-Saracino, 2010; Zukin, 1982). One group that, until recently, has received little attention in the conversations about gentrification is that of middle-class parents (Cucchiara & Horvat, 2009; Karsten 2003) and the role they play in engaging in, or, as has been the typical story, remaining disengaged from local public schools.

In their study of gentrifiers in several neighborhoods in London, Butler (2003) and Butler and Robson (2003) described how many parent-gentrifiers choose to bypass the public school route entirely and instead send their children to private school, reflecting a "circuit of schooling" (Ball, Bow, & Gewirtz, 1995) that maintains social exclusivity. The "circuits of schooling" model that Ball et al. (1995) suggested offers theoretical tools for understanding how parents select their school in the context of social dynamics around race and class that shape the school choices open to families (Ball & Reay, 1998; Ball & Vincent, 1998). Building on this model, a collection of case studies on parents' activism in London and Atlanta suggest that schooling is a key ingredient in the ways in which the gentrification dynamics of a neighborhood play out. Butler and Robson (2001a) followed the gentrification of five London neighborhoods and argued that the schools parents considered changed depending on the extent the neighborhood was gentrified (Butler, 2003; Butler & Robson, 2001b, 2003). Hankins (2005), examining activism among parents in Atlanta, traced the city's first grassroots charter school effort and argued that it was the White, middle-class gentrifiers' ability to donate their time, professional connections and money that made it possible for them to open and manage one of the city's first charter schools.

Charter schools in the United States have had a mixed reception in education policy circles since they were first introduced in the mid-1990s, but they have found favor in states such as Georgia. Georgia passed charter school legislation in 1997 to enable the creation of individual schools, and in 2012 it passed a constitutional amendment allowing for whole charter systems, overseen by the Georgia

Department of Education. As of 2013, Georgia hosts 110 charter schools and 14 charter systems, which include 107 schools. Ideologically, charter schools appeal to both the conservative right and the progressive left in that they enable local control, a key ideological element of neoliberalism, which celebrates private property and the efficiencies of the market (Hackworth, 2007; Harvey, 2005). This increasing focus on the privatization of public schools in the United States creates, as Lipman (2011) contended, increasingly fractured and geographically segregated cities.

In this chapter, we extend the concern of understanding the role schools play in gentrification dynamics by examining the (racialized) class remake of a collection of Atlanta neighborhoods over two time periods and the institutional frameworks that enabled these transformations. We argue that the engagement of parents, first in the traditional public schools during the postdesegregation era of the 1970s and 1980s, and then through the creation of a charter school in the 1990s and early 2000s, has resulted in the (re)inscription of White, affluent neighborhoods, in spite of the stated intentions of parent-activists for "integrated" and "diverse" schooling. The ability of parent-gentrifiers to create their own school with a charter enabled them to bypass engagement in the traditional public schools and other public institutions, as the parents in the 1970s and 1980s had, resulting in an accelerated process of neighborhood change. The institutional arrangements that facilitated the charter school have, we suggest, contributed to the exacerbation of Atlanta's landscape of inequality. Before we offer the empirical portraits of these two eras of activism and gentrification, we first situate these neighborhood dynamics in Atlanta's history and consequential geographies of race and class.

Atlanta's History and Geography of Race and Class

Having rebuilt itself following the Civil War and Reconstruction, the capital of Georgia emerged as the heart of the New South. Yet the small southern city that had been founded at the intersection of two railroads remained concentrated around the downtown. Families lived close to, if not above, their place of work, and those residential districts that existed presented a salt-and-pepper pattern when it came to race, ethnicity and class (Lands, 2009). At the end of the 19th century, there were several established African American neighborhoods, such as South Atlanta (Brownsville at that time) and Peoplestown to the south of the city, and Buttermilk Bottoms to the east, but most of the city was fairly integrated. As Lands pointed out, "Social practices and understandings may have separated classes and races, but housing location did not" (p. 2).

In 1906, the Atlanta race riot changed the city. The passage of Jim Crow laws along with innovations in transportation, starting with the streetcar and by the 1920s the automobile and bus line, sorted the city's neighborhoods along lines of race and class. African Americans clustered in the historic Fourth Ward, which was located east of the business district along Auburn Avenue, and west of the downtown business district, near the historic Black colleges of Morehouse, Spelman and what was to become Atlanta University. White, working-class families, moving to the city as cotton prices plummeted, settled in walking distance to the mills that were located on the city's northwest and south sides. White, middle-class families continued following the streetcar lines and the automobiles north along the now well-known Peachtree Street. By the early 1950s, the city's population swelled with the postwar economic boom, and the White, middle-class neighborhoods of the north side, collectively referred to as the community of Buckhead, were annexed, and the existing segregated schools merged with Atlanta public schools in 1952.

The Supreme Court's 1954 ruling in *Brown v. Board of Education* shook many of the city's White families, as its reverberations were felt across the South. While Atlanta avoided the violence of Little Rock and Birmingham, in large part due to the cooperative, biracial urban regime that governed the self-proclaimed "city too busy to hate," the next two decades brought rapid racial transition (Kruse, 2005, pp. 6–15; Stone, 1989). Ordered by the Fifth District Court to immediately integrate its public schools, in contrast to the violence and resistance across the South, nine African American children quietly and peacefully walked through the doors of the city's four public high schools in 1961. Over the next decade, despite the school board's tactics of delays and token integration, thousands of families fled the city as the court battle over metropolitan busing raged. A study completed by Orfield and Ashkinaze (1991) determined that Atlanta experienced greater White flight than any city in the nation:

> Atlanta's decline in white enrollment was among the most rapid of all large school districts. Between 1967 and 1986 the total Atlanta enrollment dropped by 41 percent. The school district was 38 percent white in 1968 but was only 15 percent white by 1974 and was down to 7 percent white in 1986 (pp. 110–111).

Hysteria and fear, fueled by racialized real estate practices, drove White families from the public schools to the suburban counties located north of the city ("City schools feel white flight's result," 1984; See Table 1). As the population of the metropolitan region swelled with the relocation and expanding number of Fortune 500 companies that were attracted by the Sunbelt's low tax rates and Right to Work laws, White families moving to the South listened to the real estate agents' assessment of the city's crime-ridden neighborhoods and poor and failing integrated public school system. Those more affluent and upper-middle-class White families who did live in the city relied almost entirely on the city's prestigious private schools, which greatly expanded in number and enrollments during the desegregation decade.

Table 1. Atlanta Public Schools Student Enrollment Data, 1950–2010.

Year	White	%	Black	%	Total
1950	31,951	62.7	18,972	37.3	50,923
1955	53,800	66.1	27,576	33.9	81,376
1960	54,530	57.3	40,582	42.7	95,112
1965	49,020	44.2	61,950	51.9	108,941
1970	39,318	35.9	70,296	64.1	109,664
1975	12,884	15.1	72,106	84.5	85,298
1980	6,722	9	67,158	90.4	74,321
1985	5,102	7.6	61,061	91.2	66,986
1990	4,161	6.7	56,499	91.7	61,657
1995	3,973	6.7	53,764	91.5	59,416
2010	4,930	10	40,919	83	49,300

Source: Atlanta Public Schools.

Gentrification and Parent-Gentrifiers' Commitment to Public Schools

Beginning in the 1970s, young, White, middle-class couples, who had experienced the civil rights movement in college and graduate school, began migrating to Atlanta. Often hired by one of the city's law or architecture firms, nearby universities or prospering Sunbelt-based companies, these self-identified "urban pioneers" were attracted to the historic architecture and proximity to cultural and shopping amenities of the intown neighborhoods, which had been built at the turn of the century as streetcar and automobiles expanded the city limits beyond the downtown district. In many neighborhoods, two decades of White flight had seen the Victorian and bungalow-style houses subdivided and converted to rental property. By 1970, the easement of 1,000 lots for the proposed expansion of metropolitan Atlanta's highway system further drove down home prices. However, falling property values made the homes affordable to those couples ready to restore the quirky bungalows and Victorians (Tyson, 1975a, 1975b).

In the midst of their renovation projects, a successful neighborhood movement emerged, which, after defeating Interstate 485 and Stone Mountain Freeway, the proposed highways that were to cut through the heart of their neighborhoods to the northeast of downtown, turned its attention to the mayoral and city council elections. The gentrifiers peeled off old "Atlanta Yes, I-485 No!" bumper stickers from their cars and replaced them with new "Maynard Yes!" stickers ("The interstate that almost was," 2003). A progressive coalition of African Americans and White gentrifiers emerged to elect the city's first African American mayor,

Maynard Jackson, and pro-neighborhood candidates in 1973 and again in 1977. As our research participants expressed to us, these same gentrifiers had children, and they began thinking that through their involvement in a grassroots movement, they had successfully taken on the Department of Transportation and political urban regime, so why not the schools? Due to White flight, enrollment in Atlanta public schools plummeted, and school after school was forced to close. A Virginia-Highland mother, Nancy Hamilton, explained as follows: "Many of us moved in here without kids. We moved in and then started to have families. If you were looking for education, you had this little bit of window of time to make a difference...you had five years" (E. Henry, Interview with Nancy Hamilton, February 16, 2009). With the 1977 school board elections approaching, the Hamiltons invited other parents to their home in Virginia-Highland. They felt that the incumbent, Angela Ionnides, whose children were enrolled in private schools, had lost touch with the community. After much discussion, the group of homeowners and young parents decided that the most significant impact it could have was to support a pro-public school candidate for the District 3 seat. After the meeting, Nancy Hamilton went with Betsy Richards, the wife of Pete Richards who was running for the District 8 at-large seat, to the Commerce Club. There, they met with Atlanta businessman and Morningside-Lenox Park resident, Joseph Martin, Jr. Martin recalled as follows:

> Our kids were born in 1973 and 1976. I had gone through public schools, Atlanta Public Schools, and my wife had, too. We lived in Atlanta, and our kids were going to go to the public schools...we could have afforded to do something else, but we didn't. There weren't that many kids in the neighborhood because people had [fled]. On our street there were very few school aged children...people were going to private schools or more frequently just leaving when their children reached school age. There was a group of us who said, we're going to stick it out. We're going to make it work. About that time the neighborhood political movement was beginning. A lot of that came out of the I-485 fight, and we said if that can happen in terms of city government, then it can certainly happen in terms of schools. We got together, and I became the candidate...I was just thirty-five, but what was important was the neighborhood roots. (E. Henry, Interview with Joe Martin, April 2, 2009)

From its beginnings in Nancy and John Hamilton's living room to the campaign itself, the 1977 election brought an entirely new type of parent activism to Atlanta. By bringing together parent activists who had met through the highway battle and their neighborhood associations, the successful grassroots campaign established a network of involved, middle-class parents, which stretched across five intown neighborhoods. So when newly elected school board member Joe Martin learned at his second meeting that Virginia-Highland's elementary school and the intown neighborhoods' middle school were to join the list of schools that were to close, he was quickly able to alert the chairs of the neighborhood associations' education committees. The Hamiltons again invited their friends and neighbors to their home. As the parents sat around Nancy and John Hamilton's living room, an idea for a neighborhood-based organization made up of the five education committees began to emerge, and the Close-In Reviving Communities Linked for Education, or CIRCLE, was founded. The group continued to meet at different members' homes and soon came up with the idea of moving the intown middle school to the Inman property, which was joining the already merged elementary Ansley Park and Morningside-Lenox Park schools in the fall. By March 1978, the informal parents' group was holding plenary meetings for what was to become a political organization, the Council of Intown Neighborhoods and Schools (CINS). The parents who founded it, after purchasing a home in the intown neighborhoods and enrolling their children in the struggling public schools, were making an overtly political statement. Now they had a council that viewed parent activism in relation to their reviving, intown neighborhoods, which, on the heels of the successful neighborhood movement, enabled these gentrifiers to become powerful advocates for the public schools.

Over the next 15 years, CINS developed grassroots strategies that linked the gentrification of the intown neighborhoods to its efforts to support the struggling public schools. By 1980, the parents had designed, printed and distributed what was to become the first in a series of brochures promoting intown living and the neighborhoods' public schools for real estate agents and area banks to distribute to families considering a move to the area. The success of CINS in bringing White, middle-class families to the city's intown neighborhoods and public schools hinged on the very existence of a critical mass of White, middle-class professionals moving back into the city's intown neighborhoods (Graham, 1983). However, it was the commitment by these early parent activists, with their socioeconomic and political influence, to send their children to their neighborhood public schools and not one of the city's private schools or move to the suburbs upon their children reaching school age that made the parents' group a powerful voice (Anderson, 1991; "Photograph: New residents have crowded the elementary school," 1991). Morningside Elementary School, which, in 1976, had seen its enrollment fall to 148 students, started the 1984 school year with 320 students. By 1990, the neighborhood elementary school had doubled its enrollment. As more and more families moved intown and stayed even after their children started elementary school, the collapsed housing market began to swing upward. As one real estate agent explained, "During the 1970s, you couldn't give [a house in Morningside] away." By the early 1980s, this had changed. Homes in the reviving, intown neighborhoods, which had sold for less than $16,000 during the highway battle, were selling for between $60,000 and $90,000 the same day they appeared on the market (Graham, 1983; Mooney & Johnson, 1980).[1]

During this same period, Mary Lin Elementary School, which was serving both Inman Park and Candler Park families, also began to experience a similar trend. Parent activists committed to supporting their reviving neighborhood and public school organized education committees. Their grassroots movement took on an even greater role when news broke that the Atlanta City Council had approved a four-lane, high-speed parkway leading to the planned Carter Presidential Library that followed the path of the defeated Stone Mountain Highway. The second highway battle had begun. Inman Park mother Midge Sweet, who would be elected to the Atlanta School Board, recalled how, in the gentrifiers' minds, the two battles were connected:

> The fight against the road and for the park was in some ways really fortuitous because it galvanized people...maybe we were high on beating the road...but at the exact same time we were fighting the road, we were working on Mary Lin [Elementary] and the schools... there was just this sense that we had to continue on all these fronts to make things better (E. Henry, Interview with Midge Sweet, October 27, 2011).

For young mother Midge Sweet, the fight by gentrifiers to preserve their neighborhood, against the road, and for public schools, was the essence of what she and other activists had come to value. Inspired by the revival of the neighborhood movement against the parkway, the BOND parents brought to the CINS a renewed commitment to grassroots activism and volunteerism.

The activism that shaped the transformation of the public schools in Atlanta's northeastern neighborhoods resulted in the demographic shift in their racial and class composition and consolidated the exclusivity of the neighborhoods. At the same time, a viable and thriving public Morningside Elementary eventually resulted in the school's overcrowding and the construction of a new one just a few miles away to serve many of the same addresses where, 30 years prior, few families were interested in public schools. Likewise, as of 2013, Mary Lin Elementary is under construction to provide classroom space to its approximately 575 students so that the fourth- and fifth-grade classes no longer have to meet in temporary trailers on the campus. The transformation of the northeastern neighborhoods occurred in part as a result of the investment by middle-class parents who were committed to public schools (Donsky, 2002a), institutions that are required by law to serve all school-age children in their attendance zones.

Gentrification and the Creation of a Charter School

By the late 1990s, the gentrification of Atlanta shifted with the rise in property values to the close-in, southeast neighborhoods of Grant Park and Ormewood Park. Similar to the intown parents' movement that had begun two decades earlier in the Hamiltons' living room, the grassroots efforts by gentrifiers started around the meeting held on the Blackmons' front porch. In an interview by an *Atlanta Journal-Constitution* reporter, Chris Newman, the president at that time of the

Grant Park Neighborhood Association, recalled the impact of gentrification on stirring up grassroots, neighborhood-based parent activism:

> Back then, the neighborhood was just beginning to turn around, and you could get a fixer-upper in the $12,000 to $20,000 range. We lived with a wonderful, diverse group of people, but you sent your kids to private schools. About five years ago, families made the commitment to make a difference in the education system rather than move out when their kids hit school age. (Raines, 2004, p. 8HF)

Having set up an education committee under the Grant Park Neighborhood Association following their initial meeting, the parents started by exploring and putting together a detailed informational list of all their educational options for the neighborhood's 185 soon-to-be elementary school-age children (Byrd, 1998).

After discovering the extensive waiting lists at each of the Atlanta-area private schools, the grassroots movement was galvanized. Michelle Blackmon, who had taken over as committee chair, explained as follows:

> [We] got more organized and started exploring public school options. We did a lot of going to meetings, presentations, and going door-to-door handing out information and talking to people. And when I look back on it, it really blows me away. It's crazy. We literally knocked on 2,500 doors. (E. Henry, Interview with Michelle Blackmon, November 8, 2007)

Familiar with the charter school concept, which had only recently been introduced as an educational reform effort in the state, the Grant Park parents envisioned creating a new kind of public school that would improve academic achievement, foster parental volunteerism and strengthen the sense of community in Atlanta's diverse, resurgent neighborhoods. After parents in this small group had learned everything they could about charter schools, they set out to engage their neighbors in the process. Volunteers went door-to-door to hundreds of houses and sponsored several community meetings. Blackmon explained, "These are real do-it-yourselfers, people who take control. They're going to make their neighborhood better, going to make their house better, going to make the park better. So of course they are going to want to be involved in making their child's school better" (E. Henry, Interview with Michelle Blackmon, November 8, 2007). With endorsements from the editorial staff of the *Atlanta Journal-Constitution,* along with backing from the Atlanta City Council and Atlanta Board of Education members representing the neighborhoods, community support for the charter school grew (Hankins, 2004). With three years of research and grassroots mobilizing efforts behind them, the homeowners and parents formally passed the charter petition at a public meeting in September 2000. However, opposition from the school board mounted. Blackmon elaborated, "There were people who were concerned that it was going to end up being exclusive and people would get left out. That it would end up being an all-white or upper-middle-class school...that's not what the school was about"

(E. Henry, Interview with Michelle Blackmon, November 8, 2007). Likewise, there was opposition from residents. The changes brought to the city by gentrification, including rising property values and fewer rental opportunities, were driving many of the longtime and poorer African American residents from the neighborhoods. These mounting fears and distrust of gentrifiers coalesced around opposition to the proposed neighborhood charter school and resulted in, among other concessions, the expansion of the schools' attendance zone to include the nearby neighborhoods of Ormewood Park and North Ormewood Park (which, according to U. S. Census data, had higher proportions of African Americans), an extensive mediation session with disgruntled neighbors and the expansion of the board to include more members of color (Hankins, 2007). After these changes and several revisions to their proposal, the neighborhood charter school opened in 2007 in the historic Grant Park neighborhood school, the 1907 W. F. Slaton Elementary. Its attendance zone overlapped with that of the nearby (and newly constructed) public school Parkside Elementary. The public school choice, so celebrated as a solution to failing schools, presented parents with options of where they could send their children. It also created friction, as parents had a choice between the neighborhood charter school, which, according to attendance data, had a much higher proportion of White students compared to the typical Atlanta public school and to that of nearby Parkside Elementary (Donsky, 2002b).

As the children of the parent activists that had first gathered on the Blackmon's porch approached fifth grade, the group again mobilized community and political support to form a middle school. In 2005, the neighborhood charter middle school opened in the adjacent neighborhood of Ormewood Park in another historic school building, formerly Anne E. West Elementary. By 2011, the schools merged into one administrative unit called the Atlanta Neighborhood Charter School, serving kindergarten through eighth grade.

By 2013, the waiting list for the Neighborhood Charter School reads like that of the city's most exclusive private schools. Furthermore, the demographic profile of the Neighborhood Charter School reflects the whitening of the student body and of the Grant Park neighborhood. According to 2010–2011 data from the National Center for Education Statistics, U.S. Department of Education, the school's student body was 17.9% African American, 67.8% White and 10% multiracial.[2] The opening year of the Neighborhood Charter School in 2002–2003 reflects a very different student body: 57% African American, 30% White and 10% biracial.

The creation of the charter school in Grant Park enabled the swift transformation of the neighborhood from one where middle-class parents left when their children became school-aged or spent thousands of dollars to send them to private schools to a neighborhood whose middle-class parents wait anxiously as the attendance lottery results are made known for a highly coveted place in the Neighborhood Charter School. The neighborhood itself and thus the catchment zone for the school changed significantly between 1990 and 2010: According to the 2010 U. S. Census, the population of Grant Park was just over 6,500, where 65.5% were White and 28.5% were African American. This contrasts to 1990, where the neighborhood was 52.5% White and 43.1% African American, with a population of just under 6,000. The role of the school, founded and managed by Grant Park and Ormewood Park residents, has been inextricably linked to the transformation of the southeastern section of the city.

CONCLUSIONS

Through the description of the activist experience in a collection of Atlanta neighborhoods, we make three main contributions to urban studies scholarship. First, schools are, quite clearly, an important part of the gentrification story—often discursively constructed as "failing" and doomed to closure, scattering low-income and minority populations around the city (e.g., Lipman, 2011), and, as we suggest, they can be objects and subjects of the "class remake" of the central city neighborhood. They are sites of potential engagement by activist parents, and when combined with the kinds of resources to which *middle-class* activist parents have access, this engagement can result in neighborhoods that are increasingly middle-class (and White).

Next, we highlight the role of parent-gentrifiers as an important group that represents a different set of needs than the bohemian artist or the stereotypical "urban pioneer." The city services that families require or desire, such as child care, schools and other child-centered amenities, reveal the importance of families and parents as subjects of gentrification. Their needs and their ability to marshal resources to satisfy them by, for example, creating child care co-ops or even charter schools, deeply shape the trajectories of neighborhood change.

Third, the institutional framework that enables the kind of neighborhood transformation such as that seen in the Grant Park and Ormewood Park neighborhoods influences what parent-gentrifiers are able to do. Participation in democratic processes, such as through elections to school boards and city hall, represent an avenue that ideally requires accountability and transparency. The parents in Morningside-Lenox Park, Virginia Highland, Ansley Park, Inman Park and Candler Park, while operating in an admittedly uneven landscape of race and class, nonetheless participated in broader conversations about the fate of the city as a whole in their engagement in public education. In contrast, the charter school option in Grant Park and Ormewood Park meant that the parent-gentrifiers were able to bypass full-on engagement with democratic processes and focus instead on their quasiprivate school to serve their neighborhoods.

In sum, the gentrification dynamics in the early 21st-century city cannot be separated from broader policy shifts that shape what gentrifiers, as agents of change, are able to do in their move to the central city. Education policy is particularly salient, as the profile of the gentrifier has changed from childless singles and couples to include families with children. And, as we suggest, the neoliberalization of education policy has catalyzed the rapid transformation of the central city from one of mixed, working-class neighborhoods to spaces produced for and by the middle class.

NOTES

1 In August 2013, the average listing price for a home in the Morningside-Lenox Park neighborhood was $792,824 (http://www.trulia.com).
2 As of the same 2010–2011 school year, Parkside Elementary, the traditional public school, was 77.7% African American, 11.8% White, 7.2% Hispanic and 2.4% biracial.

REFERENCES

Anderson, D. (1991, February 3). Morningside: Schools, parks attract a new generation of Atlantans. *Atlanta Journal-Constitution (Sunday Homefinder)*, p. 6–7.

Ball, S., Bow, R., & Gewirtz, S. (1995). Circuits of schooling: A sociological exploration of parental choice of school in social class contexts. *Sociological Review, 43*(1), 52–78.

Ball, S., & Reay, D. (1998). "Making their minds up": Family dynamics of school choice. *British Educational Research Journal, 24*(4), 431–448.

Ball, S., & Vincent, C. (1998). I heard it on the grapevine: "Hot" knowledge and school choice. *British Journal of Sociology of Education, 19*(3), 377–400.

Brown-Saracino, J. (2009). *A neighborhood that never changes: Gentrification, social preservation and the search for authenticity.* Chicago, IL: The University of Chicago Press.

Brown-Saracino, J. (2010). *The gentrification debates.* New York, NY: Routledge.

Butler, T. (2003). Living in the bubble: Gentrification and its "others" in north London. *Urban Studies, 40*(12), 2469–2486.

Butler, T., & Robson, G. (2001a). Social capital, gentrification and neighborhood change in London: A comparison of three south London neighborhoods. *Urban Studies, 38*, 2145–2162.

Butler, T., & Robson, G. (2001b). Coming to terms with London: Middle-class communities in a global city. *International Journal of Urban and Regional Research, 25*(1), 70–86.

Butler, T., & Robson, G. (2003). Plotting the middle classes: Gentrification and circuits of education in London. *Housing Studies, 18*(1), 5–28.

Byrd, K. (1998, December 31). Community; Parents explore charter school; Grant Park appraise public education options. *Atlanta Journal-Constitution*, p. 03JD.

City schools feel white flight's result. (1984, March 18). *Atlanta Journal*, p. 1B.

Cucchiara, M., & Horvat, E. (2009). Perils and promises: Middle-class parental involvement in urban schools. *American Educational Research Journal, 46*(4), 974–1004.

Donsky, P. (2002a, March 10). Public schools' popularity climbs. *Atlanta Journal-Constitution*, p. 1C.

Donsky, P. (2002b, April 23). New life for old school as charter fixer-upper: Parents roll up sleeves, renovate Slaton campus. *Atlanta Journal-Constitution*, p. 1B.

Graham, K. (1983, July 29). Welcome to Virginia-Highland, and the truth behind the trend. *Atlanta Journal-Constitution*.

Hackworth, J. (2007). *The neoliberal city: Governance, ideology, and development in American urbanism*. Ithaca, NY: Cornell University Press.

Hankins, K. (2004). *Site of exclusion? Practicing social citizenship and transforming urban space through the creation of a charter school* (Unpublished doctoral dissertation). University of Georgia, Athens.

Hankins, K. (2005). Practising citizenship in new spaces: Rights and realities of charter school activism. *Space and Polity, 9*(1), 41–60.

Hankins, K. (2007). The final frontier: Charter schools as new community institutions of gentrification. *Urban Geography, 28*(2), 113–128.

Harvey, D. (2005). *A brief history of neoliberalism*. Oxford, UK: Oxford University Press.

Karsten, L. (2003). Family gentrifiers: Challenging the city as a place simultaneously to build a career and to raise children. *Urban Studies, 40*(12), 2573–2584.

Kruse, K. (2005). *White flight: Atlanta and the making of modern conservatism*. Princeton, NJ: Princeton University Press.

Lands, L. (2009). *The culture of property: Race, class, and housing landscapes in Atlanta, 1880–1950*. Athens: University of Georgia Press.

Lees, L., Slater, T., & Wyly, E. (2008). *Gentrification*. London, UK: Routledge.

Ley, D. (1996). *The new middle class and the remaking of the central city*. Oxford, UK: Oxford University Press.

Lipman, P. (2011). *The new political economy of urban education: Neoliberalism, race, and the right to the city*. New York, NY: Routledge.

Mooney, B., & Johnson, S. (1980, February 9). Urban currents; 1980: City's pioneers find risky ventures pay. *Atlanta Journal-Constitution*, p. 1A.

Orfield, G., & Ashkinaze, C. (1991). *The closing door: Conservative policy and black opportunity*. Chicago, IL: University of Chicago Press.

Photograph: New residents have crowded the elementary school. (1991, February 3). *Atlanta Journal-Constitution (Sunday Homefinder)*, p. 6–7.

Raines, L. (2004, August 22). Neighborhood of the week: Grant Park; historic Atlanta area revitalized with new development, more kids. *Atlanta Journal-Constitution*, p. 8HF.

Smith, N. (1996) *The new urban frontier: Gentrification and the revanchist city*. New York, NY: Routledge.

Stone, C. (1989). *Regime politics: Governing Atlanta, 1946–1988*. Lawrence: University of Kansas Press.

The interstate that almost was. (2003, Fall). *MLPA Newsletter, 22*(3), 1–2.

Tyson, J. (1975a, April 22). Ansley Park, Morningside: Citizens work to keep in-city living. *Atlanta Journal*, p. 10B.

Tyson, J. (1975b, April 23). Midtown, Virginia-Highland: Suddenly new and young people started moving in. *Atlanta Journal*, pps.1C and 5C.

Zukin, S. (1982). *Loft living: Culture and capital in urban change*. Baltimore, MD: The Johns Hopkins University Press.

PART FIVE

The Groundswell

Community Responses to Urban Renewal

"And we sing." With these words, the poet Gwendolyn Brooks ends her poem on the 1967 dedication of the Wall of Respect, a public mural erected on Chicago's South Side. Composed of multiple traditions and genres of visual art, photography and even poetry, the Wall of Respect stood at 43rd Street and Langley Avenue, on the side of what Brooks tells us *Ebony* magazine characterized as a "typical slum building." The Wall was conceived of as a way to encourage and express Black dignity and presence in a pivotal moment of urban crisis and response. Musicians from the community played at the dedication, poets recited, political and community activists spoke. They all "sang" as a defiant expression of pride, of respect, of dissatisfaction, and importantly, of presence.

Today, many of the same spaces and people that were in crisis in the 1960s face another critical juncture. While urban renewal and gentrification often bring the kinds of benefits and attention for which inner-city communities had long been agitating, such changes are occurring concomitant with the removal of the people who make up those very communities. The essays in this section address different ways in which the people whose lives and spaces are being deeply affected by gentrification assert and affirm their presence in their communities. They do not oppose true renewal. Indeed, they struggle against a renewal that often seems dependent on their removal, coming together through different forms to express a simple, powerful notion—that these are their homes and their communities, and they belong to these spaces as they do to each other.

Sometimes the forces arrayed against these communities may seem inexorable. In a statement that is at once often-repeated and not repeated enough, Frederick Douglass once stated, "power concedes nothing without a demand." What the essays in this section demonstrate are the ways in which people in affected communities refuse the supposed inevitability of their fates. They demand and claim a say in what happens to their communities. If they differ, it is in form more so than in content. Whether through more traditional organizational structures, social clubs, or cultural bodies, people meet, collaborate, march, dance, and yes, sing, to declare their presence and their continued stakes in where they live.

AK

CHAPTER FOURTEEN

"History Still Matters"

Leveraging Historicity in Struggles to Control Space

MIRANDA MARTINEZ

INTRODUCTION

In late 2012, community activists on the Lower East Side of Manhattan cautiously celebrated after the New York City Council passed largely intact a community generated development plan for the area known as the Seward Park Urban Renewal (SPURA). The Seward Park area, located in lower Manhattan close to the Williamsburg Bridge, was designated as an urban renewal area in the 1960s, but large tracts of the site have remained vacant since 1967, when nearly 2,000 families were evicted as their tenements were torn down to make way for new housing. Forty-five years of stalemate followed, driven by a combination of political inattention, economic downturns and ethnic competition among the area's Puerto Rican, Jewish and Chinese residents over what kind of housing and commercial uses should be located on the site. After so much time, to have a plan clear the city's Uniform Land Use Review Process (ULURP) with some agreement was hailed as a historic achievement. More surprising was the significant incorporation of community demands for housing affordability, priority to local tenants in housing and job recruitment and oversight of the development process by the local community board No. 3 (CB No. 3). New York's community boards are volunteer advisory boards that are granted power under the city charter to review and comment on proposed development in a locality. Although their power is largely advisory, some community boards are more assertive in their attempts to intervene in redevelopment processes.

As in the rest of Manhattan, and large parts of Brooklyn, the gentrification of the Lower East Side has proceeded at such a pace that it seems to possess an evolutionary logic: Each wave of gentrification brings new residents wealthier than the last, attracted by increasingly luxe new condo and retail development. To have a plan reflecting the concerns of low-income and ethnic constituencies therefore speaks to a surprising resilience of community voice against a process that can feel inexorable. In this chapter, I look at how community housing activists use the neighborhood's historicity in surprising ways, to foster agreement among constituencies that are alienated by race and class and to contextualize social justice demands as part of a collective memory that has not been displaced, despite the pace of gentrification. The SPURA campaign reflects an emerging strategy among local activists to use local history to raise contemporary social justice issues with a new constituency that is less rooted in the neighborhood, while insisting on space and recognition for those with a strong local attachment.

Lower East Side and Gentrification

Located in the southeastern end of Manhattan, stretching from the Brooklyn Bridge north to 14th Street, the Lower East Side has been the focus of numerous studies on gentrification and community politics (Abu-Lughod, 1994; Mele, 2000; Smith, 1996). The primary arrival and settlement point for successive waves of immigration during the 19th and early 20th century, and a locus for New York's counterculture for many generations, the Lower East Side has long been (in)famous for its contentious political culture. Across generations of struggle and many different communities, Lower East Siders have made their mark in community mobilizations for labor and civil rights, educational reform and services. By the 1980s, gentrification created concerns about displacement and loss of local character. Many of the neighborhood's residents, especially those in the predominantly Puerto Rican northern section known as "Loisaida," remember the 1980s as a period of intense struggle to address poverty and abandonment during the years of the urban crisis. They participated in a vibrant artistic and political culture as they fought for community betterment through affordable housing and services. The major demand was for local control of land use decisions, captured by the slogan "This Land Is Ours" (Abu-Lughod, 1994; Patterson, Flood, & Moore, 2007). Community action resulted in a number of gains, although these have not prevented the continuing transformation of the neighborhood. Activists demanded city investment in affordable housing, especially through homesteading, a form of legalized squatting of vacant buildings (von Hassell, 1996). More recently, from 1997–1999, community gardeners successfully mobilized against a Giuliani administration plan to auction off dozens of community gardens that are much needed green space (Martinez, 2010). Garden preservationists successfully blocked the sale of land and created local land trusts to protect the gardens from development.

Since the 1990s, there have not been any major mobilizations, such as the ones that created the homesteads and the community gardens. Rather, as in many cities, community politics is in a state of relative quiescence, and activists seem to be waging a defensive struggle against the further encroachment of market rate and luxury development. In addition, socioeconomic polarization, and the exclusion of low-income and working-class people, especially those concentrated in the public housing tracts in the easternmost part of the neighborhood, are a concern of longtime activists. A white-hot rental market continues to drive lower income people out of the borough and the city. Certain areas of the neighborhood appear to have turned over almost entirely to an upper-middle-class population as a result of vacancy de-control enacted by the state legislature, as well as new "luxury-style" condominiums. No longer a neighborhood of working- and middle-class ethnic constituencies jostling for space, the Lower East Side now mirrors the socioeconomic polarization of the rest of Manhattan (Roberts, 2007), with low-income tenants concentrated in public housing and much higher incomes in the west. A report on neighborhood conditions by Angotti and Ervin (2008), which was issued during a contentious rezoning process, points out that average income and family size, when broken out by "subareas"—with public housing and working-class areas such as Chinatown and the Puerto Rican enclave known as "Loisaida" separated from gentrified areas—shows "major gaps in income, household size, and ethnicity between the East Village and Loisaida" (p. 52). This polarization is expressed in numerous contentious issues, such as school reform and the city's stop-and-frisk policies targeting non-White youths and public housing tenants. As the organs of community voice, especially CB No. 3, become more middle class and White, low-income constituencies must struggle to voice their needs on an equal footing with gentrifiers.

Doing Politics in a Gentrified Neighborhood

The lull in community mobilization springs in part from the character of neoliberal urban renewal, which has led to an embourgeoisement of formerly contentious working-class neighborhoods like the Lower East Side (Lees, Slater, & Wyly, 2008; Redfern, 2003). The influx of a more middle-class element, and the displacement of established working-class and minority communities, have dissipated the solidarity and collective consciousness of previous eras (Marcuse & van Kempen, 2000; Newman & Wyly, 2006). In addition, socioeconomic polarization can promote "revanchist" politics by the affluent to contain the poor (Smith, 1996).

In most instances, the first political acts of middle-class gentrifiers are assertions of class-based privilege, as they demarcate spaces apart from adjacent low-income minority populations, in order to protect housing values and services (Bridge, 2006; Cybriwsky, Ley, & Western, 1986; Kasinitz, 1988). However, there are broader structural obstacles beyond demographic shifts to be understood. An essential characteristic of neoliberal urban regimes is that they de-politicize issues of collective consumption in ways that curtail local mobilizations. The contemporary entrepreneurial city is based on privatization of service provision and competitive branding of urban spaces, which "hollow out the local state" (Patterson & Pinch, 1995) and leave local communities with no direct target at which to point collective action.

Locally specific realities also shape the mobilization context for New York City. As New York has been refashioned into a destination city for a global elite, local neighborhoods have struggled to maintain a voice. During three terms of office, Mayor Bloomberg promoted a political culture in tune with this transformation. Reasonably popular as a competent mayor who refurbished the city's image, Bloomberg was also criticized for circumventing or limiting citizen participation and disempowering communities in pursuit of rapid, streamlined redevelopment. (Greenberg, 2008; Moody, 2007). In education, neighborhood rezoning and land use, the Bloomberg approach was technocratic: Citizens were treated as consumers, entitled to good service, but not to decision-making power (Brash, 2011). The mayor's assumption of direct control of the school system was least popular. After a decade, it netted slim gains in student performance at the cost of ongoing turmoil from school closings, rigid testing regimes, bitter confrontations with the teachers' union and the decline of parent involvement. Even so, the combination of the mayor's personal wealth and philanthropy, with a perception of competence, muted citizen resistance, even while his administration pursued the most sweeping program of urban renewal since Robert Moses.

The widespread anger at New York's gentrification is often expressed through a rhetoric of imminent vanishing, as people track the loss of meaningful landmarks and note the city's cultural sanitization. Blogs such as *Jeremiah's Vanishing New York* (vanishingnewyork.blogspot.com) maintain a constant roll call of beloved spaces and business that were brought down by corporatization, high rents or luxury development, and they draw a regular traffic of mourners for the declining vibrancy of New York. This rhetoric of angry loss, while intended to question rapid redevelopment, also reflects a widespread sense of disempowerment, in that it expresses the erasure of the local as inevitable and uncontestable. Seeking to sustain local collective action, anti-gentrification activists and politicians on the Lower East Side emphasize continuity and persistence and will usually refute the idea that local character of the Lower East Side is gone. In an interview, New York City Council representative Rosie Mendez talked about the success of a number of historical preservation campaigns to illustrate ongoing efforts against gentrification. For Mendez, the possibility for future collective action lies in linking newcomers to the neighborhood with a sense of history:

A lot of people move here because they like the history of the neighborhood, and if you can explain what we went through here, they get interested. They like that it's a real community, and they get interested in certain issues. Even though they aren't from here, they get into the place. They put their toe in, and they find out what we're about, and then they start distributing petitions or donating.... So, the history of the Lower East Side, and our identity as a working class, immigrant community, still matters. Some of them want to preserve that. (R. Mendez, personal communication, August 5, 2009)

In fact, the Lower East Side will remain a working-class neighborhood for some time, albeit more polarized. The high concentration of public and other subsidized housing will help politicians like Mendez retain a significant political base and anchor a working-class, ethnically diverse population. However, she recognizes that making claims for space for the increasingly marginalized low-income and working-class community will not be accomplished in the traditional manner of appealing to common shared history and sentiments about place and neighborhood. There is an empowered, affluent constituency in play, who possess a more abstract and market-mediated relationship to the place they inhabit, but who she sees as approachable, if the right strategy is found.

Neighborhood history, in the hands of developers, is often used to market the authenticity of a place to gentrifiers (Zukin, 2010), but in the deliberations about the SPURA site, historicity was more evident on the side of gentrification opponents, who used it as a "bridging" strategy to mobilize across constituencies with "thick" and "thin" relationships to place. For these activists, keeping control of how the neighborhood's history was represented, insisting on the importance of social justice struggles as the lynchpin of neighborhood identity and linking past struggles to contemporary inequities were all crucially important to creating a broadly supported plan and forcing the city's planners to accept it.

The Seward Park Redevelopment Plan

SPURA was created as an urban renewal zone in the 1960s, when the city destroyed 20 acres of tenement housing and displaced 2,000 low-income families. At the time, they were promised new housing in the area, but the first units of new public housing targeted moderate-income people and limited acceptance of low-income people. Following protests and lawsuits by the displaced, the new housing was rented up according to quotas, but those first fights heralded an extended battle over the appropriate economic and ethnic mix on the site (Mele, 2000).

SPURA is now a six-acre set of lots, mostly used for parking, closely adjacent to the Williamsburg Bridge. The land remained undeveloped initially because the fiscal crisis of the 1970s prevented further development and later because low- and middle-income constituencies fought bitterly about what kind of housing the site should provide (Turner, 1984). The battles were sharply racialized and bitter, as low-income Puerto Ricans and middle-income Jews based in adjacent cooperative housing on Grand Street faced off over affordable housing. Middle-income cooperative dwellers fought for more market housing that would stabilize conditions in a decaying neighborhood, while Puerto Rican groups demanded additional new affordable housing that would move them from substandard tenement housing. In addition, a number of Chinese organizations fought for new housing that would ease the overcrowding in Chinatown. These conflicts produced gridlock through three different rounds of proposals for the site and left a legacy of mutual suspicion.

For Lower East Side Puerto Ricans in Loisaida, who now struggle to find affordable housing, the Seward Park area has been a thorny reminder of their marginalized status as a low-income constituency. The urban renewal scheme that led to the destruction of the original housing at the site was itself an artifact of planning doctrines that stigmatized as unfeasible working-class neighborhoods across the city and promoted policies of displacement. In addition, city planners and developers have resisted demands to use the site to meet the needs of the local community for low-income housing and open space. At other times, the city declined to honor the promises made to displaced tenants.

By the early 2000s, members of CB No. 3, politicians and advocates for affordable housing were seeking ways to intervene in the rapid redevelopment of the area, and they understood that the Department of City Planning (DCP) was close to pursuing proposals for SPURA. Debates reawakened by 2005, during a three-year process of rezoning the Lower East Side, negotiated between the DCP and a community coalition. The rezoning was designed to protect the area from large-scale redevelopment, by placing significant limits on the scale and income mix for new construction in the heart of the East Village and Lower East Side areas. The rezoning passed in 2008, but community advocates also understood that limiting those areas to small-scale development would encourage developers and the DCP to pursue large-scale commercial and luxury development on the nearby SPURA site. A coalition of 11 housing and community advocates created SPARC (Seward Park Area Redevelopment Coalition), through which they disseminated what they described as guiding principles for the site. These demands included permanent affordable housing, housing priority for residents of CB No. 3 in 50% of the created units, priority for displaced SPURA tenants, jobs for community members during construction and in future commercial development and a declaration that no big box stores would occupy the site. With these first principles, SPARC members began creating a community plan and a mobilization to pressure the DCP to accept a plan. Article 197-a in the New York City charter is concerned with city planning processes, and it creates provisions for community-based planning processes. However, community-based plans, designated as "197-a," are advisory, and the DCP does not have to accept them (Angotti, 2008).

The lead partner in SPARC was called Good Old Lower East Side (GOLES), a tenant and worker advocacy organization. The executive director of GOLES, Damaris Reyes, is from the Lower East Side, is versed in the history of the site, and as a long-serving member of CB No. 3, has been exposed to the outbursts of anti-poor rhetoric from opponents who claim that the Lower East Side needs more market-rate development because there is too "too much" affordable housing. She, like other housing advocates and politicians like Rosie Mendez, who are trying to exercise control over new development, appreciates the need to devise strategies that speak to multiple, and even opposing, constituencies. The demands listed here largely reflect the concerns of low-income and public housing tenants who feel the lack of affordable housing in the private market and who remembered the many years of broken promises. However, some of the other demands, especially the one against big box stores, could be seen as a reach to a broader constituency. Small business ownership is declining on the Lower East Side because of high commercial rents and competition from corporate chains, and preserving them is critical to maintaining the character of the neighborhood. But advocates understood that mobilizing long-term residents would not be enough to push a community plan. Publicizing the history of SPURA and its significance would create broad-based support so that DCP and the City Council would enact it as envisioned. In 2008, GOLES began collaborating with a number of public history organizations who used oral histories to craft informational campaigns to coincide with the 50th anniversary of the creation of the urban renewal zone. The materials, which I will discuss further below, highlighted the historical contentions over the SPURA site and contextualized it in terms of the broader social justice history of the Lower East Side. In 2010, GOLES hired a planning consultant from the Pratt Center for Community Development to facilitate a process of public consultation about the SPURA site.

SEWARD PARK INFORMATION CAMPAIGNS AND PLANNING PROCESS

Kicking Over the Traces

As the campaign about SPURA got underway in 2008, walkers around the neighborhood would find frequent mentions of public meetings and information sessions as well as other public history initiatives. One of the campaign's innovations was a text message guided tour called "Kicking Over the Traces" that was created in collaboration with GOLES. Using a printed map (or simply if one noticed the signs), a walker found places at the site that were marked by a sticker with a phone number and a code to be accessed by mobile phone. The texts offered short histories of a building and its relation to the SPURA site. The text messages were a historical primer, identifying important buildings like the Essex Street market, a covered small vendor market used by generations of immigrants, which would potentially be displaced or destroyed. The tour identified the public housing projects built on the site and described the demolished tenements and proposed projects for vacant areas. In an interview with Damaris Reyes (Personal communication, October 5, 2008), I discussed the strategy behind the Place Matters campaign, as an element of a local mobilization, and asked her to talk about it as a kind of marketing campaign.

> DR: Well—I guess it is like marketing. We want to draw people in to learn about the history.
>
> MM: Do you think that the people who've bought co-ops at Grand Street for prices up to 800,000 are going to be willing to talk about affordable housing at the Seward Park site?
>
> DR: I think in some ways they may be easier to talk to than the older crowd ....

In other words, for activists like Reyes, a gentrifier who can be made to understand what is at stake in the development of the site through exposure to its history, might be more open-minded about making room for low-income housing than an entrenched old-timer with a long memory of 40 years of ethnic and class resentment. In several weekends of watching people utilize the site, I noted that the text message campaign stimulated discussion and interest among participants. The stickers attracted a younger constituency, some of whom had seen the campaign noted in the local press, and others who were randomly pursuing the stickers as they found them.

The brief text messages of the walking tour could not offer much historical detail, but a SPURA 40th-anniversary map and brochure that was distributed by GOLES did, describing the various litigations around the site and offering a deeper narrative relating what was at stake at the site (GOLES, 2008). The main point to make about the texting tour and commemorative maps of the SPURA campaign is that they were clearly directed at a broadly progressive observer who lacks a base of strong local knowledge or attachment to the Lower East Side. The "Kicking Over the Traces" website gives background history on certain sites, offering snippets of local lore, and focuses strongly on the ongoing struggle for affordable housing. Certain entries make reference to the steady loss of affordable housing in New York City, as in the following, which describes how formerly price-controlled cooperative housing on Grand Street had, by the 1990s, become prohibitively expensive, and tenants voted to de-control it:

> The Hillman Houses were built in 1951 by the Amalgamated Clothing Workers Union who also built Amalgamated Dwellings. Similar to Seward Park Houses and Amalgamated Dwellings, Hillman Houses was part of slum clearance efforts. Four city blocks were cleared to build the project. It was built as affordable cooperative apartments, but in recent

years, the shareholders voted to get rid of resale restrictions and become a full equity market rate coop. (Field Play Inc., GOLES, & City Love, 2008)

The SPURA materials make oblique references to current controversies related to gentrification, the marginalization of the area's still-large minority populations or the city's pro-gentrification development policies, but they insist on the engagement with these questions as a central part of the area's identity and historicity. Many histories of the Lower East Side are essentially nostalgic, for example, framing immigrant history around an upward arc of mobility and incorporation into American society (Casey & Lee, 2006; Diner, 2002). In contrast, the SPURA history lessons are a resistive history of ongoing struggles for inclusion that encourage people to draw a line from the past to today. The entry for St. Augustine's Episcopal Church on the website "Kicking Over the Traces" offers a taste of this appeal to a historically informed sense of social justice:

> Built in 1828 (known then as All Saints Free Church) for the city's patrician elite; today it houses the largest African American congregation of any denomination on the Lower East Side. The congregation worships in the shadow of two "Slave Galleries": haunting, box-like rooms above the balcony where African Americans were forced to sit. *This rare artifact of racial segregation in New York stands as a stark, physical reminder of how and why boundaries of marginalization are drawn and contested.* Edgar Allan Poe attended church here. According to the Church's website, "Boss" Tweed attended his mother's funeral here and hid from authorities in the slave galleries (Field Play Inc. et al 2008). [Emphasis added]

Embedded alongside the gentle historicity in this passage, one hears the voices of people living today, who know intimately the confinement and second-class citizenship of those sitters in the segregated galleries, and who are pointedly challenging the reader to see the continuities within the gentrified landscape of today. In a way that seems hopeful, the campaigns show people trying to find a bottom line of agreement on what kind of place the Lower East Side has been, and should be, despite differences in relationship to place. To do this, the campaign seems to retool the technologies of gentrification for the purpose of sustaining a social justice agenda. The informational campaign is an appeal to people with an abstract sense of local history to join in solidarity with a more grounded community, to preserve a resistive local identity that can be commonly shared. The search for a just, livable community is piggybacking on the commodification of the Lower East Side as a place imbued with an undeniable authenticity based on a contentious immigrant past.

SPURA Visioning Sessions

If the walking tour and map materials targeted a constituency with a presumed "thinner" relationship to the neighborhood, the community meetings, billed as "visioning sessions," were about creating a conversation around what new development should look like from the point of view of local needs. The visioning sessions were a proactive mobilization that seemed to bring out Lower East Siders who knew the site's history and who arrived ready to demand affordable housing and a redevelopment process based on community consultation. The form of the visioning sessions themselves modeled a process of public consultation that New York's current planning processes rarely allow communities. Although consultants from the Pratt Institute guided the meeting, the sessions included long spaces of small group consultation during which people examined maps of the neighborhood, discussed different sites and issues and collaborated to fill in maps with visions of things on the site that would make the neighborhood more livable. Poring over maps, groups marked out local areas that mattered and consulted each other to make lists of what the site included. In my group, as with the others, affordable housing ("really affordable" for a low-income person was repeated often), with space for people with families, was the most requested, but other things came out as well. A small knot of teenagers mentioned a need for parks and recreational space, another young woman said the area needed more gardens and there was a spirited discussion of whether a big box retail store would help or hurt the neighborhood.

At an early visioning session in November 2008, planning consultants presented the long, frustrating history of SPURA to a room of 70 participants in the bingo hall of St. Mary's Church, located right in the heart of the area. During the question-and-answer period, participants immediately raised the historical claims of those who had been displaced from the site, and they challenged the consultants' assertion that community contention prevented development of the site. Several questioners insisted that there was basic harmony and agreement about what was needed in the neighborhood. One participant said the following: "I'm in Mitchell Llama [middle-income] housing, and I've been there for 30 years. Everyone knows what we need, we need affordable housing. There's no disagreement." Several women chimed in to agree: We don't have racial animus, we don't have gang violence and we get along in this community. After them came more local people who asked how local voice would influence SPURA and who spoke of the 40-year-old site as a constant and fresh source of grievance. One man in his 50s was a child when his family was evicted, an event that broke up his family and caused him to be sent to Puerto Rico, where he lived for years before he could return to the Lower East Side.

Forty years after SPURA was designated, the rancor over the displacement and the lack of adequate housing on the site had not dissipated. The Lower East Siders at St. Mary's, many of whom have lived for all that time in substandard housing adjacent to the site, still see it as a dispiriting evidence of their community's powerlessness, even as they prepared another salvo in the ongoing struggle to gain affordable housing. The visioning sessions were an effective act of community voice in the sense that they expressed a refusal to surrender a community claim to the site and laid the groundwork for a plan that would meet the needs of the local community. As Rosie Mendez expressed it, "history still matters" on the Lower East Side, despite the continuing expectation that low-income ethnic constituencies will somehow cease to influence development after a certain point of gentrification is achieved. The question is how to make that history work as part of community voice, given the marginalized status of low-income residents and the powerful interests promoting luxury development.

The community campaign to create a plan for SPURA that the city would respect had two prongs of attack based on an acknowledgment of the realities of a divided, rapidly changing neighborhood. The public meetings and visioning sessions regrouped the voices of original Lower East Siders for whom the Seward Park issue is a long-standing grievance, deeply involved with questions of race, class, dignity and local power. The second tactic reaches out to those who have arrived since the neighborhood's gentrification, for whom the undeveloped land below the Williamsburg Bridge is a mystery, or a policy failure based in now irrelevant ethnic antagonisms. Underlying the two-pronged approach is the recognition that with the decline of a "thick" experience of neighborhood, based on history, relationships and a lived sense of local culture, the place to start is with exposure to a "thin" textual experience that leads to a more authentic engagement with place. For activists and housing advocates trying to sustain a degree of neighborhood consensus for affordability and diversity, historically based assertions about character are a way of framing ongoing demands for housing affordability and neighborhood empowerment so that new people will be drawn into local campaigns.

What Will Happen to SPURA?

After the visioning sessions, GOLES launched a public exhibit of oral histories and perspectives on the site. A final report from the SPARC/GOLES process was passed to the CB No. 3 SPURA task force, which conducted a more conventional planning process in conjunction with the New York City Economic Development Corporation (EDC). Some tensions arose around those different plans, especially since the CB No. 3 task force came up with looser requirements than GOLES for affordable housing and community participation. However, the early start on organizing and a high degree of collaboration by housing advocates with CB No. 3 members, along with political representatives, forged unity around development principles for the site. In 2010, the community board voted unanimously on design principles that largely reflected the first motions by SPARC, including a higher percentage of affordable housing than envisioned by most conventional plans.

During the next few years, the planning process wound on, resulting in the CB No. 3 plan that was approved by the New York City Council in 2012.

The 2013 request for proposals from developers would allow 1,000 units of housing, half of which would be permanently affordable "to a range of incomes" (20% low income, 10% moderate, 10% middle-income households and 10% senior housing), and encourage community facility space. The developer selection criteria favor developers who partner with local housing organizations and who guarantee local hiring in jobs and retail spaces. In addition, community residents will receive preference for new housing. The plan also addresses community concerns about displacement. A CB No. 3 task force will monitor the developer selection process and compliance with the eventual agreement. GOLES and other advocates were unhappy with provisions such as the confidentiality requirement that prevents community representatives in the monitoring body from reporting on their deliberations.

Press coverage of the call for proposals quoted developers who were doubtful whether to bid because of the relatively high number of nonmarket units but were considering finding parcels adjacent to the site that could gain value from the development (Kaysen, 2013). The lukewarm reaction from developers is a good indicator that the aggressive, proactive public campaign and visioning process did force DCP to concede more to community control than it would have without community intervention. However, it also uncovers the many uncertainties that surround the development of SPURA. Despite the praise of the rigorous community process, once a selected developer is in control of the site, ongoing monitoring and vigilance will be the only way to maintain compliance with the original plan.

For those looking to understand how community-driven politics can persist given the quiescent and socially polarized realities of a gentrified neighborhood, new campaigns like SPURA offer some answers. The community-led planning process fostered a sense of unity and purpose around a direction for the site, driven by the idea and a sense that future development cannot be a zero-sum game where someone (usually the poor) has to lose. The two-pronged strategy I have described purposefully links new and old, as well as thin and thick relationships to place in order to create the room for a much deferred, difficult discussion. In a time of deep uncertainty that a common basis for action can arise in such a polarized neighborhood, the SPURA campaign suggests that historicity, and place attachment can be harnessed to promote action towards a shared vision of a just, livable neighborhood.

REFERENCES

Abu-Lughod, J. L. (Ed.). (1994). *From urban village to east village: The battle for New York's Lower East Side*. Oxford, UK: Blackwell.

Angotti, T. (2008). *New York for sale: Community planning confronts global real estate.* Cambridge, MA: The MIT Press.

Angotti, T., & Ervin, K. (2008). *Analysis of draft environmental impact statement: East Village/Lower East Side rezoning.* Retrieved from Hunter College Urban Affairs and Planning website: http://hunteruap.org

Brash, J. (2011). *Bloomberg's New York: Class and governance in the luxury city.* Athens: University of Georgia Press.

Bridge, G. (2006). Perspectives on cultural capital and the neighborhood. *Urban Studies, 43*(4), 719–730.

Casey, M., & Lee, J. J. (Eds.). (2006). *Making the Irish American: History and heritage of the Irish in the United States.* New York, NY: New York University Press.

Cybriwsky, R. A., Ley, D., & Western, J. (1986). The political and social construction of revitalized neighborhoods: Society Hill, Philadelphia, and False Creek, Vancouver. In N. Smith & P. Williams (Eds.), *Gentrification of the city* (pp. 92–120). Boston, MA: Allen & Unwin.

Diner, H. (2002). *Lower East Side memories: A Jewish place in America.* Princeton, NJ: Princeton University Press.

Field Play Inc, Good Old Lower East Side, & City Lore. (2008). *Kicking Over the Traces: An interactive tour of Seward Park.* Retrieved from http://www.kickingoverthetraces.com/stories.html

Good Old Lower East Side, Pratt Center for Community Development & City Lore. 2008. *SPURA matters: A 40-year commemorative map of the Seward Park urban renewal area.* Retrieved from http://www.placematters.net/files/pdfs/spura.pdf

Greenberg, M. (2008). *Branding New York: How a city in crisis was sold to the world.* New York, NY: Routledge.

Kasinitz, P. (1988). The gentrification of "Boerum Hill": Neighborhood change and conflicts over definitions. *Qualitative Sociology, 11*(3), 163–182.

Kaysen, R. (2013, February 26). After years of delay, a Lower East Side gap is ready to be filled. *The New York Times.* Retrieved from http://nytimes.com

Lees, L., Slater, T., & Wyly, E. K. (2008). *Gentrification.* New York, NY: Routledge/Taylor & Francis Group.

Marcuse, P., & van Kempen, R. (2000). *Globalizing cities: A new spatial order?* Oxford, UK: Blackwell.

Martinez, M. (2010). *Power at the roots: Gentrification, community gardens and the Puerto Ricans of the Lower East Side.* Lanham, MD: Rowman & Littlefield.

Mele, C. (2000). *Selling the Lower East Side: Culture, real estate, and resistance in New York City.* Minneapolis: University of Minnesota Press.

Moody, K. (2007). *From welfare state to real estate: Regime change in New York City, 1974 to the present.* New York, NY: New Press.

Newman, K., & Wyly, E. (2006). The right to stay put, revisited: Gentrification and resistance to displacement in New York City. *Urban Studies, 43*(1), 23–57.

Patterson, A., & Pinch, P. L. (1995). "Hollowing out" the local state: Compulsive competitive tendering and the restructuring of the public sector services. *Environment and Planning A, 27*(9), 1437–1461.

Patterson, C., Flood, J., & Moore, A. (2007). *Resistance: A radical political and social history of the Lower East Side.* New York, NY: Seven Stories Press.

Redfern, P. A. (2003). What makes gentrification "gentrification"? *Urban Studies, 40*(12), 2351–2366.

Roberts, S. (2007, August 29). New York's gap between rich and poor is nation's widest. *The New York Times.* Retrieved from http://nytimes.com

Smith, N. (1996). *The new urban frontier: Gentrification and the revanchist city.* London, UK: Routledge.

Turner, J. A. (1984). *Building boundaries: The politics of urban renewal in Manhattan's Lower East Side* (Doctoral dissertation). Available from ProQuest Dissertations and Theses database. (UMI No. 303289623).

von Hassell, M. (1996). *Homesteading in New York City: 1978–1993: The divided heart of Loisaida.* Westport, CT: Bergin & Garvey.

Zukin, S. (2010). *Naked city: The death and life of authentic urban places.* Oxford, UK: Oxford University Press.

CHAPTER FIFTEEN

Gentrification as Class Politics

JUDITH N. DeSENA

INTRODUCTION

This chapter analyzes the process of gentrification as a form of class politics. The concept of politics is defined from a feminist perspective that views everyday actions as engaging in politics. As Martha Ackelsburg stated, "Political life is community life; politics is attending to the quality of life in households, communities, and workplaces" (1988, p. 308). The working-class community in Greenpoint, Brooklyn, is in the midst of an economic crisis. It was historically damaged by economic restructuring in the United States whereby many industries lining the waterfront and employing residents closed. More recently, the cost of living in Greenpoint continues to escalate, outpacing the economic resources of its working-class residents. Gentrification is driving substantial increases in housing costs for renters and also homeowners whose property taxes are calculated based on market value. Food prices have also risen, with supermarkets stocking organic items, which cost more than nonorganic products. Greenpoint had been a neighborhood that residents from surrounding areas or local employees would shop because the prices were comparatively reasonable. Presently, many of Greenpoint's working-class residents shop in other neighborhoods. Mom-and-pop retail establishments are increasingly replaced by corporate chain stores like Starbucks, Rite Aid, Sleepy's and Brooklyn Industries. Commercial banks have multiplied, and gone are the savings and loan institutions that invested in the neighborhood by granting traditional, direct mortgages. Some independently owned stores and businesses have closed because of significant commercial rent hikes. This briefly describes the present economic predicament that Greenpoint's working-class confronts. With a focus on maintaining their place and their quality of life in Greenpoint, the working-class exercises its politics through mobilization and activism.

This chapter examines the divergent positions and strategies between a working-class community being de-stabilized by increasing gentrification and New York City policymakers regarding a rezoning and redevelopment plan for the East River waterfront in Greenpoint. It indicates the community's response to a proposed plan and the challenges posed by global capital by way of New York City government. It reveals the mobilization of the working-class community's assets in dealing with redevelopment. Through this process, the working-class community set into motion the social capital it amassed over many years. This is an attempt by ordinary people to confront the development efforts and bring attention to their consumption practices. Asserting their position, they made their voices heard regarding plans for their community of residence. Through their activism, they underscored the social and economic injustices proposed by policymakers. The chapter arrives at theory, which argues that global capitalism, through the actions of the state in accelerating gentrification on a community level, is not always successful in realizing its objective. The goals of the political economy are not entirely insurmountable. Local communities can mediate and mitigate structural and global forces on behalf of ordinary people. In the end, the working-class community managed to move beyond obstacles presented by gentrifiers and policymakers and had an impact on the future of the community by organizing through traditional, local institutions and demanding affordable housing.

Data were collected for this chapter through the use of participant observation. This researcher moved around the neighborhood as a resident and engaged in informal activities such as talking to neighbors, merchants, fellow diners at local establishments and parents with children at playgrounds and during family activities. One also took part in the community's more formal structures, through attendance at community meetings and events. In addition, one followed the rezoning and development plan discussed here through various newspaper accounts. This chapter is part of a larger, qualitative study of gentrification in Greenpoint, Brooklyn (DeSena, 2009).

THE SOCIOLOGICAL CONTEXT

The concept of gentrification is being defined as "the conversion of socially marginal and working class areas of the central city to middle-class [and elite] residential use" (Zukin 1987, p. 129). Research on gentrification focuses primarily on the causes and consequences of this urban process. Analyses fall into two major theoretical perspectives, ecological theory and critical theory (Wittberg, 1992). The ecologists examine the needs, tastes and desires of populations, which are responsible for precipitating neighborhood change in the form of gentrification (Friedenfels, 1992; Laska & Spain, 1980). Studies using an ecological framework argue that gentrification has numerous causes. Gentrification can stem from a greater need for housing by baby boomers, low vacancy rates and the relative high cost of suburban housing (Berry and Cohen 1973). Additional causes of gentrification, according to ecological theorists, include competition among social classes for urban space (Cybriwsky, 1978) and the cost of new housing and commuting as a cause for people remaining in the city (Laska & Spain, 1980), as well as lower fertility rates, since less living space is needed relative to larger families (Long, 1980).[1] Also included in this perspective are studies that examine the strategies used to create and produce gentrification and preserve upper-middle-class status in specific neighborhoods.[2] A more recent analysis of gentrification in Harlem and Clinton Hill, Brooklyn (Freeman, 2006), which examines how Black residents regard gentrification, finds that on some level it is welcomed because local services and amenities are upgraded. At the same time, there is little socializing between long-term residents and newcomers, and there are clashes between these groups over public practices, such as barbecues in the park and drinking on the corner. What exists is "strength of weak ties" (Granovetter, 1973).

Critical theorists, on the other hand, view the causes of gentrification as an outcome of the actions of the political economy, namely the investments of capital and the policies of the state (Abu-Lughod, 1994; Fitch, 1993; Smith, 1996; Zukin, 1982, 1987). The political economy takes on a "cultural" strategy that supports art and historic preservation (Zukin, 1982). In the case of New York's SoHo neighborhood, for example, the ultimate goal is not only gentrification but also the eventual attraction and residence of an elite community. Clashes between the political economy and ordinary people are also analyzed (Smith, 1996). In one conflict, city government imposed a curfew for Tompkins Square Park. Ordinary people attempted to "retake the city" by rioting against the police and engaging in other protest activities. Some residents believed that the city's claim to "clean up" the park of homeless individuals and drug traffic was actually a move to facilitate even more gentrification on Manhattan's Lower East Side.[3]

In response to the 1987 stock market crash and recession that followed, a change in the process of gentrification is documented by scholars. There was "a much reduced rate and impact of gentrification in favor of a more unevenly developed, polarized, and segregated city...the revanchist city is becoming a powerful reality" (Smith, 1996, p. 94).[4] Another major change is that the state is investing in the process more directly than in the past. In addition, corporate developers are the initial migrants prior to gentrifiers more so than before when they followed (Hackworth 2002).⁵ Moreover, actions against gentrification by ordinary people have been marginalized and ultimately successfully dismissed by those in power.

Unlike the research discussed previously, which primarily examines the causes and consequence of gentrification, this chapter investigates and analyzes the *process* of gentrification. In other words, it focuses on the period in between causes and complete change to a gentrified place.

There are clear consequences of gentrification for working-class communities. It is well documented that gentrification increases the cost of rental housing and homeownership. Property taxes and the general cost of living (food, services) also escalate with gentrification. Moreover, development of the built environment is augmented with the identification of gentrifiers seeking a lifestyle equipped with disposable income. At the same time, rental housing protections in the form of rent control and rent stabilization diminish, and rental housing in general is replaced with condominiums. Moreover, policymakers support the interests of the affluent over the working class and poor. This chapter bridges the ecological and critical views. It discusses the actions taken by both working-class residents in response to development and growing gentrification, and New York City government in promoting development and furthering gentrification. This chapter indicates that in Greenpoint, during the process of rezoning the waterfront, the working-class community surprised and outsmarted the power brokers by their ability to mobilize and unify behind the need for affordable housing.

Prior to the evolution of the issue of rezoning, there had been plans to expand garbage transfer stations and applications to build a power plant on Greenpoint's waterfront. The community was against these possibilities, since Greenpoint has its share of environmental problems resulting from toxin-producing industries and the polluted Newtown Creek. Therefore, rezoning away from manufacturing and toward residential and/or recreational use was thought to handle the threat of additional polluters locating in Greenpoint. A member of the community board was quoted in a *New York Times* article regarding rezoning:

> There's a pretty large number of people in the community that are opposed to anything above five or six stories…But 30 years of saying no to housing proposals on the waterfront is what brought us the proposal for a garbage transfer station, which we were very lucky to defeat, and it's what brought us the proposal for the power plant, which we hope to defeat. So I hope people realize that they've got to say yes to something here (Bahrampour, 2003, p. B4).

A viewpoint became prominent that rezoning and development were the "lesser evils" as opposed to more toxic waste and potential disease.

Instead of rigidly opposing development and increasing gentrification, this chapter analyzes how the working-class community attempted to carve out a place in the waterfront development plan and moderate the forces of capitalism.

THE NEIGHBORHOOD

Greenpoint is the northernmost neighborhood in Brooklyn, across the East River from Manhattan and separated from Long Island City in Queens by the Newtown Creek. Greenpoint has been a working-class community since its development, largely by immigrants from Europe, with factories lining the waterfront and manufacturing interspersed among homes. Many locals were employed at these factories, some of which were union shops. During the 1950s, Greenpoint experienced immigration from Puerto Rico. In the 1960s and 1970s, as city neighborhoods experienced White flight because of an increase of minority residents and suburbanization, Greenpoint maintained itself as a primarily White community with a relatively smaller proportion of Latinos and Latinas (DeSena, 1990, 2005). The 1980s brought another wave of Polish immigrants, as martial law was instituted in Poland in response to the Solidarity movement. Gentrification represents the latest transition for this community. The neighborhood never experienced racial transition but segregated ethnic diversity. It is presently experiencing a transition of social class.

Greenpoint is presently a community with three major populations: long-term working-class residents, immigrants and gentrifiers. The earlier gentrification of Williamsburg, a neighboring community, has had a spill-over effect into Greenpoint. Gentrifiers have been exploring Greenpoint for more living space at more affordable prices than in Williamsburg. In Greenpoint, one can now observe factory loft conversions to residences and art galleries, and sidewalk eateries on what were once desolate, industrial streets. Greenpoint has been a sought-after community with a de-industrialized waterfront. It was historically White and stable, affordable, physically well-kept, with high occupancy rates, low crime, high levels of social capital and a high degree of neighboring relative to other neighborhoods. Ironically, it is these very characteristics that attracted gentrifiers and developers. Its authenticity, as seen through the low scale of buildings and attendant small town charm, was appealing as a place of residence. Increasing gentrification has forced out many working-class residents and small businesses—the very people responsible for the maintenance of these attributes. In 2010, the total population of Greenpoint was 36,091 (U.S. Bureau of the Census, 2010). There were few changes in the racial profile of the neighborhood in that a large majority of the neighborhood remained White. The formal education of Greenpoint's residents has increased from 2000 to 2010. By 2010, the largest group of residents held bachelor's degrees. This is a dramatic change since 1980. By 2010, the median household income in Greenpoint was $56,143, indicating an increase of income. In terms of occupation, in 2010, management and business, science and arts professionals constituted the largest group, followed by sales and office occupations, and service (U.S. Bureau of the Census, 2010). This is in contrast to 1980, when the largest occupational group were technicians and salespersons, followed by operators and laborers (DeSena, 1990).

In terms of housing in 2010, 80% of households in Greenpoint are renters, and about a quarter of these are not rent regulated (Scott, 2003). Furthermore, "The percentage of affordable housing in Greenpoint is dropping faster than it is in New York City as a whole. The percentage of rental units in the most expensive category is skyrocketing" (Scott, 2003, p. 7). The notion of affordability is defined as an affordable rent burden in that no more than 30% of a household's income goes toward rental costs. In Greenpoint in 2000, "about 40% of households paid more than 30% of their income on rent" (Scott, 2003, p. 9). Market-rate rents for a two-bedroom apartment in Greenpoint in 2002 ranged from $1,500 to $1,900 monthly (Scott, 2003, p. 16). This represents a 50% increase since 1997. In 2013, a similar apartment was advertised in the window of a local realtor for $2,800 a month. Additional indicators of gentrification within the housing sector are featured in articles in the local newspapers, discussing "skyrocketing rents" and "loft regulations urged" ("Loft Regulations Urged", 2001; "Skyrocketing Rents", 2001). Houses are selling for $600,000 and upward, especially in the historic district (Mooney, 2009).

REZONING AND REDEVELOPING GREENPOINT'S WATERFRONT

Greenpoint's waterfront overlooking Manhattan's East Side lay dormant for years, quieted by de-industrialization. Brick buildings, some with steel connecting bridges, and a number of piers jutting into the East River, were left to deteriorate, but they served as reminders of a vibrant past. In some cases, private warehousing companies located on the river closed off public piers. Such action served the public, since the piers were dangerously corroded and the city was not about to address it. From the late 1970s through the 1980s, waterfront utilization ideas by residents and plans by the local community board were developed. It was not until a sustained real estate boom in New York City continued into the new millennium, and Michael Bloomberg became mayor, that Greenpoint's waterfront became a desirable prospect to city planners and private developers. This administration took the position that if sufficiently developed, the waterfront could become a lucrative investment.

The Bloomberg administration can be characterized by the corporatization of government, which is executed through private development and revenue production. All of these actors are brokers of global capitalism. Luxury development, through rezoning, redevelopment and tax abatements, is believed by this administration to generate all kinds of economic activity for New York City. And much of the way that this administration produces revenue is on the backs of ordinary New Yorkers. In general, this administration acts as a facilitator of private, luxury development as opposed to performing on behalf of working-class residents.

The Bloomberg administration has also gained a reputation regarding housing development. In 2002, the administration proposed a housing program for New York City to include "roughly 60,000 housing units throughout the five boroughs over the next four years" (Steinhauer, 2002, p. B1). Greenpoint's large waterfront was already a development priority of the Bloomberg administration, since there is little space left in New York City in which to develop new housing, and the waterfront consists of huge parcels of land. The administration's initial plan for housing was revised by the negotiations around the development plan for Greenpoint's waterfront. The modification arrived at includes a tactic untried in New York City, a commitment to affordable housing through inclusionary zoning. This means that the specific parameters regarding affordable housing units are part of the written redevelopment plan. Inclusionary zoning is also discussed by Freeman (2006) in his study of Harlem and Clinton Hill. This approach has since been integrated into the mayor's housing plan citywide.

Ultimately, through a thoughtful reconceptualization of the city's plan by Greenpoint residents, coupled with an unrelenting activation of the community's social capital, Bloomberg's commissioners were taken by surprise. The Bloomberg administration was confronted by a community injured by gentrification, one that was struggling in the face of it as the cost of living spiraled upward and out of reach of many long-term and lifelong residents. The administration was informed that its plan would only serve to accelerate gentrification and problems for Greenpoint's working class. In fact, after a community rally, an assistant commissioner commented to his colleague, who was also a former resident of Greenpoint, "Well, they're done now." His statement referred to the rally marking an end to the community's response. The planner, who had more of an intimate knowledge of the community, responded, "No…they've only just begun!" The administration underestimated Greenpoint's tradition of social activism and the power of traditional institutions, in this case, the Catholic Church and the local political club.

The Catholic churches in Greenpoint have a long history of serving working-class residents. For many, the churches are central to life's milestones. People are baptized, attend school, are married and buried all in the same church or a neighboring one. Through the years, these churches have experienced a crisis of maintenance. There are fewer priests to serve fewer parishioners and less revenue to support church facilities and schools. A number of Catholic schools in Greenpoint have closed in an attempt to contain costs. Another strategy to deal with their crisis is the creation of parish clusters. In Greenpoint, the local Catholic churches are grouped in an attempt to coordinate and plan their work within the community. The cluster realized a need to address a housing crisis in the community, since clergy were confronted with many congregants facing substantial rent increases and displacement.

The Greenpoint parish cluster was assisted by a student of urban planning who carried out a housing needs assessment in the area. The study documented the loss of affordable housing in the community, a dramatic increase of total family spending on housing and the purchase of homes in Greenpoint becoming more unattainable (Scott, 2003). The parish cluster organized meetings inviting community leaders, local politicians and community board members. Subsequently, Greenpoint's parish cluster and the adjacent Williamsburg parish cluster joined forces to mobilize around securing affordable housing for working-class and low-income residents as part of the waterfront rezoning package.

THE REZONING PROCESS

Rezoning land use in New York City is a legal process referred to as ULURP (Uniform Land Use Review Process) that is voted on by various levels of government in the city before it becomes law. The final vote rests with the city council.

Greenpoint's Waterfront Rezoning Task Force began its work with visions about the potential future of this parcel of Brooklyn's waterfront. As the task force got underway, it divided itself into various subcommittees. Those most active were focused on parks and open space, housing, height and bulk (of new residential towers), and economic development (which included industrial retention and jobs). In the course of meetings, a member asked the task force to make affordable housing its priority. For the most part, the task force rejected this position, claiming that all the areas were equally important, and affordable housing was just one issue.

AFFORDABLE HOUSING BECOMES NUMBER 1

Affordable housing emerged as the primary issue of the working-class community in Greenpoint, not parks or open space or aesthetics. As one resident expressed, "Aesthetic concerns are a luxury of affluence." The community mobilized through the tradition of the local Catholic churches. The parish clusters of Greenpoint and Williamsburg, which constitute 17 churches, organized their parishioners. A rally was held to "kick off" their work and to publicly state that they would fight for affordable housing. Over 500 residents were in attendance and were addressed by local clergy, the chairperson of the housing subcommittee of the Waterfront Rezoning Task Force, and local, elected politicians. In general, remarks criticized Bloomberg's "New Marketplace Housing Plan" for insufficient attention to the housing needs of low- and moderate-income New Yorkers.

The community's slogan and chant, "40% GUARANTEE" of affordable housing in development, evolved from this collective. The keynote speaker for the evening's event was Brooklyn's Catholic Auxiliary Bishop, Joseph Sullivan. Sullivan was well-known from his decades of service as the head of Catholic Charities in Brooklyn-Queens. A local newspaper reported Sullivan's address to the rally:

> "There's a crisis in housing in this city," the prelate thundered to the 800 or so people—middle class, poor, black, brown, white, Latino, Caribbean, Italian, Polish and every conceivable gradient in between.... This was a night for the people, not the great and grand. (Ryan, 2004, p. 1)

This was a power that the mayor and his commissioners had not foreseen. Thus, negotiations about affordable housing ended up being between the Catholic Church of Brooklyn-Queens through the parish clusters on behalf of the community, and New York City's government bureaucracy, with local politicians on the side of the Church and community. The issue of affordable housing became the major concern of rezoning and redevelopment for the working-class and adjacent poor community.

THE PLAN

The Waterfront Rezoning Task Force rebuked the original, proposed redevelopment plan presented by the Bloomberg administration. First, there was no guarantee of affordable housing units included in rezoning the waterfront or outside the rezoning site within the larger neighborhood. In addition, the plan lacked a statement defining income categories. The plan was also void of provisions giving priority of affordable housing units to current residents of Greenpoint. The call by the housing subcommittee was for "inclusionary" zoning, to make affordable housing part of the redevelopment plan, not an afterthought to an approved plan. In addition, the plan fell short of protecting existing manufacturing, which would be displaced by development. Thousands of industrial jobs would be at risk of being lost. Furthermore, open space was inadequate, and waterfront access was piecemeal. Finally, the proposed towers were too tall and bulky, resulting in the loss of sight corridors, and out of character with buildings in the neighborhood (Lander, 2004, pp. 6–8). Discussions between the task force and various city bureaucrats/specialists ensued, but ultimately, in line with the Waterfront Rezoning Task Force's recommendation, the community board voted to accept the plan with modifications.

The borough president of Brooklyn, Marty Markowitz, was next to vote. He "opposed an enormous rezoning proposal for Greenpoint and Williamsburg because it did not guarantee enough inexpensive housing, open space or economic stability for people already living in the area" (Cardwell, 2005a, p. B1). His vote mimicked the opinion of the task force.

In accordance with a requirement of the New York City Charter, the city planning department held a public hearing at a high school in the community in which residents chanted and held banners, "40% GUARANTEE!" while many residents staged a walkout as the City Planning Commission voted in favor of their plan.

In the meantime, in the New York State Assembly, local representatives Joseph Lentol and Vito Lopez sponsored a bill requiring that for developers to receive 421A benefits (tax abatements) in the Greenpoint-Williamsburg rezoning site; they must include a minimum threshold of affordable housing units. These local politicians, acting on a state level and working on behalf of the community, were attempting to pressure the city administration to revise the redevelopment plan for the waterfront. At this juncture, local elected politicians, like Lopez, were in daily contact with Church coalition leaders.

The plan moved forward to the city council. Testimony given by David Yassky, the council member representing Greenpoint, called for rejecting the plan. His testimony laid the groundwork for negotiations between the city council, on behalf of the community, and the Bloomberg administration. The night of the city council's vote, key community leaders sat with Vito Lopez, who talked by telephone with council members. Through the night and into the early morning, council members mediated between the community (Lopez and leaders) and the Bloomberg administration. Negotiations were tense and emotions were high on all sides. The administration wanted to complete the rezoning process and was willing to reconsider open space, parks and industrial retention. The most stressful part of the negotiations on all sides was over affordable housing.

The administration proposed housing vouchers to be used for affordable units. Lopez vehemently disliked this plan since it would pressure long-term and lifelong residents to move outside the neighborhood. He argued for the inclusion of apartments for ordinary people on the waterfront and within the larger neighborhood. Lopez could not be dissuaded and held firm to his position. In the end, the administration acquiesced, motivated more by politics than social justice for ordinary people. First, as a state assemblyman, Lopez chaired New York State's housing committee, making and altering legislation dealing with New York City's housing programs. The Bloomberg administration did not need to turn Lopez against it. Second, Lopez was leader of the Kings County Democratic Club. Gifford Miller, then city council speaker, in his plan to run for mayor, supported Lopez in the hope of gaining his and the Democratic club's endorsement at election time. All parties finally agreed upon a revised plan, and the city council voted to approve the modified plan based on the community's terms.

THE APPROVED PLAN

The final plan for the waterfront "rezone[s] a 175-block area of Greenpoint and Williamsburg" (Cardwell, 2005b, p. 1). It contains 54 acres of parks, including a publicly run esplanade that spans the entire waterfront. The plan also specifies an industrial business zone, protecting businesses within its confines, and creates a $4 million fund to preserve manufacturing jobs.

Housing development is the most elaborate part of this rezoning package. In order for developers to build to maximum height on the waterfront, which ranges from 30–40 stories, depending on the site, 20% to 25% of the units must be affordable to ordinary New Yorkers. Affordability is defined as 80% or below the median family income of the New York Metropolitan Area (which, for this purpose, includes the five boroughs and Putnam County) for low-income units and 125% or below for moderate income units. This would include maximum incomes (for 2005) from $35,150 up to $91,056, depending on family size and the number of bedrooms needed (Greenpoint-Williamsburg Inclusionary Housing Program, 2005, p. 6). In exchange for affordable units, developers would be eligible for a 25-year tax exemption. Those developers who choose not to include affordable units on their site would have their buildings limited in height and bulk to about 23–33 stories and would be ineligible for the tax exemption. Affordable units could also be located in buildings off-site but within the neighborhood and retain tax exemptions. This final version is "among the most ambitious such programs in the nation…and is the scene of the city's broadest test of inclusionary zoning" (Cardwell, 2005b, p. B6). The revised 421A program and inclusionary zoning are the latest tools used by the Bloomberg administration and the New York City Council to create affordable housing.

CONCLUSIONS

Ultimately, the "40% GUARANTEE" turned out to be more like 33%. The final outcome for the waterfront is bittersweet for Greenpoint. On the one hand, the working-class community strategically repositioned the issue of housing, and in the process, contributed to unburdening the community of the constant threat of toxic polluters locating in its midst.

The community also achieved the greatest amount of affordable housing units ever negotiated in one redevelopment project, and the plan makes future inclusion of them attractive to developers. In addition, this waterfront plan included the working class in the concept of affordability. The focus was not on the currently stigmatized notion of "low-income housing" but also housing for those who make this city operate "by the sweat of their brows" and work to preserve it every day. These are the Greenpointers, the Brooklynites and the New Yorkers who find themselves in the precarious position of being unable to afford life there. The redevelopment plan gives them some hope of managing, and it also gives priority for affordable units to residents.

Ultimately, the working-class community of Greenpoint serves as a model for others, illustrating that ordinary people can mitigate and mediate the forces of the state and capitalism in the processes of gentrification. In this case, the working class exercised its power through the traditional institutions of the Catholic Church and the local political club. This amounts to a huge, grassroots constituency with political clout in New York City. The actions of the working class in Greenpoint suggest the possibilities for reproduction elsewhere.

In the years ahead, Greenpoint will no longer be the same place where many residents grew up and lived. Gentrification and development will continue. The plan for the waterfront development may be stalled by the current economic crisis, but it will eventually commence, making neighborhood change by social class irreversible. The slow, multiyear, decade-long timetable for actually bringing affordable housing units to fruition will be too long for many current Greenpoint residents who are struggling to hold on to their homes, their neighborhood and their way of life.

NOTES

1. Additional demographic factors that focus on a changing family are also cited as contributors to gentrification. These include nontraditional living arrangements, increase of age at first marriage, delayed birth of first child, increasing rate of single and married women into the paid labor force and an increase of dual wage earner families. These changing demographics would account for lower fertility rates and smaller household size.
2. Kasinitz's study of Boerum Hill in Brooklyn (1988) indicates various strategies used by some residents to manufacture gentrification while at the same time their intention is to prevent deterioration. According to Kasinitz, these strategies include constructing a local history that could be advertised, renaming the neighborhood or portions of it, producing house tours to "show off" the neighborhood and participating in historic preservation by the creation of an historic district. Similarly, Krase's study of Lefferts Manor in Brooklyn (1982) discusses the use of a restrictive covenant in maintaining housing "single family only" in order to keep the community middle class.
3. Like Zukin, Smith believes that the culture industry adds to the growth of gentrification. Smith (1996) views the notion of "back to the city," a phrase associated with gentrification, as meaning that capital investment, not people, moved back to the city. For him, gentrification represents postmodern urbanism.

4. For Smith, upper-middle-class and elite gentrifiers found themselves with lowered property values and forced interactions with minority groups, immigrants and women as they competed for the urban terrain.
5. In other words, there is a tendency toward "If you build it, they will come."

REFERENCES

Abu-Lughod, J. L. (1994). *From urban village to east village: The battle for New York's Lower East Side.* New York, NY: Blackwell.

Ackelsberg, M. A. (1988). Communities, resistance and women's activism: Some implications for a democratic polity. In A. Bookman & S. Morgen (Eds.), *Women and the politics of empowerment* (pp. 297–313). Philadelphia, PA: Temple University Press.

Bahrampour, T. (2003, June 18). City seeking to rezone Brooklyn waterfront. *The New York Times.* Retrieved from http://www.nytimes.com/2003/06/19/nyregion/city-seeking-to-rezone-brooklyn-waterfront.html

Berry, B., & Cohen, Y. (1973). Decentralization of commerce and industry: The restructuring of metropolitan America. In L. Mazotti & J. Hadden (Eds.), *The urbanization of the suburbs* (pp. 431–455). Beverly Hills, CA: Sage.

Cardwell, D. (2005a, January 15). Brooklyn rezoning plan assailed. *The New York Times.* Retrieved from http://www.nytimes.com/2005/01/15/nyregion/15marty.html?_r=0

Cardwell, D. (2005b, May 3). City is backing makeover for decaying Brooklyn waterfront. *The New York Times.* Retrieved from http://www.nytimes.com/2005/05/03/nyregion/03brooklyn.html

Cybriwsky, R. (1978). Social aspects of neighborhood change. *Annals of the Association of American Geographers, 68,* 17–33.

DeSena, J. N. (1990). *Protecting one's turf: Social strategies for maintaining urban neighborhoods.* Lanham, MD: University Press of America.

DeSena, J. N. (2005). *Protecting one's turf.* Lanham, MD: University Press America.

DeSena, J. N. (2009). *Gentrification and inequality in Brooklyn: The new kids on the block.* Lanham, MD: Lexington Books.

Fitch, R. (1993). *The assassination of New York.* New York, NY: Verso.

Freeman, L. (2006). *"There goes the hood": Views of gentrification from the ground up.* Philadelphia, PA: Temple University Press.

Friedenfels, R. (1992). Gentrification in large American cities from 1970–1980. In R. Hutchison (Ed.), *Research in urban sociology: Gentrification and urban change* (Vol. 2, pp. 63–93). Greenwich, CT: JAI.

Granovetter, M. (1973). The strength of weak ties. *American Journal of Sociology, 78,* 1360–1380.

Greenpoint-Williamsburg Inclusionary Housing Program. (2005). The City of New York.

Hackworth, J. (2002). Postrecession gentrification in New York City. *Urban Affairs Review, 37,* 815–843.

Kasinitz, P. (1988). The gentrification of "Boerum Hill": Neighborhood change and conflict over definitions. *Qualitative Sociology, 11*(3), 163–182.

Krase, J. (1982). *Self and community in the city.* Washington, DC: University Press of America.

Lander, B. (2004, January 12). *Rezoning Greenpoint-Williamsburg: What does it really mean for the community?* Unpublished presentation for community leaders.

Laska, S., & Spain, D. (Eds.). (1980). *Back to the city: Issues in neighborhood renovation.* New York, NY: Pergamon Press.

Loft regulations urged. (2001, March 1). *Greenline,* p. 1.

Long, L. H. (1980). *The city-suburb income gap.* Washington DC: Special Demographic Analyses, U.S. Bureau of the Census.

Mooney, J. (2009, May 31). Polish is still spoken, but industry is history. *The New York Times.* Retrieved from htttp://www.nytimes.com/2009/05/31/realestate/31living.html?pagewanted=all

Ryan, M. (2004, May 6–31). Affordable housing: The people come out in force. *Greenline,* p. 1.

Scott, R. (2003). *Inclusionary zoning: A proposal for North Brooklyn and New York City.* Unpublished report.

Skyrocketing rents. (2001, March 1). *Greenline,* p. 1.

Smith, N. (1996). *The urban frontier: Gentrification and the revanchist city.* New York, NY: Routledge.

Steinhauer, J. (2002, December 9). Mayor envisions housing revival unmatched since the 80s. *The New York Times.* Retrieved from http://www.nytimes.com/2002/12/09/nyregion/mayor-envisions-housing-revival-unmatched-since-the-80-s.html

U.S. Bureau of the Census. (2010). *Census of population and housing* (generated by J. DeSena). Retrieved from http://factfinder.census.gov

Wittberg, P. (1992). Perspectives on gentrification: A comparative review of the literature. In R. Hutchison (Ed.), *Research in urban sociology: Gentrification and urban change* (pp. 17–46). Greenwich, CT: JAI.

Zukin, S. (1982). *Loft living.* Baltimore, MD: The Johns Hopkins University Press.

Zukin, S. (1987). Gentrification: Culture and capital in the urban core. *Annual Review of Sociology, 13,* 129–147.

CHAPTER SIXTEEN

Foreclosure Crisis AND THE Role OF Community Organizing IN A U.S. Latino Community

JAMES JENNINGS

INTRODUCTION

This chapter examines how a predominantly Latino and low-income community organized in response to a wave of foreclosures threatening to destroy their community. The social, cultural and economic infrastructure of U.S. urban Latino communities has been considerably weakened as a result of the recent and continuing foreclosure crisis. Latino "neighborhoods have sustained comprehensive and lasting damage," according to some observers (Ricks, 2009, p. 1). In addition to a reduction in property values and loss of homes, neighborhood consequences of spatial concentrations of foreclosures resulted in an aggravation of problems like overcrowding and decline in public safety. The spatial concentration in the location of foreclosures has adverse effects on the social and economic well-being of Black and Latino neighborhoods (Li & Morrow Jones, 2010). Small businesses and microenterprises in low-income and working-class communities have been closed as a result of foreclosures and loss of disposable income within those communities. Because tax revenues are reduced due to loss of business and homes, it means government services have also been impacted adversely.

The pace and numbers of foreclosures in Latino communities have produced adverse effects on spousal relationships and children's relationships with each other, weakened extended family and social networks and many times have resulted in decline in access to medical care, according to a sobering report published in

2010 by the National Council of La Raza (Bowdler, Quercia, & Smith, 2010). The physician and urban scholar, Mindy T. Fullilove (2001), described how the era of urban renewal in the 1950s and 1960s negatively impacts Black health conditions today. Massive numbers of foreclosures that are concentrated in urban areas are having similar disruptive effects on urban communities today.

The story of one small grassroots organization, Arlington Community Trabajando (ACT), shows the importance of political activism and community organizing as response to the foreclosure crisis wrought by corporate de-regulation and excesses. ACT works in a neighborhood with the highest number and rates of foreclosure petitions in the entire state of Massachusetts (Massachusetts Foreclosure Monitor, 2009). While this small Latina-run organization has not worked alone, it has played a significant role in organizing the community in its neighborhood; its work offers important lessons for similar communities in other parts of the nation. The author utilized agency documents and participant observations for this study and served as a strategic and organizational volunteer to ACT for more than five years. While retained by the U.S. Department of Housing and Urban Development's (HUD's) Community and Enterprise Development Center, based at Northern Essex Community College (2005–2008), the author also provided technical assistance and capacity-building services to other Latino and grassroots organizations in Lawrence, Massachusetts, in the area of neighborhood revitalization and economic development.

ACT's mission is to empower residents with community development initiatives and activities such as affordable housing, foreclosure prevention, first-time homebuyer education, family financial literacy and business and youth development.[1] It is the belief of the leadership of this organization, including its executive director, Ana Luna, and its board, that antiforeclosure activities will not be successful unless planned and pursued in ways that strengthen the social and economic fabric of the neighborhood, and this happens most effectively through community organizing. ACT has a small staff of three or sometimes four individuals, depending on the state of its budget, along with a dedicated cadre of community volunteers. Since its founding, this organization has provided a range of financial literacy workshops for low-income families. In addition, the organization has sponsored and organized youth services and summer employment and has worked with numerous small businesses and microenterprises in the development of business plans.

ACT's antiforeclosure work is integrated with these kinds of activities and is driven by community mobilization and raising the quality of civic engagement and participation on the part of residents. ACT emerged as an early leader in the struggle against the high number of foreclosures in the city. Over the last three years, it has worked with hundreds of individuals and families in the area of foreclosure prevention, credit counseling and financial literacy. Since 2007, ACT has prevented more than 150 foreclosures by directly negotiating with lenders and helping to modify loans. The organization has been involved with a broad range of community actors, including faith-based leaders and public health agencies, and has also worked with local businesses.

Unfolding Foreclosure Crisis

Corporate-driven policy preferences leading to deregulation of the banking and financial sectors opened the door to massive predatory lending in Latino and Black communities (Bowdler, 2005; Calhoun & Bailey, 2005). The deregulatory context included adoption of laws like the Financial Institutions Reform, Recover, and Enforcement Act (1989) and the Gramm-Leach-Bliley Financial Services Modernization Act (1999). Both laws were aimed at weakening the regulatory power of the Glass-Steagall Act (1933) under Roosevelt's New Deal, a law that prohibited the merging of commercial, insurance, investment and savings institutions.[2] Under a deregulated housing market, low-income and communities of color easily became prey for financial manipulation and voracious lending on the part of nonconventional financial institutions that had minimal state and federal oversight.[3]

As a result of deregulation and predatory lending, the loss of Latino-owned homes within a space of a few years is striking in many U.S. cities. Working-class communities, but especially Latino and Black communities of color, have been devastated by the loss of thousands of homes due to the inability to maintain mortgage payments, usually after high-priced loans began to trigger higher adjustments in mortgage interest rates. Janet Murguia, president of the National Council of La Raza, noted that, "An estimated 1.3 million Latino families will lose their homes to foreclosure between 2009 and 2012" (Bowdler et al., 2010, p. ii). Zip codes in Massachusetts with the highest proportion of African American and Latino families have been hit particularly hard with foreclosure petitions, the first step toward home loss (Massachusetts Foreclosure Monitor, 2009, p. 5). Researchers for the Federal Reserve Bank of Boston identified several causes for the recent wave of foreclosures, including the fall in housing prices, job losses and increases in interest rates (Reade, 2007).[4] But, they add that it is likely that "Blacks and Hispanics [have received] a disproportionately high share of higher-cost loans" (Reade, 2007, p. 17).

Rick Cohen, a journalist for the *NonProfit Quarterly* magazine, argues that the crisis reflected structural racism. It was partially a consequence of active targeting of low-income Black and Latino communities for predatory and unjustified higher-cost loans (Cohen, 2008).[5] The Center for Responsible Lending reported that Blacks and Latinos were easy targets for subprime lending and mortgage manipulation: "High levels of segregation create a natural market for subprime lending and cause riskier mortgages, and thus foreclosures, to accumulate disproportionately in racially segregated cities' minority neighborhoods. By definition, segregation creates minority-dominant neighborhoods, which, given the legacy of redlining and institutional discrimination, continue to be underserved by mainstream financial institutions" (Rugh and Massey, 2010, p. 632).

ACT and the Foreclosure Crisis in Lawrence, Massachusetts

ACT is located in the city of Lawrence with a population of 76,377 in 2010.[6] In the Arlington neighborhood, a community within Lawrence where ACT focuses its work, more than 80% of the residents are Latino/a. Almost three quarters of the residents five years and older in this neighborhood, or 73.1% compared to 8.8% for the entire state, speak Spanish at home as their primary language. The neighborhood is relatively small (10,637 persons in 2009) and very low income, where the per capita income in 2009 was $11,770, compared to the state's per capita income of $32,738. More than half of the 3,499 households in this neighborhood had incomes of less than $25,000, compared to 19.5% of all households in Massachusetts at this same income level.[7]

Over several years, this city and its poorest neighborhood were targeted as lush for economic exploitation through predatory lending and questionable profit-making on the part of financial and real estate interests. Real estate activity became especially intense after 2003 and 2004. Since 2003, the overall city had the third-highest subprime lending activity in the entire state, with 318 subprime loans representing 19.3% of total loans/mortgages made (2,985). This was a similar case for other predominantly Black and Latino urban neighborhoods, also hit with disproportionately high levels of shaky and unaccountable mortgage arrangements. Subprime mortgage lending was targeted to communities of color regardless of the income levels of Black and Latino applicants for home loans. Blacks and Latinos received considerably higher proportions of subprime loans when compared to Whites in comparable categories (Rugh & Massey, 2010, p. 631).

The Arlington neighborhood represented 13.4% of the city's total population in 2006 but 35% of all foreclosure petitions in this same year. Residents faced numerous "opportunities" from lenders and brokers to cut quick deals in order to become homeowners. Luna described one case where a particular broker "used God a lot in his advertising" in order to emotionally and ethnically manipulate potential homeowners to sign highly fallible mortgages.[8] At a November 2007 meeting, the author was told by administrators of Lawrence's Housing Authority that even residents in public housing, a sector with the lowest of household or family incomes, were being presented with "opportunities" to become homeowners. By 2008, Lawrence had the state's highest proportion of housing units affected by foreclosure petition activity at 48.9 housing units per 1,000, according to a highly respected housing organization, the Massachusetts Housing Partnership. The statewide figure for this same time period was 12.0 housing units per 1,000 (Massachusetts Foreclosure Monitor, 2009, p. 4). One small area in this neighborhood registered 63.7 housing units affected per 1,000, by far the largest ratio in the entire state of Massachusetts.

The effects of the high numbers and spatial concentration of foreclosures threatened the well-being and very existence of this neighborhood. The bunching of foreclosures has displaced many families in the Arlington community. According to observers, some of these families returned to the Dominican Republic or Puerto Rico; others simply relocated to live with relatives or friends, contributing to overcrowded housing. The foreclosure crisis has also resulted in a major reduction in wealth assets in terms of the savings held by working-class residents. A key economic sector in this neighborhood included hundreds of local and small businesses and microenterprises. These businesses, many with five employees or less, include beauty salons, barber shops, general contractors, furniture stores, convenience stores, dry cleaning and laundry facilities, clothing stores, grocery stores, auto repair shops and restaurants—collectively, a critical economic and employment base for the neighborhood. In addition to increases in vandalism, city services for the neighborhood were threatened to be reduced drastically. In 2007 and 2008, school closings and firefighter and police layoffs were discussed as a way of balancing the city's budget, a situation created partially by foreclosures and a drop in real estate taxes collected by the city.[9]

This neighborhood's situation is reflected nationally in other urban Latino communities. The Pew Research Center conducted an analysis and found that Latino households were more adversely affected by the mortgage industry crisis than those of White homeowners: "The bursting of the housing market bubble in 2006 and the recession that followed from late 2007 to mid-2009 took a far greater toll on the wealth of minorities than whites. From 2005 to 2009, inflation-adjusted median wealth fell by 66% among Hispanic households and 53% among black households, compared with just 16% among white households" ("Twenty-to-one: Wealth gaps," 2011, p. 1). The report noted the following:

> The net worth of Hispanic households decreased from $18,359 in 2005 to $6,325 in 2009. The percentage drop—66%—was the largest among all groups. Hispanics derived nearly two-thirds of their net worth in 2005 from home equity and are more likely to reside in areas where the housing meltdown was concentrated. Thus, the housing downturn had a deep impact on them. ("Twenty-to-one: Wealth gaps," 2011, p. 5)

Small business owners in this neighborhood rely heavily on the availability of disposable income on the part of local residents. One long-time beauty salon owner, for instance, attributed her decline in business not only to the bad economy but also to clients who are struggling to make ends meet as a result of the massive waves of foreclosures. Regarding this development, Luna (2010) noted that, "many clients are going to the salon less often and have also decreased the number and types of services they pay for." These are very small and localized businesses that must count on the expenditures of local residents who, whether homeowners or renters, have some degree of housing stability. According to Luna, many of these small businesses have closed as a result of the foreclosure crisis. One subsector of these businesses includes home-based child care centers, which represented an important support for working families who did not have access to bigger or more expensive facilities. These home-based centers, like other microenterprises, represented a significant employment base for residents. In 2010, the staff at ACT counted numerous child care providers closing their home-based businesses as a result of foreclosures. One of ACT's organizers, Anabelle Roldan, noted that child care providers experienced a major decline in service demand as foreclosures increased, which was due to clients' unemployment status and the fact that as unemployed parents, they were home taking care of their children.

ACT Responds with Community Organizing

Luna and other members of ACT continually highlight the importance of community building and mobilization as a critical part of an effective neighborhood-saving strategy. ACT has approached community organizing as an effort to help build a sense of bonding and collaboration among low-income and working-class residents and also as a way to organize social and political mobilization as a mechanism to respond to neighborhood problems and challenges. Under this framework, ACT focuses on small businesses as part of antiforeclosure efforts in these kinds of neighborhoods. Luna (2010) observed the following:

> Strategies such as civic involvement and technical assistance for businesses are key when rebuilding these affected communities. These strategies not only help re-engage community members to improve our neighborhoods, but also help support local businesses that may be struggling to continue. Once businesses are stabilized, their focus turns to helping the community and improving the services they provide.

There are several "big picture" tools utilized by ACT in pursuing community organizing: facilitating cultural empowerment, expanding civic discourse about spatial inequalities, participating and supporting coalitions and building community equity.

Facilitating Cultural Empowerment

ACT seeks to respect the cultures of residents as a key part of community organizing. All the activities planned by ACT are conducted in bilingual settings, and there is emphasis on recruiting and training residents to assist with workshops and activities. One of the most effective organizing tools utilized by ACT and its board is its annual Christmas toy drive. Beginning in 2007, ACT started to solicit donations and contributions from businesses, primarily to sponsor a free toy giveaway each Christmas season. These events were first held at a place called Rita Hall, but due to the hundreds of families that have come to participate in this initiative, the annual event was moved to a larger gymnasium at a local Catholic high school. The event has grown significantly over the past five years, from hundreds of individuals to close to a thousand. In 2012, the event drew approximately 1,100 people and children who received free toys. Many families attending the event are impoverished, and the opportunity to receive toys for children is most appreciated. As important, the event brings neighborhood residents together in an environment that is culturally comfortable and generates a sense of bonding.

This is also evident at another annual event utilized to mobilize residents. "National Night Out," a national anticrime, pro-neighborhood initiative, takes place the first Tuesday of every August. This event allows residents and local business leaders to participate in a host of street activities. ACT is a major organizer of this event and mobilizes youth to encourage residents to participate. ACT uses this event to present youth leaders with opportunities to practice newly learned community organizing and fund-raising skills.

Expanding Civic Discourse about Homeownership

ACT has triggered and expanded civic discourse about the meaning of homeownership as a neighborhood resource and not just along an individual dimension. In other words, homeowners and renters have similar interests in preserving the neighborhood and working together on a range of issues. This can be difficult, since both sectors have seemingly different economic interests. But residents are encouraged to understand the "big picture," as sociologist C. W. Mills once explained, and see their immediate economic interests as part of a broader network affected by political and economic decisions.[10] This requires that they understand the "scalar" politics and private and public decision making that created the housing crisis in the first place (DeFilippis, Fisher, & Shragge, 2010; Fraser, Lepofsky, Kick, & Williams, 2003). ACT has sought to do this through popular education workshops and neighborhood forums. It has also utilized a local radio program, *La voz del pueblo*, where related issues are discussed and listeners have an opportunity to ask a range of questions. Cable episodes have been sponsored by ACT not just to provide technical assistance information to residents but also as a platform for discussion of broad public policy issues. Through utilizing these popular mediums, as well as organizing numerous community meetings, ACT has provided information and talks to impress on residents that it is not just homes that are in jeopardy but the entire neighborhood, including businesses, nonprofit organizations and the very health of children and families.

Participating and Supporting Coalitions

ACT has worked with other organizations in order to encourage community coalitions to respond to foreclosures as a community-wide issue. An example of this kind of work includes ACT's leadership in triggering one of the first coalitions aimed at the foreclosure crisis engulfing the city. Even before foreclosures were on the radar screen of some public and civic officials, ACT and community development practitioner Mayte Rivera, then director of the Community and Enterprise Development Center, a HUD Community Outreach Partnership Center, helped spearhead a coalition of organizations in 2007, which included Neighborhood Legal Services, International Institute of Greater Lawrence, AMEDAL (Asociación de Ministros Evangelicos del Area Lawrence), Merrimack Valley Project, Office of Planning and Community Development, Lawrence Community Works, Bread and Roses Housing (a community land trust) and local business and religious leaders.

The broad aim of this coalition was to develop collaborative strategies aimed at immediately minimizing foreclosures and loss of homes. One action on the part of ACT involved community mobilization to petition the Lawrence City Council to pass a resolution on March 12, 2007, declaring the Arlington neighborhood a "foreclosure watch zone" for a period of five years. ACT and its supporters believed that passage of even an essentially symbolic resolution would be useful for planning and developing antiforeclosure strategies that should be "community-wide" and based on strengthening the social and economic networks and assets of this neighborhood, as well as helping individual homeowners. This may have been the first instance in which the city council officially considered the growing deleterious effects of foreclosures in Lawrence.

As stated earlier, the resolution did not carry any resources or enforcement capacity but was a significant acknowledgment that the problem of foreclosure was a neighborhood, and citywide, issue. ACT and its supporters, including Bread and Roses Housing, the Lawrence Housing Authority and various religious and business leaders, believed that such a designation could facilitate community outreach and future funding to prevent foreclosures. One of the early elected officials supporting the resolution, former city council member Nilka Alvarez-Rodriguez, also hoped that such a resolution could emerge as a civic engagement strategy for other cities in Massachusetts.

With planning and support from Rivera and the Community and Enterprise Development Center, a legislative breakfast to highlight concerns about the growing foreclosure crisis in Lawrence and other places was held. Along with a visit to Lawrence from then Governor Deval Patrick of Massachusetts, in which he lauded the work of Luna and others, these efforts helped lay the foundation for the state's adoption of a new law in November 2007 designed to prevent future foreclosures: Chapter 206 of the Acts of 2007: An Act Protecting and Preserving Home Ownership. Under this legislation, Massachusetts now provides some limited funding for a range of anti-foreclosure activities throughout the state. These include a mandatory waiting period (90 days) before banks can file a foreclosure in order to provide time for a homeowner to renegotiate a back-due mortgage. There is now a prohibition imposed on mortgage lenders from making a subprime variable or adjustable rate mortgage loan to a first-time home loan borrower unless the borrower affirmatively opts in writing for the variable or adjustable rate mortgage loan and receives certification from a nonprofit housing counselor; There are new tenant protections, including one that establishes that the lease for a tenant whose rental payment is subsidized under state or federal law shall not be affected by a foreclosure sale. While many organizations have contributed to these legislative accomplishments, ACT was one of the earliest to mobilize residents around effective antiforeclosure policy responses.

Building Community Equity

ACT links civic awareness and community building as a key component of financial literacy. Familiarity with checking and savings accounts is certainly important for families in low-income neighborhoods. At the same time, however, there should be a focus on helping residents understand how their quality of life is being compromised by financial decision making at political and economic levels to which they may have little direct access. This is critical because, as noted by Lydia Lowe and Doug Brugge in their 2007 case study of the Chinese Progressive Association in Boston, the raising of civic awareness about invisible political and economic decision making can possibly move to some "immediate improvements in the community as well as to alter the terms of public debate and raise popular consciousness toward building a longer-term social change movement" (p. 46). Information that makes individuals aware of how they can be exploited by real estate companies, check-cashing outlets or credit card companies is vital in low-income urban communities in order to prepare and buffer residents from the kind of economic exploitation that helped create the foreclosure crisis in the first place. Broader approaches to financial literacy, which go beyond simply what an individual or family can do to balance its own checkbook, can elevate the discourse about how a community and neighborhood can be economically exploited and also show how a community can use its fiscal and financial assets and resources to expand economic opportunities for residents.

ACT takes this latter approach by working with individuals and families but also helping small neighborhood businesses and microenterprises learn more about the accumulation of wealth. One workshop in 2008, for example, brought a small group of business representatives together and exposed them to the market characteristics of the neighborhood that they were not aware of, for instance, the amount of disposable income, the particular consumer expenditure patterns and the aggregated accumulated wealth of residents. The purpose was not only to enhance the capacity of these businesses but also to show how linked the well-being of their business is to that of the neighborhood. Financial literacy workshops have been organized for residents and businesses, reaching more than a hundred residents and families. These workshops have triggered a few efforts on the part of residents to investigate and implement ideas for local business development and entrepreneurship. The latter have been provided with additional workshops focusing on the development of business plans, credit and debt management, understanding of local regulations and opportunities to network with larger enterprises.

More recently, ACT entered into a collaborative project with a housing developer, HOPE Companies, and another grassroots organization, the Lawrence Youth Team, and with *Youth Build*, they collaborated to build a small number of affordable housing units where families and youth would be linked to a range of human services.[11] Under this venture, ACT will also be working with the Lawrence Youth Team to provide job training placement opportunities for young adults in construction. The four housing units will be affordable to area residents who will be connected to a network of supportive services to ensure that they understand how to hold and grow with the property. Given the number of abandoned housing units in Lawrence, ACT hopes that while small in scope, this emerges as a model for other organizations to work collaboratively on cooperative homeownership.

CONCLUSION

In recent years, the foreclosure crisis in the United States has destroyed homes and neighborhoods. National responses on the part of the U.S. government have been limited in that the framing of the foreclosure crisis is presented and treated as a problem for individuals and families rather than entire communities. The federal government has committed approximately $6 billion under the American Recovery and Reinvestment Act (February 2009) and the Housing and Economic Recovery Act of 2008's "Neighborhood Stabilization Program" to assist local and state governments in responding to the significant jumps in foreclosed and abandoned properties. The Homeowner Affordability and Stability Plan adopted by the U.S. Congress in February 2009 addresses, in part, concerns of holders of subprime loans, and the decline in property values, where mortgages and equity are lower than the market value of the home.[12]

These are significant policy efforts, to be sure. But low-income and working-class residents in places wrecked by foreclosures require much more focus on understanding the "big picture" and knowing how to mobilize collectively on behalf of the well-being of their neighborhoods. This is noted by urban planner Alan Mallach (2009):

The wave of home mortgage foreclosures that began in 2006 continues to surge, greatly destabilizing neighborhoods, towns and cities across the United States. Without robust, carefully-targeted federal policies to mitigate the community-level impacts of foreclosures, local and state efforts will invariably fall far short of what is needed. (p. 2)

Mallach further observed, "Since the mortgage crisis erupted in 2006, the federal government has played only a limited role in attempting to mitigate its effects on families and communities" (p. 3). A similar emphasis is made by the Latino Policy Forum: "Helping those harmed by this crisis, whether by foreclosure, eviction, or depreciation in property value, requires a comprehensive understanding of the roots of this crisis.... Policies and community initiatives should address the potential increases in overcrowding and homelessness amongst Latinos" (Feliciano & Hernandez, 2008, p. 16).

Grassroots organizations like ACT, although small, are key for community-building strategies. More broadly, sociologist Susan Ostrander (2013) highlighted the critical importance of smaller nonprofit organizations as a support for shared governance in a democracy. She examined a "voluntary association convened in summer 1998 with the specific aim of establishing a public process for future development of the 145-acre flat piece of land directly bordering the city of Boston..." (p. 51). This case study indicates that community-based nonprofit organizations are an important piece for efficacious democracy and representation but also that this sector works parallel to the state, especially when the state does not represent interests of more vulnerable or less powerful groups. This reflects part of the role played by ACT in Lawrence.

At least two observations or lessons emerge for Latino leadership at the local level based on the work of ACT. First, strategic responses to the foreclosure crisis must be based on strengthening the social, economic and cultural fabric of communities. The community, in a sense, must be the unit of analysis in terms of evaluating ideas in this area. Second, Latino urban and grassroots organizations must work to elevate and mobilize the voices of people and families affected adversely by corporate-driven policies on local land. Doing this helps guarantee that the strategies and policy changes advocated will be comprehensive and equitable.

We should be reminded that ACT's community organizing and related actions represent a tradition of activism reflected in earlier struggles on the part of Latino working-class communities in the United States. These lessons, part of a tradition of community organizing in Latino urban history, are not new. As a result of community-building and mobilization, for instance, Latinos in Boston were able to establish *Inquilinos Boricuas en Acción,* the first resident-owned Puerto Rican cooperative housing establishment in the nation, to ensure the availability of affordable housing in the South End neighborhood (Small, 2004). In the 1960s and 1970s, community building and street protests in New York City were utilized to eliminate or reduce toxic environmental conditions in mostly Puerto Rican neighborhoods and to ensure that disenfranchised parents could provide input regarding the education of their children (Gandy, 2002; Jennings & Chapman, 1998). And, earlier than these examples, in Los Angeles circa 1930s and 1940s, Latinos pursued community and labor organizing to ensure improvement in working conditions, which benefited all workers (Sanchez, 1993).

A final thought about ACT and community organizing. I believe that struggle and conflict between powerful corporate interests and local communities over the ownership, control and utilization of urban land is one of the most important political issues today.[13] An example of such a struggle is being played out in many low-income and working-class communities seeking to preserve homes and neighborhoods in the face of the current foreclosure crisis, although this is being played out as well in conflict regarding strategies for local economic development.

Some scholars have proposed that community building and organizing are limited in their capacity as a tool to effectively challenge corporate power or rectify its excesses in these areas. In short, powerful interests can exploit community building to expand the globalization of cities, making them more attractive to capital and corporate elites. Fraser et al. (2003) noted, ironically perhaps, "the basic foundation of the community-building field function to mitigate the responsibility of extra-neighborhood, public and private institutions whose (in)action has played a major role in the creation of de-valorized neighborhoods" (p. 421). There certainly is some validity to this concern in that relatively powerless residents are pitted against much more powerful and wealthy interests. And the former can succumb to divide and conquer strategies, as we have seen in many instances.

At the same time, however, grassroots organizations like ACT are providing examples of how low-income and working-class residents can try to take some control of their communities' space and resist disempowering developments that point to a loss of social and economic well-being. Community organizing is not a panacea, but it can help lay a foundation for residents to understand why and how they, and their homes and urban spaces, are being targeted for economic exploitation. This is not a universal remedy, of course. But ACT's community organizing is critical, nevertheless, as an example of struggle and resistance, and some victories—for low-income and working-class communities in other places seeking to protect their communities against the excesses of corporate power and agendas.

NOTES

1. See the graduate thesis by Amy E. Kuykendall, "Assessment of Latino/a Community-Based Organizations: The Case of Arlington Community Trabajando, Inc., in Lawrence, Ma," Tufts University, Urban and Environmental Policy and Planning Department, (2012), for a case study of the ACT organization.

2. See Financial Institutions Reform, Recover, and Enforcement Act 1989: FIRREAf. Pub. L. 101–73.103 Stat. 183 (August 9, 1989); Gramm-Leach-Bliley Financial Services Modernization Act, 1999: Pub. L. 106–102, 113 Stat. 1338.
3. See Final Report of the National Commission on the Causes of the Financial and Economic Crisis in the United States, *The Financial Crisis Inquiry Report,* The Financial Crisis Inquiry Commission (2011, January).
4. A potential explanation for the dramatic increase in foreclosures is the high level of unemployment in some places. At least one national study, however, disputes this suggestion. By reviewing historical trends in the relationship between unemployment and foreclosures, the study found that, "the connection between unemployment and foreclosures has been weak. During previous periods of high unemployment, while delinquency levels did rise, foreclosure numbers remained essentially flat" (Bocian, Li, & Ernst, 2010, p. 14).
5. Predatory lending can include a range of nefarious practices, including charging higher interest rates without relationship to credit worthiness and targeting communities of color and the elderly with deceptive and high-pressure marketing of mortgages and loan packages (see Squires, 2003).
6. U.S. Census (2010).
7. See http://quickfacts.census.gov/qfd/states
8. K. Cooper interview for *New England Ethnic News*, A. Luna, personal communication, November 12, 2007.
9. Email correspondence, Anabelle Roldan, March 30, 2013.
10. C. W. Mills (1959).
11. Press Release, September 18, 2012: "ACT and Charles HOPE Companies announce a collaboration to provide new affordable housing for Lawrence, job training for youth."
12. See Homeowner Affordability and Stability Plan, Executive Summary, THE WHITE HOUSE, Washington (February 18, 2009).
13. See James Jennings and Julia Jordan S. Zachery's *Urban Spaces: Planning and Struggles for Land and Community* (Boston: Lexington Books, 2009).

REFERENCES

Bocian, D. G., Li, W., & Ernst, K. S. (2010). *Foreclosures by race and ethnicity: The demographics of a crisis.* CRL Research Report. Retrieved from http://www.responsiblelending.org/mortgage-lending/research-analysis/foreclosures-by-race-and-ethnicity.pdf

Bowdler, J. (2005). *Jeopardizing Hispanic homeownership: Predatory practices in the homebuying market* (Issue Brief, No. 15). Washington, DC: National Council of La Raza.

Bowdler, J., Quercia, R., & Smith, D. A. (2010). *The foreclosure generation: The long term impact of the housing crisis on Latino children and families.* Washington, DC: National Council of La Raza. Retrieved from http://www.nclr.org/index.php/publications

Calhoun, M., & Bailey, N. (2005). Predatory lending: Undermining economic progress in communities of color. *Poverty & Race Research Action Council, 14*(1), 17–19.

Cohen, R. (2008). *A structural racism lens on subprime foreclosures and vacant properties.* The Ohio State University, Kirwin Institute for Study of Race and Ethnicity. Retrieved from http://www.racialequitytools.org/resourcefiles/cohen.pdf

DeFilippis, J., Fisher, R., & Shragge, E. (2010). Contesting community development: Promoters and critics. In J. Jennings & J. Jordan-Zachery (Eds.), *Urban spaces: Planning and struggles for land and community* (pp. 117–134). New York, NY: Lexington Books.

Feliciano, R. D., & Hernandez, R. P. (2008). *Nuestro hogar: Addressing the foreclosure crisis in the Latino community.* Chicago, IL: Latino Policy Forum.

Fullilove, M. T. (2001). Root shock: Consequences of African American dispossession. *Journal of Urban Health, 78*(1), 72–80.

Fraser, J. C., Lepofsky, J., Kick, E. L., & Williams, J. P. (2003). The construction of the local and the limits of contemporary community building in the United States. *Urban Affairs Review, 38*(3), 417–445.

Gandy, M. (2002). Between Borinquen and the barrio: Radical political activism in the New York City Puerto Rican community 1969–1972. *Antipode, 34*(4), 730–761.

Jennings, J., & Chapman, F. (1998). Puerto Ricans and the community control movement: An interview with Luis Fuentes. In A. Torres & J. E. Velázquez (Eds.), *The Puerto Rican movement: Voices from the diaspora* (pp. 280–295). Philadelphia, PA: Temple University Press.

Li, Y., & Morrow-Jones, H. A. (2010). The impact of residential mortgage foreclosure on neighborhood change and succession. *Journal of Planning Education and Research, 30*(1), 22–39.

Lowe, L., & Brugge, D. (2007). Grassroots organizing in Boston Chinatown: A comparison with CDC-style organizing. In S. Ostrander & K. Portney (Eds.), *Acting civically: From urban neighborhoods to higher education* (pp. 44–71). Medford, MA: Tufts University Press.

Luna, A. (2010, June;). Interviews by J. Jennings.

Mallach, A. (2009). *Metropolitan Policy Program: Stabilizing communities: A federal response to the secondary impacts of the foreclosure crisis.* Washington, DC: The Brookings Institution.

Massachusetts Foreclosure Monitor. (2009, January). Table 3: Foreclosure petition activity, top 20 municipalities; Table 4: Foreclosure petition activity, top 20 Massachusetts zip codes. Boston: Massachusetts Housing Partnership.

Mills, C. W. (1959). *The sociological imagination.* London, UK: Oxford University Press.

Ostrander, S. A. (2013). Agency and initiative by community associations in relations of shared governance: Between civil society and local state. *Community Development Journal, 48*(4), 511–524.

Reade, J. (2007, November). *Foreclosures in New England.* Boston, MA: Federal Reserve Bank of Boston.

Ricks, N. W. (2009). *Factsheet: Arrested development: Foreclosures eroding the Latino community.* Washington, DC: National Council of La Raza.

Rugh, J. S., & Massey, D. S. (2010). Racial segregation and the American foreclosure crisis. *American Sociological Review, 75*(2), 629–651.

Sanchez, G. (1993). *Becoming Mexican American: Ethnicity, culture and identity in Chicano Los Angeles, 1900–1945.* New York, NY: Oxford University Press.

Small, M. L. (2004). *Villa Victoria: The transformation of social capital in a Boston barrio.* Chicago, IL: University of Chicago Press.

Squires, G. D. (2003). The new redlining: Predatory lending in an age of financial service modernization. *Sage Race Relations Abstracts, 28*(3&4), 5–18.

Twenty-to-one: Wealth gaps rise to record highs between whites, blacks and Hispanics. (2011, July 26). Pew Research Center, Social and Demographic Trends. Retrieved from http://www.pewsocialtrends.org/files/2011/07/SDT-Wealth-Report_7-26-11_FINAL.pdf

CHAPTER SEVENTEEN

Community Parading AND Symbolic Expression IN Post-Katrina New Orleans

DIANE GRAMS

POST-KATRINA AND POST-BIG EASY

The process of gentrification in New Orleans, as in other urban areas, has been one of revalorization of neglected inner cities, competition for the most valued properties and class-based transition from lower- to middle- and upper-class residents. The disaster following Hurricane Katrina's landfall on August 29, 2005, in which 80% of the city flooded, stimulated a real estate boom for property above sea level throughout the region. As competition over revalorized high ground heated up, poor and working-class residents living in communities facing post-Katrina transformation took to the streets.

Amid the destruction, Black New Orleanians turned to their historic parading practices to make their presence known. Public parading for African Americans in the city has long been a way to celebrate the living and honor the dead. After Katrina, parades were an important way for African Americans to continue this cultural practice, while also asserting a spiritual presence of resilience and claiming ownership of increasingly valued space. Returning to participate in parades provided locals living in the post-Katrina diaspora with opportunities to reengage and even reestablish life back in New Orleans. The neighborhood-based parades referred to as "Second Lines"[1] were a vehicle for returnees to rebuild social ties. While these parades were a nourishing ritual for body and soul, they also provided a forum through which to resist unjust rebuilding policies that enabled gentrification of working-class neighborhoods and disinvestment in those people most devastated by the post-Katrina flood. During this research, these parades were the subject of increased permit fees (Blumenfeld, 2007; Troeh, 2006), increased regulation (Reckdahl, 2007), and surveillance (Barrios, 2010). Such efforts by civic authorities to constrain these parades were as persistent as community efforts to maintain them.

In this essay, I recount a series of racist city policies that unfolded in the aftermath of the disaster and that sought to suppress the parades of the predominantly Black Social Aid and Pleasure Clubs (SAPCs) and Mardi Gras Indian tribes. As Sakakeeny (2013) acknowledged, contemporary noise and permit ordinances have been inconsistently enforced to constrain impromptu revelry of Blacks in their own historic neighborhoods, which have been increasingly invaded by middle-class Whites. I focus on how the resurgence of this historic cultural practice helps shape strategic social action in the time of recovery and how these practices are integral to the rejuvenation of the city's historic Black communities.

BACKGROUND OF A DISASTER

From the 17th century, the city's most advantaged residents—historically a merchant class augmenting the power of royal appointees and governing officials—have been situated on the high ground above sea level along the crescent-shaped curve of the Mississippi River; the largely uninhabitable low lying "swamp" areas were left to an array of "noncitizen" groups, including Native Americans and "maroon"[2] communities of runaway slaves. Through two centuries of industrial growth, the city expanded as a patchwork of ethnically and racially segregated neighborhoods: first expanding on the high ground along the river from the French Quarter, then through swamplands drained by an elaborate system of pumps and levees allowing the patchwork of neighborhoods to extend from the Mississippi River all the way to Lake Ponchartrain. Mid-century highways and bridges traversed urban poverty, resilient swamps, the Mississippi River and lakes surrounding the city, stimulating White flight to suburbs north and west, leaving most of the lowland historic city to a predominantly Black population.

When Hurricane Katrina made landfall, it did not select its victims by race or class, but the organization of the city was such that the most valued property was also on the safest ground. So when breached levees allowed the 80% of the city below sea level to flood, it was, as experts predicted, the "most disadvantaged…[who were] most likely to be located in harm's way" (Erikson, 1976/2006, p. vi). Indeed, the worst cases of victimization and greatest percentage of those displaced in the aftermath of Katrina were disproportionately elderly, and poor and working-class African Americans.[3]

Although most of the city properties—even those along the crescent of the Mississippi—sustained some sort of damage from the storm and from the weeks of neglect in the postdisaster shutdown, those homes that were not flooded and those homes constructed with stone or brick with plaster interiors were habitable in the months after the storm. But the vast majority of surviving wood-framed homes with drywall interiors would have to be fully gutted and refurbished before they could be inhabited. This meant that more than half the city's population already evacuated from the city would exist in a post-Katrina diaspora for as much as a year or two.

In the fall 2005, rumors began to emerge of plans for a "new" New Orleans, one that was smaller and turned large portions of African American neighborhoods into nonresidential parkland. The trauma of the destruction was replaced by fear that the city would permanently disinvest in the neighborhoods owned by generations of African American families. Such rumors were compounded by the difficulty some residents faced returning home, including selective demolition of neglected but habitable properties in neighborhoods intended for rebuilding, selective enforcement of tax code violations leading to property repossessions and a rental market excessively inflated by the influx of private insurance reimbursements and government housing vouchers worth as much as five times the prestorm rental rates. Together, these policies ominously pointed to an intentional rebalancing of the city's racial proportion from predominantly Black to White. Indeed, a sense of Confederate resurgence filled the air, as was eloquently summarized by Clyde Woods (2005) in his post-Katrina volume of *American Quarterly:* "There is an omnipresent fear, rarely openly discussed in the literature, that the tortured past of Alabama, Louisiana, and Mississippi is reasserting itself. No longer content to haunt the American psyche, it aspires to be resurrected" (pp. 427–428).

RESURGENCE OF PARADING TRADITIONS

The importance of public parades[4] in New Orleans became apparent when residents who did not evacuate the city began parading, even though many neighborhoods remained flooded and highways into the city were cordoned off, blocking evacuees' return. On October 9, 2005, as the New Orleans City Council took up consideration of plans for an abbreviated Mardi Gras the coming year (Koenig, 2005a, 2005b), a jazz funeral was staged to celebrate the life of a well-known chef, Austin Leslie, who died in Atlanta shortly after the storm. And then in late November, there was an abridged Second Line parade by the Black Men of Labor (BMOL) Social Aid and Pleasure Club (SAPC), whose annual parade date on Labor Day weekend was canceled because of Katrina. Contemporary Sunday Second Line parades are a 20th-century incarnation of historic jazz funerals, but they take place on the anniversary of the inaugural parade of a particular SAPC. Second Line parades are sponsored by a particular club on the same Sunday each year[5] to celebrate its members both living and those who have passed. The rescheduled 2005 BMOL parade, reportedly staged and funded in part by Spike Lee, was the centerpiece of the cultural section of Lee's (2006) four-hour HBO series, *When the Levees Broke*.

By December 18, 2005, when the Original Prince of Wales mounted its annual Second Line parade, recovery of the city came to be gauged by the recovery of its parading culture.[6] The storm delayed the club's 77-year-long practice of staging its anniversary parade on the second Sunday in October. One of the city's oldest parading SAPCs, its members nonetheless proceeded amid the city's dysfunction and destruction. They did so, according to its president, Joe Stern, to cheers and tears of just about everyone remaining in the city (Joseph Stern, interview with Alexandra Moon, 2007). The parade signaled for many that this important element of New Orleanian culture might return. Moreover, it motivated cultural advocates to take the lead toward recovery so that the second lining would remain part of the "cultural gumbo" that characterizes life in the city.

BMOL and Original Prince of Wales are among the 45 different SAPCs that sponsor and participate in weekly parades during the 38-day parade season extending from the end of August to mid-June. The total number of parades in the post-Katrina city has grown from approximately 30 parades in the season following Katrina's landfall (September 2005 to mid-June 2006) to an estimated 150 annual parades in 2012.[7]

Participatory Forms of Parades

Participatory forms of New Orleanian parades have been traced back through the 19th century from a "circum-Atlantic"[8] funerary tradition (Roach, 1996, pp. 56–63), colonial street masking and informally organized public revelry (Kinser, 1990, p. 8) and the modern form of Mardi Gras spectacle parade first staged by the Mistick Krewe of Comus in 1857 (Kinser, 1990; Roach, 1996).

"Second line" anniversary parades exemplify the participatory character. In such parades, a lead group of dancers followed by a brass band of 10–15 musicians moves through the city attracting joiners. As the parade progresses between destinations, honoring the living and remembering those who have passed, it reconstitutes its sense of community through ritual celebration.

The participatory parades of the Black community take place on the backstreets of the city. Just as the Second Lines wind their way through neighborhoods on Sundays from Labor Day to Father's Day, the same community of participants dress as Mardi Gras Indians and meander through the backstreets of the city while traditional Mardi Gras parades roll down the main streets. As members of SAPCs and Mardi Gras Indian tribes are predominantly poor or working-class African Americans, their communities were among those most vulnerable to the damage

Divide and Conquer: Segregated Ordinances for Segregated Parades

My interest in the city's recovery began as an outsider living in Chicago and conducting research on local culture and cultural policy when Hurricane Katrina hit. I had interviewed local cultural producers in New Orleans for a book on nonprofit organizations and had featured discussion of community-based activities sponsored in conjunction with the 2003 Louisiana Purchase bicentennial. I followed the post-Katrina recovery effort, including development of a cultural plan in late 2005 under the leadership of famed trumpeter Wynton Marsalis and the development of the Unified New Orleans Plan by late 2006. I read reports on arts blogs and sociology list-serves as plans for 2006 Mardi Gras and then the city's efforts to suppress the parades in the Black neighborhoods unfolded. For most outsiders like myself, the postdisaster recovery unfolded as a case study of structural racism at work.

The event that gave the New Orleans Police Department (NOPD) the leverage it needed to act to limit the parades of Black New Orleans was an incident of gun violence on Martin Luther King, Jr., Day, January 15, 2006, in the week before Mardi Gras. Thirty-two different SAPCs joined in a Second Line parade under the theme "ReNew Orleans." The collective parade was simultaneously a celebration in honor of Dr. King, a protest of disinvestment in communities and a call for evacuees' right to return. Near the end of this massive Second Line parade, gunfire sprayed though the crowd from one group of bystanders to another, killing one person and injuring two others (Michelle Longino, interview with author, 2009). While opportunist shooters not involved in any SAPC in the parade perpetrated the violence, the police took action against the parading groups. Two weeks after the incident, NOPD Superintendent Warren Riley released a memorandum effectively pricing SAPCs off the street. The parade permit fee, once in line with the $750 cost to parade during Mardi Gras, was raised to over $7,000. The memo required each club sponsoring a Second Line parade to pay $4,000–$7,000 as well as provide a $10,000 bond to receive a permit to parade. According to the SAPCs, the new fee amounted to cultural racism (Scheets, 2007), as most clubs would be unable to afford such fees.

Although such news was prevalent and available to outsiders, I would soon learn just how culturally segregated New Orleans was. When I was interviewed for a faculty position in early 2007 at Tulane University, I was aware of the lawsuit filed in federal court against the city and the state by SAPCs claiming violation of both First and the Fourteenth Amendment rights. However, I would learn that my soon-to-be new colleagues were unaware of the lawsuit, and few had ever been to a Second Line parade. Local press did not cover the parades. In fact, parades were left unmentioned in the *Times Picayune* unless violence was somehow associated.

The suit, filed in late 2006 by the American Civil Liberties Union on behalf of the Social Aid and Pleasure Club Task Force and with 17 representatives of different clubs (*Social Aid and Pleasure Club Task Force et al. vs. City of New Orleans et al.*, 2006), claimed the fees established by the police department were disproportionate to the efforts required by the police, amounting to unconstitutional infringement on the First and Fourteenth Amendment Rights of the SAPCs. The suit compared Mardi Gras parades (which extend several miles in length, featuring as many as 30 floats and requiring hundreds of police) to Second Line parades (which require as few as 10 officers to manage traffic). As Second Lines are a mobile form of parade, police officers ride on horseback, motorcycle or in patrol cars and move with the second liners, blocking traffic only as the parade moves through a neighborhood. Furthermore, Second Line parades are not random private events, but rather they are historically sanctioned, having occurred for more than 100 years. Moreover, competition by SAPCs for parade dates has led to a routinized schedule of parades every Sunday from the last Sunday in August through mid-June. Established clubs have an annually designated parade date; new clubs are able to secure a Sunday parade date once a previously established club abandons it.

In a deal arrived at before the courts on April 27, 2007, the NOPD agreed to cut the fee in half (Meeks, 2007). The deal was heralded for helping to pave way for a resurgence of weekly parades. However, the deal did not result in a ruling for treatment of the SAPCs equal to the Mardi Gras krewes. Although the new base permit fee of approximately $2,000 was substantially less than what the NOPD initially wanted to charge, it was still three times the cost of a Mardi Gras parade permit ($750).[9] NOPD could still charge a fee greater than other parades, but the fee structure now had to be based on what it claimed were its actual costs. Rather than creating cultural equity by requiring an equitable fee structure for all kinds of parades, the deal created a new intermediary structure in which the Social Aid and Pleasure Club Task Force was empowered to act as an intermediary between SAPCs and the NOPD (Wynoka Boudreaux, interview with author, 2010; Tony Hookfin, interview with author, 2009; Michelle Longino, interview with author, 2009). The clubs were able to pay the increased fees through a fund established by Quint Davis, the producer and director of the New Orleans Jazz and Heritage Festival.[10] With an annual budget of $60,000–$100,000, the Norman Dixon Sr. Annual Second Line Parade Fund would contribute an average of $2,000 to each SAPC to cover a portion of the city permit and band fees for their annual Second Line parade.

As an outsider who had never attended a Second Line parade, it was unclear to me exactly how the city had succeeded in isolating the parades staged by the predominantly Black SAPCs with its restrictions. State laws regulating Second Line parades are general ordinances for parade permits. They specifically exempt Mardi Gras and other parades on officially sanctioned holidays, funerals, parades sponsored by government, labor unions and the military from the same provisions (LA RS 14:326; NO MC Sec. 154). Local ordinances gave the NOPD the authority to establish fees for nonexempted parades as if the police department were a private escort service. By contrast, 50 pages of municipal code (MC Chapter 34) set Mardi Gras apart from all other holiday parades by defining what constitutes a Mardi Gras krewe, requiring affidavits of nondiscrimination, limiting the number of Mardi Gras parades to 34 and establishing the permit fee of $750 for each of the parades it permits, among other provisions. Similar protections do not exist for the Second Line form of parade, one that is as historically present in the city, yet popular in predominantly Black neighborhoods. With this ruling, Second Lines remained "unrecognized" within municipal code and susceptible to arbitrary restrictions privileging local cultural events on state-sanctioned holidays tied to a tourist economy being promoted as central to a "new" New Orleans.

Parade Participation

My first Second Line experience happened on September 30, 2007, barely a month after I moved to New Orleans. I learned of the parade from Nasha, a security guard at our apartment off St. Charles Avenue. She later told me about the Ladies of Unity, a new club founded by her and her sister Nokie, whose inaugural parade was scheduled in December (Nasha Pleasant, interview with author, 2007). Nasha would become my entrée into much of the city's Black cultural production, a nexus of Second Line parades and Mardi Gras Indians, and their music and dance.

As an ethnographer, I participated in 64 parades, out of an estimated 228 parades staged by the Black community in the city over six parade seasons (August 2007–June 2013). These parades, which proceeded from neighborhood to neighborhood, provided a ground level view of recovery. On parade days, I would interact with parade participants with digital camera and camcorder in hand, capturing both the activities of the parade and the neighborhood context. I paraded with other second liners as the parade briskly moved through trashed and abandoned side streets, wound around dumpsters or stopped for food and refreshments. As participants learned of my ethnographic interest in public parades, I was invited to more events beyond a club's annual parade. In my first year, I documented the annual parades of 17 SAPCs.[11] I became an honorary member of the Ladies of Unity, as I documented all of their activities over five years.

Besides taking photographs and video, I conducted informal interviews at parade stops and in-depth interviews before and after the parades with 48 parade participants. Through such activity, I came to understand how parades are planned and organized, how and why relatively poor people invest so much every year for their annual parade and how funds are raised to cover the $10,000–$20,000 estimated to stage such events. More importantly, I came to understand and share the centrality of parading in the lives of New Orleanians. As part of my role, I provided club members with copies of pictures and video on DVDs and later began uploading low-resolution versions to YouTube and a Facebook page available to my "friends" only. This sharing of images led to me being considered by some club members as "their personal photographer."

At the weekly Second Line, I joined local residents who actively asserted what it means to inherit this incredible tradition of Black culture. These practices emerged in a city where historic local ordinances, following Catholic theological tradition, deemed slaves "free" on Sundays; slaves were "free" to gather, make music, dance and buy and sell their own goods in what was known as "Congo Square," today's Louis Armstrong Park, located between the French Quarter and the historic Black neighborhood of Tremé. It was here, among a rich variety of African and Caribbean men and women gathered on Sundays, where the performative traditions of communal dance[12] emerged. From the Civil War era on, an array of private social clubs, segregated by class, race and gender, emerged to maintain the cultural traditions of the city's groups. These clubs built their activities around a tradition of dancing, balls and public parades, while shaping the repertoires and rituals of social life that continue today.

WHY PROTECT THESE PARADES?

As music and collective performance have been the tools through which Black Americans have historically resisted domination and stimulated social change, it is not surprising that music and parades provided means through which Black New Orleanians faced the resurgent injustices in the post-Katrina city.

The social action, the networking, the raising of money, interaction with police and other civic entities all required to stage parades were all part of the apparatus through which residents negotiated their survival and rebuilt their everyday lives. Parades involved the gathering of people to plan, create and implement the parade, the municipal regulations that simultaneously constrain and enable the parade, and a distinctive alternative economy that included people who made the outfits and parade accessories, including fans and baskets, musicians, and venders selling refreshments along the parade routes.

Interviewees focus on the culture involved in parading, that is, the yearlong collective efforts to plan a parade and the parade activities itself, as the way in which the community combats street violence. Young people involved in street violence are said to have been "lost to the streets" or that the "streets got 'em." Every Sunday, the community gathers to collectively take back the streets in celebration. As accounted by Aaron Atkinson, "It's knowledge that we try to spread. And for the young kids [we do it] so they can follow the family traditions and values in the families. Children are socialized into the parading culture as a way to pass on the knowledge of the culture" (Aaron Atkinson, interview with author, 2008). As Wynoka Boudreaux, a mother of a preteen boy who is a spy boy for his grandfather's tribe, the Golden Eagles, claimed, "It's our way of keeping the world from eating them up. After school, they might play a little sports and do their homework, but then they have to get to their beadwork" (Wynoka Boudreaux, interview with author, 2007).

Paraders might best be considered "cultural activists" who mobilize the processes of cultural production to construct and maintain their social networks and social lives. In doing so, they tap the resilience that has been a product of these performances. Such leadership acts in the character of the "Second Line," that is, setting activity in motion so that the rest would follow. In the era of post-Katrina recovery, they provide the momentum to rebuild homes, neighborhoods and the city while creating a stable reality within a complex, fragmented and unpredictable postdisaster environment that increasingly threatens their removal.

Although Mardi Gras parades are seen as cultural symbols of the city and are widely understood as important to its cultural milieu, the routine activity of weekly parading among the Black working-class New Orleanians remained largely invisible. Parade schedules, routes and information about parading clubs were not publicized; moreover, local or national media did not portray these parades as cultural symbols of the city. Not only are they as important to the cultural milieu of the city as other parades, but these parades produce a symbolic world; they create and perpetuate the symbols and the ideas through which the participants define themselves (Alexander & Smith, 2001; Durkheim, 1912/1995; Geertz, 1973; Turner, 1982).

There is little disagreement among scholars that these parades are rooted in African and Caribbean religious and festive practice (Gaudet & MacDonald, 2003; Gill, 1997; Harrison-Nelson, 1996; Lipsitz, 1990; Regis, 1999; Roach, 1996) and have been important to the development of improvisational jazz (Berry, Foose, & Jones, 1986; Lomax, 1990; White, 2008). Although this New Orleanian parading practice has developed over two centuries for most practitioners of Western, White American culture, it is often misunderstood. The difficulty lies in the fact that outsiders approach the activity as a spectacle parade, akin to, for example, watching the Macy's Thanksgiving Day Parade. However, these New Orleanian parades are better understood as participatory ritual. In short, they are not meant to be watched but to be joined. Through such collective participation, the line separating the observer and observed dissolves into shared experiences of "collective effervescence" (Durkheim, 1912/1995, p. 224). In such moments of euphoria, the sights, sounds and experiences of unification with a community become embedded in individual memory. In New Orleans, these moments are recalled, repeated and modified each week when one participates in another parade sponsored by yet another club whose creativity is evident by how members modify the ritual and its associated symbols. Their meanings are built and rebuilt through the communicative interaction and through the shared experience of collective effervescence.

Such performance has a long history in Black culture and is widely viewed as part of the continuum of African cultural practices "retained" (Herskovits, 1958; Southern, 1971/1997) through slavery and helps to explain what has been maintained in the postdisaster city. Contemporary scholarship of cultural forms in Black America (Baraka, 1999/2002) focuses on how Black cultural forms and narratives shift in response to historical context. It is not surprising that the culture of a people who emerged from slavery into poverty and disadvantage is conveniently designed to accommodate disaster, even the kind that followed Katrina. The form of the Second Line parade is based upon the premise of a lead club and its band using music and dance to set in motion a communal celebration. Similarly, as Lipsitz (1990) argued, the reoccurring theme of the popular counternarrative enacted by Mardi Gras Indians is one of resistance, that is, of heroic warriors resisting domination (p. 236). The ongoing expectation of the Indian remaking his suit each year using patches from previous suits and even from others' suits undermines Western artistic traditions of permanence and ownership in favor of a cultural experience made anew each and every year. This practice is experiential rather than object-oriented and therefore incorporates a process in which artisans are quite able to pick up the pieces from a disaster and start anew. The postdisaster environment provided new context for parading practice to begin anew, while reusing from the old that which was relevant.

HAVE TIMES CHANGED?

A violent shooting on Mother's Day, May 12, 2013, mirrored the 2006 shooting, but a change occurred in the city's response. In a brazen act of violence, two shooters sprayed the crowd at a Second Line parade, injuring 19 people. Coverage of the shooting broadcast internationally framed it as the senseless violence it was, rather than using the incident to quash Second Line parading or suppress Black culture. The day after the shooting, comments by White city officials acted from the playbook of the recent Boston Marathon tragedy and sought to mobilize the city in support of the parading tradition and against street violence, rather than tap into the long history of racial tension.

At a press conference called the day after the violence, Mayor Mitch Landrieu stated, "Everybody on this street knows that what happened yesterday has nothing to do with the cultural beauty of New Orleans. It happened during a sacred event." Similarly, NOPD Superintendent Ronal Serpas declared, "There is no question that Second Lines are a culturally important factor and a fact of life in the city of New Orleans. This [shooting] we believe had nothing to do with the Second Line, except it occurred near where the Second Line was happening." Both took the lead in what was ultimately a successful effort to identify, locate and arrest the perpetrators.

This moment seemed to mark a turning point in the cultural rejuvenation of post-Katrina New Orleans. But was it a moment when civic authorities began to value the city's vastly diverse parading culture? Did city authorities recognize Second Line parades as a conduit for a shared experience of joy, exuberance, healing, and even violence prevention? Or were authorities simply performing well in the international spotlight? Or does it mark a new era of regulation and incorporation of Second Lines into a tourist economy for a newly gentrifying New Orleans?

As I argue in this essay, public parades have been central to the post-disaster recovery of the city, not because of their capacity to stimulate economic development or for their exotic appeal to tourist interests, but instead for the meaning they have for participants. It is time for the parades of Black New Orleans to be recognized and protected by municipal ordinance and state law, not simply regulated into a tourist economy. Paraders produce and maintain a festive culture that defines a mythopoeic space rich with reinvented symbols of the past and makes possible distinctive individual and group identities for a present in which the assertion of their presence is crucial. Such heightened moments of experience renew ideas that define a group and recreate memories important to individual and collective identity and are worth rebuilding a city for.

NOTES

1. "Second Line" or "Second Line parade" refers to a unique form of neighborhood-based parade indigenous to black communities in New Orleans. The term also refers to the act of participating in such a parade and the form of dance movement typical of participating, that is, "to second line." It also refers to the people who participate, that is, "second liners."
2. See Midlo-Hall (1992) for discussion of Native Americans and Maroon communities living on the margins of the colonial city.
3. According to a Congressional Research Service report to Congress, "one-fifth of those displaced by the storm were likely to have been poor, and 30% had incomes that were below 1½ times the poverty line. African Americans are estimated to have accounted for approximately 44% of

the storm victims. An estimated 88,000 elderly persons (age 65 and older), many with strong community ties, may have been displaced, along with 183,000 children, many of whom were just starting the school year when the storm struck" (Congressional Research Service, 2005, p. 1). Gotham (2007) and Molotch (2005) each provided accounts of the racial tension in post-Katrina New Orleans. Elliott and Pais (2006) provided an analysis of survey data from evacuees, illuminating the influence of race and class on individual preparations and responses to the storm.
4. By "public parade," I refer broadly to processional activity, sponsored by a nongovernmental social or cultural organization, taking place on city streets and involving costumed performers and live music from marching bands, brass bands or percussionists with chanting choruses. See Grams (2013) for a detailed discussion of the types of parades in post-Katrina New Orleans.
5. Such as, for the BMOL, the first Sunday in September, which falls on the day before Labor Day.
6. Joel Dinerstein (2009) provided a moving account of these first post-Katrina parades.
7. This figure is calculated from the weekly emails from the Backstreet Museum list of Second Line parades published by Social Aid and Pleasure Club Task Force, the Mardi Gras parade schedule in New Orleans Municipal Code (MC Chapter 34-1) and published announcements of additional parades in association with festivals, sporting events, political events, holidays and funerals.
8. Joseph Roach developed a theory of circum-Atlantic culture as "geohistorical locale" that "insists on the centrality of the diasporic and genocidal histories of Africa and the Americas, North and South, in the creation of the culture of modernity" (1996:4).
9. These efforts to suppress the parades by the poor and working classes by regulating them out of existence have been part and parcel of modernity since early 19th-century France, when local festivals were suppressed in order to centralize culture and its power within the state (Hunt, 1984, pp. 205–206).
10. The New Orleans Jazz and Heritage Festival is one of the city's largest nonprofit organizations with a 2011 budget topping $27 million (http://www.Guidestar.org).
11. These included Goodfellas, Ladies of Unity, Men of Unity, Original Prince of Wales, Men of Class, Nine Times, New Orleans Bayou Steppers, Uptown Swingers, Lady Buck Jumpers, Sudan, Women of Class, Original Four, Mahogany Ladies, Black Men of Labor, Versatile Ladies of Style, the Young Men Olympian and the Original CTC Steppers.
12. For descriptions of bamboula and calinda in Congo Square, see Samuel Kinser (1990, pp. 36–39).

REFERENCES

Alexander, J. C., & Smith, P. (2001). The strong program in cultural sociology. In J. Alexander (Ed.), *The meanings of social life*. New York, NY: Oxford.

Baraka, A. (1999/2002). "Blues people: Looking both ways." New introduction, to *Blues people: Negro music in white America*, 1963 by LeRoi Jones, aka Amiri Baraka. New York, NY: Harper Collins.

Barrios, R. E. (2010). You found us doing this, this is our way: Criminalizing second lines, super Sunday, and habitus in post-Katrina New Orleans. *Identities*, *17*(6), 586–612.

Berry, J., Foose, J., & Jones, T. (1986). *Up from the cradle of jazz: New Orleans music since World War II*. Athens: University of Georgia Press.

Blumenfeld, L. (2007, April 17). Not wash away: The fight for New Orleans' culture continues, one parade at a time. *The Village Voice*. Retrieved from http://www.villagevoice.com/2007-04-17/music/not-wash-away/

Congressional Research Service. (2005). *Hurricane Katrina: Social-Demographic: Characteristics of impacted areas*. Washington, DC: Library of Congress.

Dinerstein, J. (2009). Second lining post-Katrina: Learning community from the Prince of Wales social aid and pleasure club. *American Quarterly, 61*(3), 615–637.

Durkheim, É. (1912/1995). *The elementary forms of religious life* (K. E. Fields, Trans.). New York, NY: Free Press.

Elliott, J. R., & Pais, J. (2006). Race, class and Hurricane Katrina: Social differences in human responses to disaster. *Social Science Research, 35*, 295–321.

Erickson, K. (1976/2006). *Everything in its path: Destruction of community in the Buffalo Creek flood*. New York, NY: Simon & Schuster.

Gaudet, M., & MacDonald, J. C. (Eds.). (2003). *Mardi Gras, gumbo, and zydeco: Readings in Louisiana culture*. Jackson: University of Mississippi Press.

Geertz, C. (1973). *The interpretation of cultures*. New York, NY: Basic Books.

Gill, J. (1997). *Lords of misrule: Mardi Gras and the politics of race in New Orleans*. Jackson: University of Mississippi Press.

Gotham, K. F. (2007). *Authentic New Orleans: Tourism, culture and race in the Big Easy*. New York, NY: New York University Press.

Grams, D. (2013). Freedom and cultural consciousness: Black working class parades in post-Katrina New Orleans. *Journal of Urban Affairs, 35*(5) 501–529.

Harrison-Nelson, C. (1996, September 8). Guardians of the flame: Upholding community traditions and teaching with art in New Orleans. *In Motion Magazine*. Retrieved from http://www.inmotionmagazine.com/flame.html

Herskovits, M. J. (1958). *The myth of the Negro past*. Boston, MA: Beacon Press.

Hunt, L. A. (1984). *Politics, culture, and class in the French Revolution*. Berkeley: University of California Press.

Kinser, S. (1990). *Carnival American style: Mardi Gras at New Orleans and Mobile*. Chicago, IL: University of Chicago Press.

Koenig, R. (2005a, October 12). *Carnival continues to roll*. Retrieved from http://www.nola.com

Koenig, R. (2005b, November 23). *City gears up for Mardi Gras 2006*. Retrieved from http://www.nola.com

Lee, S. (2006). *When the levees broke: A requiem in four acts*. New York, NY: Forty Acres and a Mule Productions.

Lipsitz, G. (1990). *Time passages: Collective memory and American popular culture*. Minneapolis: University of Minnesota Press.

Lomax, A. (1990). *Feet don't fail me now* [Film]. Association for Cultural Equity at Columbia University, New York, NY. Retrieved from http://www.folkstreams.net

M. C. Chapter 34. Code of Ordinances, City of New Orleans, LA. Retrieved from http://library.municode.com/index.aspx?clientID=10040&stateID=18&statename=Louisiana

Meeks, D. (2007, April 25). NOPD halves fees for second-line parades. *Times Picayune*. Retrieved from http://blog.nola.com/times-picayune/2007/04/nopd_cuts_parade_fees_for_seco.html

Midlo-Hall, G. (1992). The formation of Afro-Creole culture. In A. R. Hirsh & J. Logsdon (Eds.), *Creole New Orleans: Race and Americanization* (pp. 58–90). Baton Rouge: Louisiana State University Press.

Molotch, H. (2005). *Death on the roof: Race and bureaucratic failure.* Social Science Research Council. Retrieved from http://forums.ssrc.org/understandingkatrina/death-on-the-roof-race-and-bureaucratic-failure/

Reckdahl, K. (2007, April 24). Parade fee fight rolls to court. *Times Picayune.* Retrieved from http://blog.nola.com/times-picayune/2007/04/parade_fee_fight_rolls_to_cour.html

Regis, H. A. (1999). Second lines, minstrelsy, and the contested landscapes of New Orleans Afro-Creole festivals. *Cultural Anthropology, 14*(4), 472–504.

Roach, J. (1996). *Cities of the dead: Circum-Atlantic performance.* New York, NY: Columbia University Press.

Sakakeeny, M. (2013). Music versus noise (pp. 36–37). *Louisiana Cultural Vistas.* New Orleans: Louisiana Endowment for the Humanities.

Scheets, G. (2007, March 29). Permit fees raining on second-line parades. *Times Picayune.* Retrieved from http://blog.nola.com/topnews/2007/03/permit_fees_raining_on_secondl.html

Southern, E. (1997). *The music of Black Americans: A history* (3rd ed.). New York: W. W. Norton. (Original work published 1971).

Troeh, E. (2006, May 6). 4/6—Second line fee increase puts strain on tradition. *WWOZ Street Talk.* Retrieved from http://wwozstreettalk.blogspot.com/2006/05/46-second-line-fee-increase-puts.html

Turner, V. W. (1982). *From ritual to theatre: The human seriousness of play.* New York, NY: Performing Arts Journal Publications.

White, M. (2008). New Orleans African American musical traditions: The spirit and soul of a city. In M. Marable & K. Clarke (Eds.) *Seeking higher ground: The Hurricane Katrina crisis, race and public policy reader* (pp. 87–106). New York, NY: Palgrave MacMillan.

Woods, C. (Ed). (2005). Do you know what it means to miss New Orleans? *American Quarterly, 57*(4), 1005–1018.

Afterword

Things Have Fallen Apart but We Are Planning to Stay

MINDY THOMPSON FULLILOVE

Loïc Wacquant opens this volume with the challenge that we move past the "gentrification of gentrification research," the too-easy celebration that making a neighborhood wealthier and prettier has solved a social problem. The chapters that follow delineate the seriousness of gentrification as a symptom of the state of American capitalism, as a threat to human communities and as a process whose inexorability is hard to deny. I want to discuss two issues raised or implied in this book. First, gentrification represents the latest of many upheavals that have undermined communities and led to a serious social fracture in the United States. Second, we, as human beings, need stable communities. We must shift from acquiescence to policies of displacement to affirmation of our plan to stay.

WHAT DO I HAVE IN COMMON WITH A FOOT SOLDIER OF THE TEA PARTY?

In the fall of 2013, the U.S. government was shut down by a confrontation between the two parties—the Republicans, driven by their radical Tea Party fragment, and the Democrats, led by President Barack Obama. Ostensibly about preserving "freedom" from the "slavery" of obligatory health insurance, the Tea Party argued that it was better to have no government than to have the modest enhancement in health insurance promised by "Obamacare." This confrontation exemplifies what is really at stake in our understanding of gentrification: the chasms that have developed in a society that has ferociously employed divide-and-conquer strategies to immobilize the working class.

Let us start with a simple definition of "working" class, using the old-fashioned concept that one's relationship to the means of production defines one's class. By this definition, "working" class includes all of us who have to work for a living, who do not own means of production or have a trust fund that provides a source of income. By this simple definition, the people who live in the neighborhoods, and those arriving to "gentrify," are all working people, albeit some are paid better than others. The title of "gentrifier," by its association with the aristocracy, names and reifies a difference that undermines the ability of the class to name and assert its interests. Indeed, the stories of resistance to gentrification in this volume are stories of *communities* resisting change, rather than the *working class* resisting change or, better yet, putting forward some demands. What I have in common with a foot soldier of the Tea Party is that we both work for a living. We probably also have in common flat wages eaten away by rising costs, anxiety for our children's futures, fear of minimal security for our old age and even uncertainty about that old standby *The Weather*. Yet we stand staunchly divided by *Issues*, like Obamacare. It is hard to imagine that she and I might get to a conversation about our common interests, but we surely have them.

One might pose a different question, then, and ask this: How did the working class fall apart? Of course, the divisions of the working class are many and go back to the birth of capitalism, so it is more to the point to consider the divisions that catch our attention in the present moment. I think the answer emerges with great clarity throughout this volume: The American industrialized economy was replaced by a financialized economy. Suddenly, "working" class came to refer to old-style industrial workers, including construction trades and civic workers from teachers to sanitation workers, while "middle to upper" class became the referent for people who work in white-collar positions in, education, healthcare (often abbreviated as "eds and meds" [Dubb, 2013]), and finance, insurance, and real estate–often shortened to "FIRE" [Fitch, 1996]). These are the people who have enough money and privilege to "gentrify," and they emerge, at least in the way we identify them, as a distinct sector, no longer seen as "working" class. As industry leaves the cities, and FIRE/EdsMeds enter, the industrial working class is pushed out and the residential space is given over to the so-called gentry. Yet we must acknowledge that these individuals work very long hours in the sweatshops of their own industries.

The story of the transition from industry to finance is intertwined with the story of gentrification in another way, having to do with the commodification of housings, such that "homes" replace "widgets" in the production of profit. Investment in neighborhoods is stripped of connection to "use value." People are moved willy-nilly because large organizations can make money from creative destruction and new development.

This is a terrifying moment in human history, the moment at which, in pursuit of money, we start cannibalizing the roots of our own habitat, the bedrock of our society. Making habitat a source of profit has introduced a destabilizing factor, similar to the hyper-exploitation of the fields and forests and oceans. The key to making money from habitat appears to be its reorganization for new use, not its maintenance for existing use. This puts the stability of human communities squarely in the crosshairs of profit. Because the consequences for human health are most flagrant among the poor, the capitalists are able to use "health disparities" to justify new displacements, whether it is clearing slums or dispersing concentrated poverty. Despite the stated "good intentions" of helping the poor to a "better place," in fact, these clearances follow, and even intensify, the well-established lines of division by class and race, aggravate the "health disparities" and create the forerunners of new clearance in decades to come.

The even more terrible product of this constant upheaval is a deep fracture among social groups. As Rodrick Wallace has explained, upheaval breaks social bonds, both strong and weak ties (Wallace, Fullilove, & Flisher, 1996). Groups crystallize around the strong ties of family, ethnicity, religion and political organization. As they differentiate, the groups take on characteristic identities and behaviors that operate as part of an inextricable behavioral language of self-affirmation. The behaviors are necessary, even when they are self-destructive, because they assert the viability of the individual and the individual's group within that fracture.

In the United States, upheaval has followed upheaval, and economic transformation has undermined the security of many. We can list, at a minimum, the federal urban renewal program of the 1950s and 1960s; the de-industrialization that has nearly eliminated the American manufacturing sector; planned shrinkage and other forms of disinvestment; unmitigated epidemics of violence, addiction and infectious disease; the development of the prison-industrial system and explosion of incarceration; and the "new urban renewal" of HOPE VI and gentrification (Fullilove & Wallace, 2011). Each of these programs has led to the rupture of social bonds and the dispersal of populations. None has been adequately mitigated and some have impeded recovery. Mass incarceration, for example, by linking a felony conviction to denial of housing, education, employment and voting rights, has created a permanent "criminal sector" with little hope for a return to preconviction life (Alexander, 2010).

The fractures caused by these many upheavals have diminished or eliminated connections that used to exist. Congresspeople from the two sides of the aisle used to socialize, Americans used to read the same books and listen to the same music and many people ascribed to the value of the nation supporting the common welfare. This interaction has been eroded by decades of unmitigated upheaval and community dispersal. Shutting down the government is only the tip of the iceberg of damage that the commoditization of habitat has caused in our nation.

Somehow, we must find common cause. I thought that one of the most helpful points in this volume was in the chapter by Grams, which discussed the Second Line parades that occurred in New Orleans, where African Americans affirmed something about the city post-Katrina that, eventually, the leaders of the city agreed was its cultural core and essential to its future. It is that that all of us must find, the thing that gets past the corrosive difference and allows us to visualize a common future. Whatever I have in common with the foot soldier of the Tea Party, she and I must find it so that all of us may survive.

WE ARE PLANNING TO STAY

Whether people think it's good or bad, they do agree that gentrification replaces one population (poorer) with another (richer) one. It is that depopulation that leads to social fracture, to alienation and collapse of the collective. Whatever we think gentrification buys us, we can't afford it any more. We have to plan to stay.

> Morrish and Brown (2000) opened *Planning to Stay* with the following thought:
>
> Planning a neighborhood is a participatory act of community membership and an expression of belief about the future of one's community. Before residents and merchants can begin planning for their neighborhood, participants must make a sincere declaration about themselves: "We're planning to stay." Planning to stay in a neighborhood can be a transforming experience in which participants discover new dimensions of being a good neighbor and a good citizen. When coupled with planning, "to stay" becomes an active verb. Once this active role is embraced, two questions need to be addressed:
>
> 1. What is it about this place that draws us here?
>
> 2. What could we add to this place that will keep us here?

"Planning to stay" was what the great urbanist Jane Jacobs called "unslumming." It is a pivotal decision, and I have identified it as one of nine key elements of urban restoration that people are using to make new connections among themselves, in their neighborhoods and cities (Fullilove, 2013). In my observations, people love the questions Morrish and Brown have posed, taking particular delight in adding something. One neighbor of mine adds yarn to neighborhood fences and light poles, which softens the area's post-apocalyptic look. The director of the local community development corporation found a 1950s trailer that he added to the yard of a small youth center. It rouses trailer envy among men who wander into the yard, and they dream of what they would do with such a find. An artist, as part of a sculpture show, made a figure riding a lion. The rest of the sculptures are gone, but that one has stayed, riding the range of the neighborhood's dreams and possibilities.

It is easy to add delight and to envision potential. It is another thing to be allowed to keep what one has made. The sad possibility that ameliorating environmental hazards could be a trigger for displacement is one my neighbors and I face, saved only by a long recession. We do not know what will happen when the market heats up—our protections are too few. What are the tools we can use to ensure that we stay? As this anthology makes clear, there are some, but none are very good, and the real problem lies in the boardrooms of Wall Street, not anywhere near here.

What we—all of us—need is a new approach to habitat, one that respects the biological imperatives of living in a place over many generations, with access to resources, in concert with other citizens of all political persuasions and with our eyes on the needs of the Seventh Generation. Richmond, Virginia, was, as of this writing, poised to tackle its high rate of poverty with a program of considerable interest. Timothy Williams (2013), writing for *The New York Times*, reported the following:

> Emphasizing programs the city can afford on its $760 million budget, Richmond is considering pairing every at-risk child 11 to 15 with an adult mentor; razing public housing and replacing it with mixed-income housing units with the option of homeownership; settling up an intensive manufacturing training program that would give the unemployed the skills needed to work at one of the many local companies looking for workers; starting a farm-to-school program to promote adequate nutrition; and establishing an assistance program to help pay water and wastewater bills for low-income households. (p. A15)

This program reflects the endorsement of the mayor's office and other city leaders; it is a multidimensional approach covering education, jobs, housing and access to fundamental resources like water and waste removal and is conceived as possible within the existing funds of the city. Such an approach opens doors to new and creative thinking about the best uses of the human and other resources that already exist and that might be used to better effect in solving seemingly intractable problems.

In this complex of interventions, I would argue that the demolition of public housing is potentially troublesome; there are no neighborhood-scale demolitions that do not cause at least as many problems as they solve. The need for care in altering human habitat is poorly understood, particularly where the poor are concerned. People say, "Well, anything would be better than what they have now." This turns out not to be true. People need to keep whatever they have and add more.

The soundness of much of the Richmond program, with this glaring exception of the respect for stable residential communities, is the state of our struggle for a new approach to human habitat. This volume, by its careful consideration of the threats of gentrification, will push the debate in useful ways.

Now is the time to plan the conversation on how we shall stay.

REFERENCES

Alexander, M. (2010). *The new Jim Crow: Mass incarceration in the age of color-blindness*. New York, NY: The New Press.

Dubb, Steve (2013). Why Eds and Meds Matter for Community Economic Development. Rooflines: The Shelterforce Blog, Posted 12/3/13. http://www.rooflines.org/3530/why_eds_and_meds_matter_for_community_economic_development/

Fitch, Robert (1996). *The assassination of New York*. Brooklyn, NY: Verso Press.

Fullilove, M. T. (2013). *Urban alchemy: Restoring joy in America's sorted-out cities*. Oakland, CA: New Village Press.

Fullilove, M. T., & Wallace, R. (2011). Serial forced displacement in American cities, 1916–2010. *Journal of Urban Health, 88*(3), 381–389.

Morrish, W. R., Brown, C. R., with Robinson, M., Quimby, C., & Brown, S. (2000). *Planning to stay: Learning to see the physical features of your neighborhood*. Minneapolis, MN: Milkweed Editions.

Wallace, R., Fullilove, M. T., & Flisher, A. J. (1996). AIDS, violence and behavioral coding: Information theory, risk behavior and dynamic process on core-group sociogeographic networks. *Social Science & Medicine, 43*, 339–352.

Williams, T. (2013, October 15). Richmond awaits a bold antipoverty plan: Anxiety rises amid a mayor's bid to revive one of the nation's poorest areas. *The New York Times*, pp. A15, A16.

Contributors

Tabia Henry Akintobi, PhD, MPH, is an associate professor at Morehouse School of Medicine in the Department of Community Health and Preventive Medicine and director of the Prevention Research Center. Her research and publications center on collaborations with institutional, local, national and international leaders seeking to facilitate effective translational research and evaluation between researchers and communities through community-based participatory research designed to reduce chronic health disparities.

Antwi Akom is an associate professor of environmental sociology, public health, and STEM education at San Francisco State University and the founding director of I-SEEED, an innovative nonprofit organization based in Oakland, California. Professor Akom is nationally and internationally recognized for his work building sustainable cities and schools.

William Ayers, formerly distinguished professor of education and senior university scholar at the University of Illinois at Chicago, has written extensively about social justice, democracy and education, and teaching as an essentially intellectual, ethical and political enterprise.

Donald Barr is a professor of pediatrics in the Stanford University School of Medicine and a professor (by courtesy) in the Stanford Graduate School of

Education. He received his MD from the University of California, San Francisco, and his PhD in sociology from Stanford University. In addition to more than three decades as a practicing physician, he has taught health policy and health disparities at Stanford for nearly two decades.

Sue Books, a professor of secondary education at SUNY New Paltz, is the editor of *Invisible Children in the Society and Its Schools* and author of *Poverty and Schooling in the U.S.: Contexts and Consequences.* She teaches courses in comparative education, education and poverty and teacher research.

Ronald L. Braithwaite, PhD, is a professor in the Departments of Community Health and Preventative Medicine, Psychiatry and Family Medicine and director of the Center of Excellence on Health Disparities at Morehouse School of Medicine. His research involves HIV intervention studies with juveniles and adults in correctional systems, social determinants of health, health disparities and community capacity building. He has published over 125 manuscripts and authored or coedited seven books on health issues.

Kevin Burke examines curriculum theory and teacher education, most particularly the ways in which gender and religion come to inform, reform and deform student possibility and teacher training. His first book, *Masculinities and Other Hopeless Causes in an All-Boys Catholic School* (2011), is available from Peter Lang.

Judith N. DeSena, PhD, is a professor of sociology at St. John's University. Her scholarship focuses on community, neighborhoods and gender studies. Her present research investigates gentrification and its effect on community relationships in Brooklyn, New York. She has authored, edited and coedited numerous books and research articles.

Anika Dodds, BS, has a background in quantitative measurement and biomedical sciences. She has a Bachelor of Science degree in bioinformatics from Claflin University and a graduate certificate in health information technology from Southern Polytechnic State University.

Daniel Faber is a professor of sociology at Northeastern University and director of the Northeastern Environmental Justice Research Collaborative (NEJRC). His most recent work is concerned with problems of environmental injustice and equity in America and includes *Capitalizing on Environmental Justice: The Polluter-Industrial Complex in the Age of Globalization* (Rowman & Littlefield, 2008). In 2010, he received the Environmental Sociology Practice and Outreach Award from the Environmental Sociology and Technology Section of the American Sociological Association.

Mindy Thompson Fullilove, MD, is a professor of clinical psychiatry and sociomedical sciences at Columbia University. She has conducted research on epidemics in poor communities, with a special interest in cities. Her most recent book is *Urban Alchemy: Restoring Joy in America's Sorted-Out Cities*.

Henry A. Giroux holds the Global TV Network Chair in English and Cultural Studies at McMaster University in Canada. His most recent books include *America's Education Deficit and the War on Youth* (Monthly Review Press, 2013) and *Youth in Revolt* (Paradigm, 2013).

Diane Grams, PhD, conducts research on urban culture. Her book, *Producing Local Color: Art Networks in Ethnic Chicago* (University of Chicago Press, 2010), investigates art producers in Chicago's Bronzeville, Pilsen and Rogers Park communities. She is currently working on New Orleanian parading cultures and on a book comparing New Orleans to Chicago.

Stuart Greene is an associate professor of English with a joint appointment in Africana Studies at the University of Notre Dame. His new book is *Race, Community, and Urban Schools: Partnering with African American Families* (Teachers College Press, 2013).

Katherine B. Hankins, an associate professor in the Department of Geosciences at Georgia State University, examines urban politics and neighborhood change in Atlanta around issues of schooling, housing and community development.

Elizabeth Egan Henry is an academic advisor at East Georgia State College. Her dissertation work examined school activism in Atlanta between 1972 and 2012.

Adrienne Holloway, PhD, an assistant professor at DePaul University, is an accomplished professional with over 20 years of experience in housing, community development and academia. Her research interests include housing policy, poverty studies and race and ethnic issues.

James Jennings received his PhD from Columbia University in American government. He is a professor of urban and environmental policy and planning at Tufts University. His teaching and research focus on neighborhood revitalization, race and poverty. He has published widely in these areas and is involved in numerous local community-building initiatives and civic activities.

Russell Lopez has a master's in city and regional planning and a doctorate in environmental health. His research focuses on the social, physical and built environment influences on health, including the effect of segregation, income inequality and urban sprawl.

Shelley McDonough Kimelberg is an assistant professor of sociology at Northeastern University. She teaches and conducts research in the areas of urban sociology, urban education, economic development and inequality.

Miranda Martinez is an associate professor at Ohio State University. She researches community-based movements and the Latino experience of urban space.

Her 2010 book, *Power at the Roots: Gentrification, Community Gardens, and the Puerto Ricans of the Lower East Side,* was published by Lexington Books. She is currently researching issues of financial access in low-income communities.

Maria McKenna is an assistant professor of the practice in Africana Studies and the Education, Schooling, and Society program. Her recent work on culturally sensitive parent engagement can be found in the *School Community Journal.*

Aaron Nakai is currently the program director of health equity and community engagement at I-SEEED. He is an expert facilitator whose work includes designing health and environmental equity curriculum, implementing community-based action research projects and training youth and youth-serving adults in YPAR, e-PAR and B-TEC.

Emily Rosenman is a PhD candidate in geography at the University of British Columbia. Her research focuses on financialization, community development and housing.

Aekta Shah is the program director of technology and youth engagement at the Institute for Sustainable Economic, Educational and Environmental Design (I-SEEED). Shah's work and research focus on developing culturally and community responsive tools and approaches that link GIS mapping, social media, education and sustainable community development.

Loïc Wacquant is a professor of sociology at the University of California, Berkeley, and a researcher at the Centre Européen de Sociologie et de Science Politique, Paris, France. A MacArthur Foundation Fellow and recipient of the Lewis Coser Award of the American Sociological Association, his research spans urban relegation, ethnoracial domination, the penal state, incarceration and social theory and the politics of reason.

Samuel Walker is a doctoral student at the University of Toronto in the Department of Geography. His research interests include North American shrinking cities, political ecology and the alternative food movement.

William H. Watkins, PhD, is professor of education at the University of Illinois at Chicago. He is the author of *The White Architects of Black Education* (2001) and lead editor of *Race and Education* (2001). His numerous articles, chapters, and essays have appeared in scholarly journals, books, and encyclopedias. His research and writings are informed by critical inquiry and Marxian theory that challenges conventional beliefs about the system of education in the U.S.

Elvin Wyly is an associate professor of geography and chair of the Urban Studies Coordinating Committee at the University of British Columbia. He is editor-in-chief of *Urban Geography.*